PROGRESS IN COLLOID & POLYMER SCIENCE

Editors: H.-G. Kilian (Ulm) and G. Lagaly (Kiel)

Volume 83 (1990)

Interfaces in Condensed Systems

Guest Editor: G. H. Findenegg (Bochum)

Springer-Verlag
Berlin Heidelberg GmbH

ISBN 978-3-662-15063-4   ISBN 978-3-7985-1686-1 (eBook)
DOI 10.1007/978-3-7985-1686-1
ISSN 0340-255 X

© 1990 by Springer-Verlag Berlin Heidelberg
Originally published by Dr. Dietrich Steinkopff Verlag GmbH & Co. KG, Darmstadt in 1990
Softcover reprint of the hardcover 1st edition 1990

Chemistry editor: Dr. Maria Magdalene Nabbe; English editor: James Willis; Production: Holger Frey.

Type-Setting: Graphische Texterfassung, Hans Vilhard, D-6126 Brombachtal

Preface

The 34th General Meeting of the Kolloid-Gesellschaft e.V. was held at the Ruhr-Universität-Bochum, 1—4 October 1989. It was organized jointly by the Kolloid-Gesellschaft and the Fachausschuß "Grenzflächen" of the VDI-Gesellschaft Verfahrenstechnik und Chemie-Ingenieurwesen. The meeting was attended by about 200 participants from 12 countries. In the opening session of the meeting J. Th. G. Overbeek was awarded the Wolfgang-Ostwald-Price of the Kolloid-Gesellschaft for his outstanding fundamental contributions to the field of pure and applied colloid science.

The main topic of the meeting was "Interfaces in Condensed Systems. Equilibrium and Dynamic Phenomena" and covered adsorption and wetting phenomena at interfaces solid-liquid, liquid-liquid, and gas-liquid, as well as colloidal phenomena related to interfacial behaviour. Invited lectures were presented by nine eminent scientists in this field: J. Th. G. Overbeek (Utrecht, NL), P. G. de Gennes (Paris, F). H. Hoffmann (Bayreuth, D), D. Horn (Ludwigshafen, D), D. Langevin (Paris, F), J. Lyklema (Wageningen, NL), C. A. Miller (Houston, USA), M. J. Schwuger (Jülich, D), and Th. F. Tadros (Bracknell, GB). In addition, more than 60 contributed papers dealing with fundamental and applied aspects of interfacial chemistry in condensed systems were presented either orally or as posters.

This Progress Volume contains a selection of the papers presented at the Kolloid-Tagung 1989. The papers are grouped together somewhat arbitrarily into the sections General (Invited Papers), Interfaces, Monolayers, Surfactant Systems, and Dispersed Systems. These contributions cover both the fundamental and technological aspects, as well as modern experimental techniques of the field.

I thank the members of the Organizing Committee of the meeting, K. Kosswig, J. Lyklema, M. J. Schwuger and A. Zembrod, for their help and advise in the preparation of the scientific program. I would also like to thank my coworkers in Bochum for support in the organization of the meeting, and the sponsors for financial support. Last not least, I wish to thank the authors and referees of the papers submitted for this volume.

Gerhard H. Findenegg

Contents

Dispersed Systems

Progress in Colloid & Polymer Science   Progr Colloid Polym Sci 83:1—9 (1990)

Microemulsions: theoretical estimates of droplet sizes and size distributions

J. T. G. Overbeek

Van't Hoff Laboratory, University of Utrecht, The Netherlands

Abstract: After a brief introduction on general properties of microemulsions, this paper presents an expression for the Gibbs energy of a water in oil (W/O) microemulsion containing a range of droplet sizes. Droplets are partitioned in categories, j, where j is the number of surfactant molecules per droplet. The Gibbs energy takes into account: the amounts and chemical potentials of the constitutents of the dispersion medium and of the droplets, including the adsorption layers, the interfacial tension, the bending stress and the concentration of droplets of type j. The droplets of all sizes are in a mass-action equilibrium with single molecules in the water and/or oil regions. — A few necessary parameters are taken from experiments with the system: water, cyclohexane, sodium dodecylsulphate (SDS), 1-pentanol, and NaCl. — The concentration of droplets of type j can then be expressed in their radius, a_j, their surface tension and its dependence on curvature, and known constants (such as kT and parameters pertaining to the electrical double layer) both for saturated (Winsor II equilibria) and unsaturated microemulsions. — The size distributions are found to be fairly wide, the width increasing with increasing average droplet size.

Key words: Microemulsions; droplet sizes; size distribution; Winsor II equilibrium; curved interfaces; capillary pressure

Introduction

Microemulsions are mixtures of water, oil, relatively large amounts of one or more surfactants, and often electrolytes, all present in one thermodynamically stable phase. Their properties depend on their composition, but not on the method of preparation. They form spontaneously when the ingredients are brought together. In many cases microemulsions contain small droplets (diameter of the order of 20 nm) of one medium, e.g. water, W (or oil, O), dispersed in the other medium, O or W. There is a continuous transition from swollen micelles to microemulsion droplets. But, whereas in swollen micelles direct contact and interaction between surfactant and swelling agent exist, microemulsion droplets may be so large that most of their content is not in direct contact with the surfactant at the oil-water interface.

Systems in which microemulsions occur have complicated phase diagrams [1]. In this paper we shall concentrate on water in oil (W/O) droplet-type microemulsions, containing an ionic surfactant and a non-ionic "cosurfactant" (e.g., a medium chain length alcohol) [2—4] and on their phase equilibria with a non colloidal aqueous phase (Winsor II equilibrium) [5, 6].

In microemulsions the interfacial area between W and O is very large; therefore, the interfacial tension must be very low, so that the interfacial free energy can be compensated by the negative free energy of dispersion of the droplets in the medium. As a rule a sufficiently low interfacial tension is not obtained by the use of single chain ionic surfactants, since they form micelles before the interfacial tension is low enough. Addition of the cosurfactant, which lowers the interfacial tension in addition to the effect of the surfactant, allows the interfacial tension to drop to zero before micelles are formed [7].

By measuring the interfacial tension, σ, of a macroscopic oil-water interface as a function of the activities, a_i, of surfactant and cosurfactant their surface excess concentrations, Γ_i, can be calculated from the Gibbs adsorption equation,

$$-\frac{\partial\sigma}{RT\,\partial\ln a_i} = \Gamma_i . \tag{1}$$

It is found that, at cosurfactant and electrolyte concentrations at which microemulsions can be formed, surfactant and cosurfactant show saturation adsorption, i.e., their surface concentrations are virtually independent of their bulk concentrations. Since, below the CMC the surfactant concentration is low and, assuming that the surface concentration at the curved surface of the droplets is not too different from that at a flat surface, the total area of the droplets can be found from the total amount of surfactant. Their total volume, V_d, is nearly equal to the total amount of water (for W/O microemulsions) and thus the radius a of the droplets, assumed to be of uniform size, is equal to

$$a = 3 \times \frac{\dfrac{4\pi a^3}{3}}{4\pi a^2} = \frac{3V_d}{A}$$

$$\cong \frac{3\bar{V}_w n_w \Gamma_{sa}(a)}{n_{sa}} \approx \frac{n_w}{n_{sa}} \text{ Å} . \tag{2}$$

In these equations A is the total area of the droplets, n_w and n_{sa} are the amounts of water and surfactant respectively, $\bar{V}_w$ is the molar volume of water and the last near equality implies that $\Gamma_{sa}(a) \cong 1.8\,\mu$ mol m^{-2}, as has been found [7] for sodium-dodecylsulphate (SDS) at a flat ($a = \infty$) water-cyclohexane interface.

Equation (2) suggests that it should be possible to prepare thermodynamically stable emulsions with large droplets, e.g., with a radius of 1 μm, by choosing a large water-to-surfactant ratio. This, however, does not succeed. At given salt and cosurfactant activities, a given amount of surfactant can only solubilize a certain maximum amount of aqueous phase into a microemulsion. If more aqueous phase is offered it forms a separate phase in a so-called Winsor II equilibrium (W/O microemulsion + aqueous phase). This behavior can only be understood if the interface has a preferred curvature and a definite stiffness [6, 8—10]. At least two factors contribute to this stiffness, the crowding of the hydrocarbon chains of surfactant and cosurfactant in the interface, which tend to bend the inter-

face around the water, and the electric double layer at the water side, which tends to bend the interface around the oil.

These facts lead to the following model for microemulsions.

Model

We describe a W/O microemulsion as a suspension of spherical droplets, having a certain size distribution. The droplets contain all water and salt in the system and a small concentration ($\leqslant$ CMC) of the ionic surfactant. The interfacial adsorption layer contains all the ionic surfactant minus the small amount dissolved in the droplets, a sizable fraction of the cosurfactant, and the negative adsorptions of salt and oil. The Gibbs surface is chosen so that water is not adsorbed ($\Gamma_{water} = 0$). Since most of the ionic surfactant is in the interface, its concentration in the water can vary within fairly wide limits (zero to CMC) without much change in volume or interfacial area of the droplets. With any change in surfactant concentration the interfacial tension, σ, changes and thus we have the somewhat paradoxical situation that σ must be quite low to allow the existence of a microemulsion; however, σ is a parameter that adapts itself to variations in volume and area and, in particular, to changes in the concentrations of salt and cosurfactant [3].

Free energy of a microemulsion

In a previous paper [3] a thermodynamic analysis was given for a microemulsion with uniform droplet size. In this paper we follow the same line of thought, but introduce a particle-size distribution. The Gibbs free energy, G^M, of a droplet-type microemulsion can be written in two equivalent ways:

$$G^M = \sum_i n_i\mu_i = \sum_i n_{im}\mu_i + \sum_j n_{dj}\mu_{dj} . \tag{3}$$

n_i and μ_i are the total number of moles and the chemical potentials respectively of type i. n_{im} are the number of moles of type i in the continuous medium, n_{dj} and μ_{dj} are the number of droplets of category, j, and their chemical potential per droplet, respectively. The droplets are considered to also include the adsorption layers. j is defined as the number of surfactant molecules in the adsorption layer of a single droplet with radius, a_j. This leads to the following relation between a_j and j:

$$j = 4\pi a_j^2 \Gamma_{sa,j} N_{Av} , \tag{4}$$

where N_{Av} is Avogadro's constant and $\Gamma_{sa,j}$ is the surface excess concentration at a droplet of radius a_j. It has been argued in [3] that the surface of constant (close) packing of surfactant and cosurfactant is a small distance, ξ (a few Å), away from the Gibbs surface for $\Gamma_w = 0$ and thus,

$$\Gamma_{sa,j} = \left(\frac{a_j + \xi}{a_j}\right)^2 \Gamma_{sa} , \tag{5}$$

where Γ_{sa} is the surface excess concentration at a flat surface.

Since the pressure inside and outside the droplets are, in general, different, we first formulate the Helmholtz free energy, F^M, of the microemulsion.

$$F_m = \sum_i n_{im}\lambda_{im} + \sum_i \sum_j n_{ij}\lambda_{ij} - pV_m$$

$$- \sum_j (p + \Delta p_j)V_{dj} + \sum_j \sigma_j A_j + F_{mix} , \tag{6}$$

where λ_{im} and λ_{ij} are the chemical potentials of i in the continuous medium and in the droplets of size j, respectively, when they are considered separately (i.e., not yet mixed, the medium at pressure, p, equal to the ambient pressure, the droplets at $p + \Delta p_j$). n_{im} and n_{ij} are the amounts of i in the continuous medium and in the droplets, j, respectively, n_{ij} containing both bulk and interface of the droplets. V_m is the volume of the medium, V_{dj} the volume of all droplets of type j. σ_j and A_j are the interfacial tension and the interfacial area of all droplets of type j. F_{mix} is the free energy of mixing of droplets and medium. It also contains a contribution of the preparation of the droplets from a macroscopic phase of suitable composition.

With the total volume of the microemulsion $V^M = V_m + \Sigma V_{dj}$ and assuming that $G_{mix} = F_{mix}$, we obtain

$$G^M = F^M + pV^M$$
$$= \sum_i n_{im}\lambda_{im} + \sum_{ij} n_{ij}\lambda_{ij} - \sum_j \Delta p_j V_{dj}$$
$$+ \sum_j \sigma_j A_j + G_{mix} . \tag{7}$$

In [3] an equation for G_{mix} was used that was valid up to quite high concentrations of uniform droplets. Modifying this equation for nonuniform droplet size is easy for low concentrations (ideal behavior), but gets complicated for higher concentrations because the interaction between the droplets depends on the concentrations of all droplets categories. Since essential effects of a droplet size distribution are already observed at low concentrations, we shall limit the treatment here to ideal behavior, and we write for G_{mix}

$$G_{mix} = \sum_j n_{dj}kT$$

$$\times \left(\ln \frac{n_{dj} \times \frac{4}{3}\pi a_j^3}{V^M} - 1 \right.$$

$$\left. + \frac{3}{2} \ln \frac{v_w}{16a_j^3} \right) , \tag{8}$$

where k is the Boltzmann constant, v_w is the volume of a water molecule, and the term $1.5 \ln(v_w/16a_j^3)$ reflects Reiss's [11, 12] treatment of the formation of the droplets from a bulk solution of the same composition.

Introducing

$$\Delta p_j = \frac{2\sigma_j}{a_j} - \frac{2c_j}{a_j^2} , \tag{9}$$

$$A_j = n_{dj} \times 4\pi a_j^2 , \tag{10}$$

and

$$V_{dj} = n_{dj} \times \frac{4}{3}\pi a_j^3 , \tag{11}$$

where the bending stress coefficient, c_j, of the interfaces is defined as

$$c_j \equiv (\partial\sigma/\partial(2/a)_{a=a_j} , \tag{12}$$

we can write Eq. (7) as

$$G^M = \sum_i n_{im}\lambda_{im} + \sum_{ij} n_{ij}\lambda_{ij} + \sum_j \frac{4}{3}\pi a_j^2 n_{dj}$$

$$\times \left(\sigma_j + \frac{2c_j}{a_j} \right) + G_{mix} . \tag{13}$$

Now we want to find out at what size distribution the system is in equilibrium, i.e., under what conditions G^M has a minimum value. By combining Eqs. (3), (8), and (13) we can find an expression (Eq. (17)) for the chemical potential, μ_{dj}, of the droplets of size j, which contains the concentration n_{dj}/V^M. We can also find μ_{dj} with a mass-action approach, as the sum of the chemical potentials of all the constituents of the droplets (Eq. (18)). Elimination of μ_{dj} between the two equations (17) and (18) then solves our problem, at least in principle. Combination of Eqs. (3), (8), and (13) leads to

$$
\begin{aligned}
G^M &= \sum_i n_{im}\mu_i + \sum_j n_{dj}\mu_{dj} \\
&= \sum_i n_{im}\lambda_{im} + \sum_{i,j} n_{ij}\lambda_{ij} + \sum_j \frac{4}{3}\pi a_j^2 \\
&\quad \times \left(\sigma_j + \frac{2c_j}{a_j}\right) n_{dj} + \sum_j n_{dj}kT \\
&\quad \times \left(\ln \frac{n_{dj} \times \frac{4}{3}\pi a_j^3}{V^M} - 1 \right. \\
&\quad \left. + \frac{3}{2}\ln \frac{v_w}{16 a_j^3}\right) .
\end{aligned}
$$

(14)

The difference between μ_i and λ_{im} of the continuous medium is found by differentiation of G^M with respect to n_{im} at constant $n_{km}(k \neq i)$ and n_{dj}:

$$
\begin{aligned}
\left(\frac{\partial G^M}{\partial n_{im}}\right) &= \mu_i \\
&= \lambda_{im} - \frac{\sum_j n_{dj}kT}{V^M}\frac{\partial V^M}{\partial n_{im}} \\
&= \lambda_{im} - \frac{\sum n_{dj}kT}{V^M}\bar{V}_i ,
\end{aligned}
$$

(15)

where $\bar{V}_i$ is the molar volume of i. And thus,

$$
\begin{aligned}
\sum_i n_{im}\mu_i &= \sum_i n_{im}\lambda_{im} \\
&\quad - (\sum n_{dj}kT) \times (1 - \varphi) ,
\end{aligned}
$$

(16)

where φ is the volume fraction of the droplets. From Eqs. (14) and (16) it follows then that

$$
\begin{aligned}
\mu_{dj} &= \sum_i \frac{n_{ij}\lambda_{ij}}{n_{dj}} + \frac{4}{3}\pi a_j^2 \left(\sigma_j + \frac{2c_j}{a_j}\right) + kT \\
&\quad \times \left(\ln \frac{n_{dj} \times \frac{4}{3}\pi a_j^3}{V^M} - \varphi\right. \\
&\quad \left. + \frac{3}{2}\ln \frac{v_w}{16 a_j^3}\right) ,
\end{aligned}
$$

(17)

and from mass action,

$$
\mu_{dj} = \sum_i \frac{n_{ij}}{n_{dj}}\mu_i .
$$

(18)

φ might be dropped from Eqs. (16) and (17) with the argument that other terms also proportional to φ would appear in these equations as soon as droplet interaction is not neglected. The conditions (16) through (18) also guarantee that G^M is at a minimum for constant p, T, and all n_i.

It is awkward that in Eq. (17) each particle category, j, has its own values for λ_{ij}, σ_j, and c_j. However, in a system with a given composition of the water and the oil regions σ and c are fairly simple functions of the radius of curvature, a. As has been argued in [3], σ has a more or less parabolic shape with a minimum when plotted against $1/a$ and c is an increasing function of $1/a$ when a is called positive for interfaces curving around the aqueous medium. In [3] a linear relation between c and $1/a$ gave satisfactory results. Fig. 1 gives a sketch of the situation.

We thus write:

$$
c = -b + \frac{d}{a}
$$

(19)

and, using Eq. (12)

$$
\sigma = \sigma_\infty - \frac{2b}{a} + \frac{d}{a^2} .
$$

(20)

In these relations d was shown to be practically independent of salt- and cosurfactant concentrations in W/O microemulsions consisting of water, cyclo-

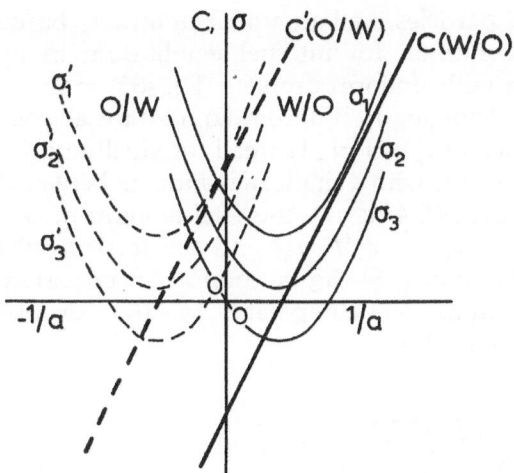

Fig. 1. Interfacial tension, σ, and bending stress coefficient, c, plotted against the curvature (= reciprocal radius, $1/a$). The drawn lines refer to a W/O situation where the preferred curvature is around the water. The broken lines refer to O/W. The three values for the interfacial tension all lead to the same line for c, showing that for fixed c, σ can still vary greatly

hexane, NaCl, sodium dodecylsulphate (SDS), and 1-pentanol. b is positive for W/O, increases with NaCl and cosurfactant concentrations, and it is zero for conditions where W/O goes over into O/W. σ_∞ is adapted to equilibrium conditions by changes in the small SDS concentration. The droplets consist of bulk, of volume $4/3\pi a_j^3$ and an adsorption layer of area $4\pi a_j^2$. The bulk contains all water, the ionic surfactant minus its adsorption, and the NaCl plus the negative adsorption of the coions (we neglected the solubilities of oil and cosurfactant in the aqueous medium). In the surface layer all adsorbed surfactant and cosurfactant are present, with the negative adsorption of the salt and, to obtain zero volume, the negative adsorption of the oil. Writing

$$x = \Gamma_{sa}\bar{V}_{SDS} + \Gamma_{Cl}\bar{V}_{NaCl} ,\tag{21}$$

(where sa stands for the surfactant ion) for the equivalent thickness of the adsorption layers of surfactant and salt, we find the components, water, surfactant, and salt in the volume

$$\sum_{\substack{\text{water}\\ \text{SDS}\\ \text{NaCl}}} \frac{n_{ij}}{n_{dj}} \bar{V}_i = \frac{4}{3}\pi a_j^3 + 4\pi a_j^2 x .\tag{22}$$

For these components, λ_{ij}, being the chemical potential at the pressure $p + \Delta p_j$ in a stagnant droplet, can be written

$$\lambda_{ij} = \mu_i' + \Delta p_j \bar{V}_i = \mu_i' + \left(\frac{2\sigma_j}{a_j} - \frac{2c_j}{a_j^2}\right)\bar{V}_i ,\tag{23}$$

where $\bar{V}_i$ is the partial molar volume of component i and μ_i' is its chemical potential at the ambient pressure p. Since the bulk composition of the droplets may be assumed not to depend on their size, μ_i' does not depend on j. It differs from the chemical potential μ_i in the microemulsion as a whole, since μ_i' does not contain a contribution from G_{mix} and μ_i does. The adsorbed cosurfactant and the negatively adsorbed oil belong to the droplet, but their chemical potential is the same as in the continuous medium at pressure p.

By elimination of μ_{dj} from Eqs. (17) and (18) and use of Eqs. (19)—(23), Eq. (24) is obtained.

$$\ln \frac{n_{dj}}{V^M} = \ln \frac{48 a_j^{3/2}}{\pi v_w^{3/2}} - \frac{4\pi d - 8\pi x b}{kT}$$

$$+ \frac{8\pi a_j}{kT}(b - x\sigma_\infty) - \frac{4\pi a_j^2}{kT}\sigma_\infty$$

$$+ \left(\sum_i \frac{n_{ij}}{n_{dj}} \frac{\mu_i - \mu_i'}{kT}\right)_{i = \text{water, SDS, NaCl}} .\tag{24}$$

The term with $\mu_i - \mu_i'$ does not contain contributions from oil and cosurfactant, since for these components μ_i and μ_i' differ by a factor proportional to the droplet concentration that should be neglected for small volume fractions (say < 0.05).

In [3] it was shown that $(\mu_i - \mu_i')/\bar{V}_i$ does not depend on i. Therefore, we may write

$$\sum_i \frac{n_{ij}\bar{V}_i}{n_{dj}} \frac{\mu_i - \mu_i'}{\bar{V}_i kT} = -\left(\frac{4\pi}{3} a_j^3 + 4\pi a_j^2 x\right) MU ,\tag{25}$$

where $MU = (\mu_i' - \mu_i)/(\bar{V}_i kT)$, ($i$ = water, SDS, NaCl).

MU, being proportional to $\mu_i' - \mu_i$ for the aqueous components, is zero for saturated microemulsions, for at the Winsor II equilibrium the composition of the bulk of the droplets and the equilibrium aqueous phase are identical, and the chemical potentials μ_i' and μ_i are both at ambient pressure.

For unsaturated microemulsions, MU is positive, because such microemulsions will spontaneously take up a solution, that has the composition of the bulk of the droplets and, thus, $\mu_i' > \mu_i$.

Values for Γ_{sa}, d, and x are known from our previous work [3]. For the system water, cyclohexane, SDS, pentanol, and NaCl, Γ_{sa} varies from 2.0×10^{-6} mol m^2 for 4.11% pentanol in cyclohexane and 0.3 M NaCl, to 1.56×10^{-6} mol m^{-2} for 19% pentanol and 0.1 M NaCl, d, the slope of the c vs $1/a$ line (Eq. (19)) is about 2.5 to 3.0 kT and x (Eq. (21)) = 0.4 nm.

On the assumption that the influence of the hydrocarbon tails on c is constant at constant cosurfactant activity, $-b$, the cut-off of c for $1/a = 0$ is obtained from the influence of NaCl on the Winsor II equilibrium and varies from $-0.51 \cdot 10^{-12}$ N for 19% pentanol and 0.3 M NaCl to zero for 19% pentanol, and 0.15 M NaCl or for 7.65% pentanol and 0.3 M NaCl.

These values have been found from an analysis in which all drops were assumed to have the same size, but they should also be good approximations if a size distribution is present.

Furthermore, as mentioned before, σ_∞ (Eq. (20)) must adapt itself by changes in the small SDS concentration to the volume (in essence the amount of water) and the interfacial area (given by the amount of SDS) of the droplets.

Finally, MU, reflecting the difference between μ_i' and μ_i that is due to the free energy of mixing, changes even at constant composition of the oil and water media by changes in the volume fraction and interfacial area and, thus, in the sizes and numbers of the droplets.

Results in terms of calculated size distributions

On the basis of Eqs. (24) and (25) a simple computer programm can be set up for calculating the size distribution, n_{dj}/V^M vs. a_j or, easier vs $j = 4\pi(a_j + \xi)^2 \Gamma_{sa} N_{Av}$ since each integer j corresponds to one size category of droplets.

Γ_{sa} and x are known from experiments. When ξ is chosen (within narrow limits, see [3]), b follows from experiments. Then MU is set equal to zero, so as to obtain a microemulsion in Winsor II equilibrium, and preliminary values of d and σ_∞ are chosen, d being slightly larger than the value found for uniform droplets (the concentrations of the categories j are much smaller than the concen-

tration of particles if all are uniform) and σ_∞ based on the condition for internal equilibrium in an emulsion with uniform droplets (Eq. (33) in [3]).

After calculating n_{dj}/V^M from Eq. (24) for all relevant values of a_j or j (n_{dj} is small for small and for large values of j with a single maximum in between) the total area ($\Sigma\, 4\pi a_j^2 n_{dj}$), the total amount of surfactant ($\Sigma\, 4\pi(a_j + \xi)^2 \Gamma_{sa} n_{dj}$) and the total droplet volume ($\Sigma\, 4\pi a_j^3 n_{dj}/3$) can be found. An important average radius, $\langle a \rangle$, comparable to "the" droplet radius of Eq. (2) is

$$\langle a \rangle = \frac{\sum 4\pi a_j^3 n_{dj}}{\sum 4\pi (a_j + \xi)^2 n_{dj}}$$

$$= \frac{3 \sum\limits_j V_{dj} \Gamma_{sa}}{n_{sa}} \cong \frac{n_w}{n_{sa}} \, \text{Å} \,. \qquad (26)$$

The values obtained can be compared with experimental data. If necessary, a better fit can be reached by adapting σ_∞ and, to a lesser extent, d, until the droplet volume and the total amount of surfactant agree with experiments.

In a next step the calculation is repeated for an unsaturated microemulsion, keeping the composition of water and oil regions constant, and thus keeping Γ_{sa}, b, d, x and ξ constant, but choosing finite positive values for MU and adapting σ_∞, so that, for example the total droplet volume is halved, the total amount of surfactant in the interface is kept constant, and thus the average droplet radius is about halved too. Good tentative values for MU and σ_∞ can again be obtained from the results in [3] for uniform droplets (esp. Eqs. (33) and (46)).

In Figs. 2 and 3 and in Table 1, we show a few typical results obtained in this way. The *size distributions are fairly wide*. They become narrower (relatively narrower) for smaller average droplet size, irrespective of whether the smaller droplets are obtained by making the microemulsions unsaturated at constant composition of the water and oil regions, or by increasing salt and/or cosurfactant concentration, but keeping at the Winsor II boundary, where MU is zero.

Another obvious application consists in checking how droplet size is affected by changing the surfactant concentration (or, equivalently, the volume fraction, φ) in Winsor II equilibria. It is found that the average droplet size decreases with decreasing concentration as might be expected on qualitative

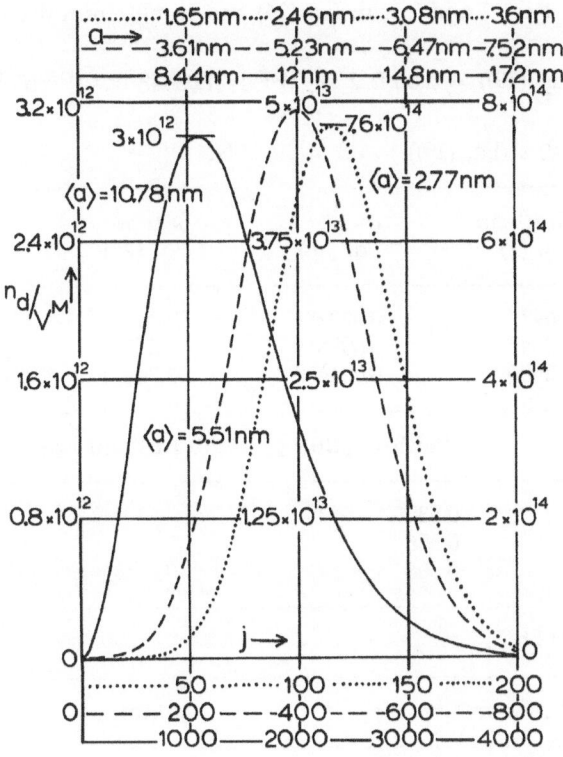

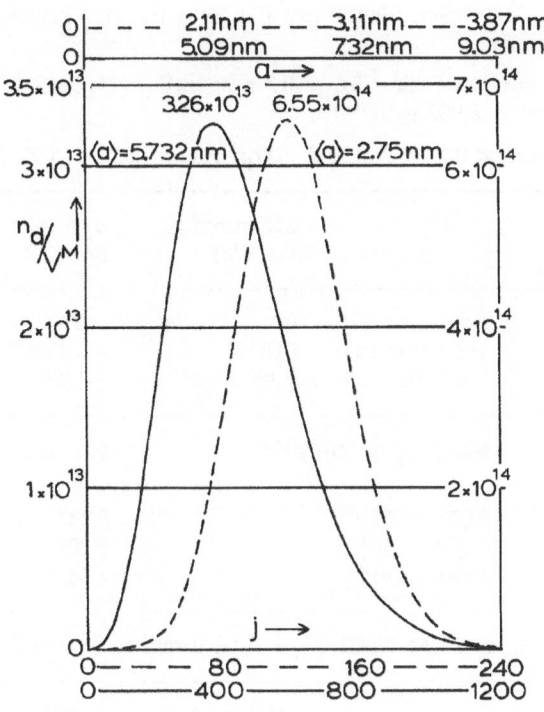

Fig. 2. Number of droplets of category j per cm³ (n_{dj}/V^M) according to Eq. (24) plotted against j (linear scale) or a_j (non linear scale). The units on the horizontal and vertical axis have been chosen so as to give the curves about the same maximum height and about the same width. The data correspond to microemulsions of 0.2 M NaCl in water in 19% (w/w) 1-pentanol in cyclohexane. Detailed data are collected in Table 1

Fig. 3. As for Fig. 2, except that the data correspond to 0.3 M NaCl

Discussion

It may be useful to discuss qualitatively the size distribution following from Eqs. (24) and (25). If the free energy of mixing is neglected the droplet size would tend to be uniform at $c = 0$, i.e., at $a = d/b$ and σ would adapt to zero, too. The free energy of mixing causes the droplets to be smaller than $a = d/b$, because that implies a gain in entropy. Very small droplets are unfavorable, since for them the surface free energy (in particular the $(d/a^2) \times 4\pi a^2 \times n_d$ term) becomes too high. Somewhat larger droplets are present in larger numbers, in the first instance due to the $a_j^{3/2}$ term, which reflects the fact that the center of mass of a larger droplet gains more freedom in being dispersed in the volume V^M than a smaller droplet. This idea is due to Reiss [11, 12] and also explained in [3]. At further increase of the droplet size the advantage given by the $a_j^{3/2}$ term is offset and more than offset by the increase in the surface free energy, in particular by the $4\pi a_j^2 \times \sigma_\infty$ term and still more strongly by the MU term, which is proportional to the droplet volume and which reflects the fact that more material has to

arguments. At high dilution the entropy of mixing becomes relatively more important and favors a larger number of smaller droplets. The effect is quite marked, especially at low concentrations, but not as pronounced as with uniform droplets.

Table 2 compares the average radius $\langle a \rangle$ (Eq. (26)) for a series of volume fractions φ, with radii a of uniform droplets at corresponding values of V_{hs}/V^M, where V_{hs} is the total volume of hard spheres. V_{hs}/V^M is larger than φ in the ratio $(1 + 24 \, \text{Å}/\langle a \rangle)$ (See [3], Eqs. (31) and (35)), because V_{hs} also contains the adsorbed layers of surfactant and cosurfactant. No attempt has been made to match $\langle a \rangle$ and a for a particular volume fraction. This could have been easily achieved by simultaneous small changes in d and σ_∞.

Table 1. Data used for the calculation of the size distributions in Figs. 2 and 3 with Eq. (24) and results from these calculations

For all curves: ξ (Eq. (5)) = 0.3 nm; x (Eq. (21)) = 0.4 nm; d (Eq. (19)) = 2.8 kT; $\Sigma\, jn_{dj}/V^M$ = 1.03443 × 10^{-5} mol SDS/cm^3 = 2.982 g SDS/l.

For curves at 0.2 M NaCl: Γ_{sa} (Eqs. (1) and (5)) = 1.73 μ mol m^{-2}; b (Eq. (19)) = 1.8 × 10^{-13} N.

		MU/nm^{-3} Eq. (25)	σ_∞/mN m^{-1} Eq. (20)	A/m^2 cm^{-3} Eq. (2)	φ Eq. (16)	$\langle a \rangle$/nm Eq. (26)
No. 1	Saturated	0	0.0225	5.643	0.02028	10.782
No. 2	1/2 Saturated	0.01516	−0.13367	5.3647	0.009856	5.512
No. 3	1/4 Saturated	0.18	−1.02758	4.8500	0.004472	2.766

	Max. $(n_{dj} \times$ cm$^3)/V^M$	at a_j/nm	Width = $(a(0.9\ \varphi) - a(0.1\ \varphi))/a(0.5\ \varphi)$
No. 1	3.003 × 10^{12}	8.837	0.532
No. 2	4.934 × 10^{13}	5.207	0.356
No. 3	7.640 × 10^{14}	2.64	0.308

For curves at 0.3 M NaCl: Γ_{sa} = 1.82 μ mol m^{-2}; b = 5.1 × 10^{-13} N

		MU/nm^{-3}	σ_∞/mN m^{-1}	A/m^2 cm^{-3}	φ	$\langle a \rangle$/nm
No. 4	Saturated	0	0.104763	5.109	0.009764	5.732
No. 5	1/2 Saturated	0.132	−0.607087	4.602	0.004218	2.750

	Max. $(n_{dj} \times$ cm$^3)/V^M$	at a_j/nm	Width = $(a(0.9\ \varphi) - a(0.1\ \varphi))/a(0.5\ \varphi)$
No. 4	3.261 × 10^{13}	4.946	0.482
No. 5	6.555 × 10^{14}	2.627	0.334

Table 2. Influence of the droplet concentration on the (average) radius of the droplets

Volume fraction φ	Average radius, Eq. (26) $\langle a \rangle$/nm	Hard sphere Volume fraction V_{hs}/V^M corresp. to φ	Radius of mono-dispersed droplets, a/nm, [3]
0.080383	6.552	0.10983	6.463
0.037114	6.246	0.051374	6.015
0.018249	5.970	0.025584	5.656
0.0097641	5.733	0.013852	5.361
0.0017292	5.096	0.0025436	4.593
0.00066199	4.756	0.00099603	4.186

be brought from μ_i to μ_i' ($\mu_i < \mu_i'$) in the building of larger droplets. Or, in other words, the standard chemical potential of larger droplets is higher than that of smaller droplets as a consequence of surface effects and volume effects, but the reverse is the case, due to the Reiss effect connected with the formation of individual droplets from a bulk phase.

Very little *experimental information* on the droplet size distribution in the system studied here (water, cyclohexane, SDS, 1-pentanol, NaCl) is available, but preliminary SAXS measurements show that the W/O microemulsions are not monodispersed, but have a size distribution with a width of at least 20% in the radius [13].

A change in the average size with the droplet concentration has not been found experimentally, but the size range tested was higher than corresponds to Table 2, and at the higher concentrations in that table hard sphere interactions are probably already not negligible.

Finally, I want to point out that after I had presented a preliminary version of this paper in April 1989, Prof. H.-F. Eicke showed me the typescript of his paper (meanwhile published [14]) treating droplet-size distribution in microemulsions. In general the two approaches are similar, but many important details are different. In [14] a fairly wide size distribution is also found.

Acknowledgement

I express my gratitude to Mrs. Marina Uit de Bulten and Margaret de Groot for preparing the typescript and to Mr. Theo Schroote for drawing the figures.

References

1. Friberg S (1977) In: Prince Leon M (ed) Microemulsions, Theory and Practice. Academic Press, New York, pp 133—146
2. Hoar TP, Schulman JH (1943) Nature 152:102—103
3. Overbeek JTG, Verhoeckx GJ, de Bruyn PL, Lekkerkerker HNW (1987) J Colloid Interface Sci 119:422—441
4. de Bruyn PL, Overbeek JTG, Verhoeckx GJ (1989) J Colloid Interface Sci 127:244—255
5. Winsor PA (1948) Trans Faraday Soc 44:376—398
6. Robbins ML (1977) In: Mittal KL (ed) Micellization, Solubilization and Microemulsions. Plenum, New York, Vol 2, pp 713—754
7. Verhoeckx GJ, de Bruyn PL, Overbeek JTG (1987) J Colloid Interface Sci 119:409—421
8. Bowcott JE, Schulman JH (1955) Z Elektrochemie 59:283—288
9. Miller CA, Neogi P (1980) AIChEJ 26:212—220
10. Mukherjee S, Miller CA, Ford T Jr (1983) J Colloid Interface Sci 91:223—243
11. Reiss H (1975) J Colloid Interface Sci 53:61—70
12. Reiss H (1977) Adv Colloid Interface Sci 7:1—66
13. van Aken G (1989), private communication
14. Borkovec M, Eicke H-F, Ricka J (1989) J Colloid Interface Sci 131:366—381

Received November 27, 1989;
accepted December 27, 1989

Author's address:

Prof. J. Th. G. Overbeek
Van't Hoff Laboratory
University of Utrecht
Padualaan 8
3584 CH Utrecht, The Netherlands

Progress in Colloid & Polymer Science   Progr Colloid Polym Sci 83:10—15 (1990)

Optical studies of fluid interfaces

D. Langevin

Laboratoire de Physique Statistique de l'Ecole Normale Supérieure, Paris, France

Abstract: Experimental investigation with optical techniques give very useful information on liquid-gas and liquid-liquid interfaces. We have developed a number of these techniques in our laboratory: — ellipsometry, which gives information about interfacial thickness, anisotropy, and roughness; — fluorescence microscopy, which gives information about phase separation in adsorbed layers; — surface light-scattering, which gives information about interfacial tension and surface viscoelasticity; — excited-wave techniques (electrocapillarity, compression) which give the same dynamic information as surface light scattering, but in a different frequency range. The potentials of these techniques will be briefly reviewed and illustrated with results obtained recently in our laboratory in surfactant systems: insoluble monolayers at the free surface of water, soluble monolayers at the free surface of water and oil-water interfaces.

Key words: Liquid surfaces; surfactant systems; monolayers; optical techniques

1. Introduction

Optical techniques are nonperturbative and thus potentially well suited for the study of liquid surfaces. A variety of such techniques has been developed or improved in recent years, and information about both interfacial structure and dynamics has been gained. Several of these will be briefly reviewed in this paper and illustrated with results obtained recently in our laboratory.

The first one is fluorescence microscopy [1—3], which allows to image the surface and is potentially interesting when large scale inhomogeneities are formed in the surface plane. This happens frequently when monolayers of amphiphilic molecules are adsorbed at the surface, because of the coexistence of different monolayer states that leads to pattern formation. The technique gives information about interfacial structure in the surface plane. In order to obtain information about the structure perpendicular to the surface plane, light reflectivity can be used [4, 5]. The resolution can be considerably improved when working at the Brewster angle and measuring the ellipticity of the reflected light; optical thicknesses below 1 nm can then be measured [6].

The surface dynamics can be investigated by studying the propagation of surface waves. Light-scattering by thermally excited capillary waves allows to determine large and ultralow interfacial tensions as well [7]. It also allows to determine surface dilational viscoelasticity when thin layers are adsorbed at the surface. Capillary [8] and longitudinal [9, 10] waves can also be excited by external sources at the surface. They allow to obtain information about the same parameters, but in different frequency ranges. This is particularly important for a number of surfactant layers that exhibit a highly viscoelastic character and for which elasticity and viscosity are frequency-dependent.

2. Fluorescence microscopy

This technique is restricted to insoluble monolayers. When amphiphilic fluorescent molecules are dissolved in a monolayer, their concentration usually varies in the different monolayer states: the concentration is small in the dilute phases or the fluorescence is quenched due to possible contact of the fluorescent part of the molecule with the water subphase; the concentration is also small in the more concentrated phases (solid, liquid condensed), because the fluorescent molecules behave like an impurity and are expelled out of these phases. As a result, when different phases coexist in the monolayer, the intensity of the fluorescent light emitted by the domains of a given phase differs from the intensity emitted by the domains of the other phases. The domains can then be visualised with an optical microscope if their size is large enough compared to the optical resolution (~1 µm). The technique has been applied to a number of interesting two-dimensional phase equilibria [1—3]. It helped to solve the issue of the order of the phase transition between liquid expanded and liquid condensed monolayer states. For a number of years the transition was claimed to be first order by several authors, second order by others. The existence of well-defined domain boundaries between the two phases proved unambiguously that the transition was a first order one.

The introduction of fluorescent labels in the monolayer can, of course, slightly affect the phase diagrams. This current problem is usually addressed by studying the influence of the label concentration and reducing this concentration to the lowest possible value compatible with good quality images. We have used a different approach in our laboratory. We have studied monolayers of purely fluorescent molecules: nitrobenzoxadiazol stearic acid [11]. We have observed a phase transition between area per molecule A of 84 and 32 Å^2 (Fig. 1).

Monolayer images in the coexistence region reveal the presence of elongated domains (Fig. 2). Observations under polarized light show that the domains are optically anisotropic, and that light absorption is maximum for an electric field perpendicular to the main domain axis. Observations of the mechanical behavior of the domains suggest that they are solid-like; they flow freely in the dark phase, which is, therefore, liquid-like. The fluorescence of the molecules in the latter phase is probably quenched by water, thus leading to a good optical contrast between the two phases. The origin of the observed elongated domain shapes remains to be understood.

3. Light reflectivity

Information about interfacial thickness can be obtained from reflectivity of either light, x-ray or neutrons. The resolution is limited by the wavelength: of the order or 1 µm for light, below 1 nm for x-ray or neutrons. However, due to practical difficulties (collimation, small reflectivities) the last two methods have only been developed recently [12, 13]. The resolution for light beams can go beyond the wavelength in a particular case, at the Brewster incidence. The method is then called ellipsometry, because the information about interfacial thickness is contained in the degree of ellipticity of the polarization of the reflected light. Interfacial optical thicknesses down to below 10 Å can be deduced from the measurements.

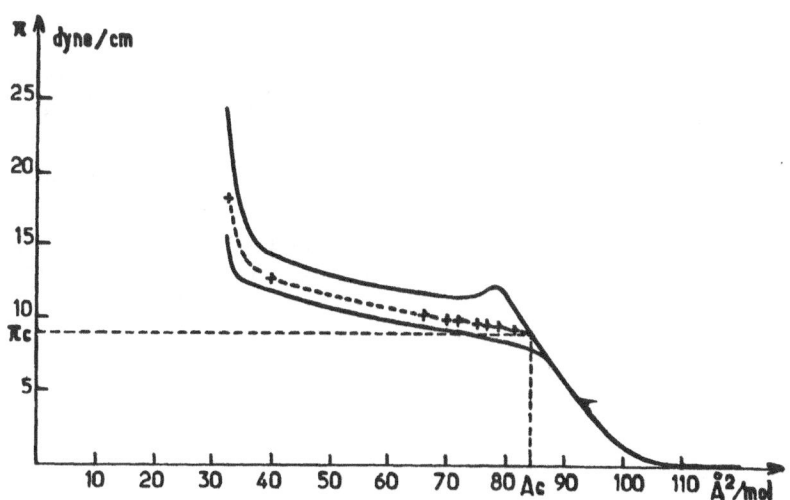

Fig. 1. Surface pressure of NBD-stearic acid vs area per molecule. Upper curve: continuous compression (2 Å^2/mol/min); lower curve: continuous decompression; dashed curve: points obtained after waiting about 30 min at a given area

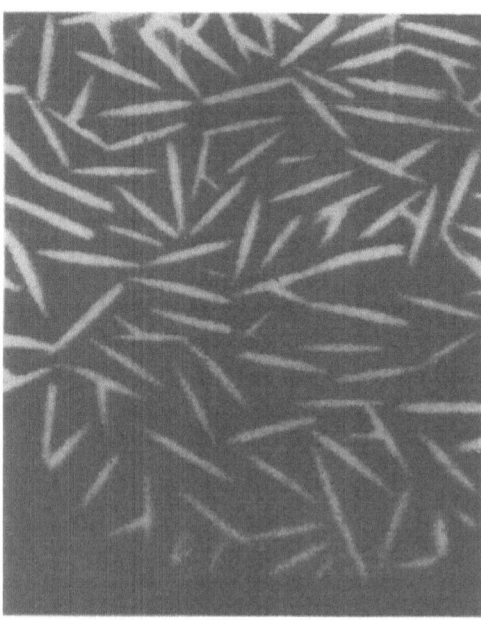

Fig. 2. Fluorescence microscopy images. The size of the picture is 200 × 250 μm

In the method, a light beam having an electric field with components E_p and E_s, respectively parallel and perpendicular to the plane of incidence is sent onto the surface. The reflected electric fields component are, respectively, $E_p^r = r_p e^{i\varphi_p} E_p$ and $E_s^r = r_s e^{i\varphi_s} E_s$. At the Brester angle, $|\varphi_p - \varphi_s| = \pi/2$ and the ellipticity degrees is defined as $\rho = |r_p/r_s|$.

There are two main physical origins for the ellipticity of the reflected light. One is the finite intrinsic thickness of the interface. It can be evaluated by the Drude's formula

$$\rho_p = \frac{\pi}{\lambda} \frac{(\varepsilon_1 + \varepsilon_2)^{1/2}}{\varepsilon_1 - \varepsilon_2} \int \frac{(\varepsilon - \varepsilon_1)(\varepsilon - \varepsilon_2)}{\varepsilon} \, dz \,, \quad (1)$$

where ε is the dielectric constant at optical frequencies (squares of the refractive index), and ε_1 and ε_2 are its values in the coexisting bulk phases.

The second origin is the roughness of the interface due to thermal motion [14]:

$$\rho_r = -\frac{3}{4\lambda} \frac{\varepsilon_1 - \varepsilon_2}{(\varepsilon_1 + \varepsilon_2)^{1/2}} \frac{kT}{\gamma} q_{max} \,, \quad (2)$$

where γ is the interfacial tension and q_{max} the cut off wave vector for capillary waves.

a) Monolayers at the air-water interface

When surfactant molecules are adsorbed at the air-water interface, the surface tension is decreased, but remains large enough so that the roughness contribution to the ellipticity is negligible.

If one assumes that the monolayer is homogeneous, and if d is the thickness at maximum coverage, Eq. (1) becomes

$$\rho_p = \frac{\pi}{\lambda} \frac{(\varepsilon_1 + \varepsilon_2)^{1/2}}{\varepsilon_1 - \varepsilon_2} \frac{(\varepsilon - \varepsilon_1)(\varepsilon - \varepsilon_2)}{\varepsilon} d \,. \quad (3)$$

If the monolayer coverage is intermediate, the measured ellipticity will be given by the above formula in which d will represent an apparent thickness proportional to the monolayer density. In other words, because ellipsometry only gives an integrated optical thickness, it does not allow to distinguish between a dilute thick layer and a compact thin one.

When the monolayer is inhomogeneous, and when the size of the inhomogeneities is not too small compared with the beam size at the surface, the ellipticity fluctuates around a mean value still given by Eq. (3). The fluctuations are produced by the motion of the domains across the boundaries of the illuminated region.

These features are illustrated in Fig. 3 for the monolayers of the fluorescent amphiphile described in § II. At small surface concentrations, one observes ellipticity fluctuations probably corresponding to the liquid-gas coexistence region. This coexistence was not evidenced in the surface pressure measurements (not accurate enough at these low surface pressures) and in the microscopy observations (absence of optical contrast between the domains).

Above $n_s = 0.9 \ 10^{-2}$ mol/Å^2 ($A \sim 110$ Å^2), the ellipticity no longer fluctuates and varies linearly with n_s up to $n_s = 1.19 \ 10^{-2}$ mol/Å^2; the fluctuations appear again in the liquid-solid coexistence region. The ellipticity variation with concentration is still linear, but the slope is larger. We have interpreted this feature by taking into account the optical anisotropy in the monolayer. For an optically uniaxial layer, the optical axis being in the surface plane, the ordinary and extraordinary refractive index being n_0 and n_e, respectively, the corresponding contribution to the ellipticity is [6]

$$\rho_A = \frac{1}{2} (n_e^2 - n_0^2) d \,. \quad (4)$$

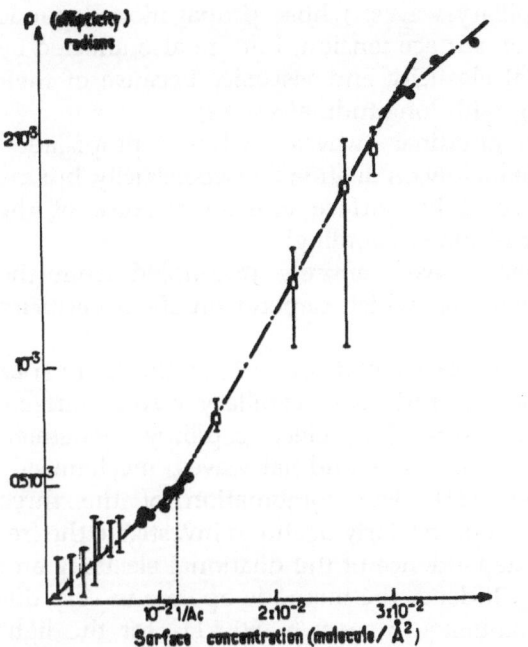

Fig. 3. Measured ellipticity vs surface concentration. The open squares are mean values

The values of n_0 and n_e obtained for this monolayer are 1.56 and 1.72, respectively.

Above $n_s = 3.1 \ 10^{-2}$ mol/Å^2 ($A \sim 32$ Å^2) one enters the solid region. The fluctuations disappear and the slope drops to its initial value (because the optical anisotropy remains constant).

b) Monolayers at the oil-water interface

When surfactants are adsorbed at the oil-water interface, they can decrease the interfacial tension γ to ultralow values in certain cases. This is generally accompanied by the formation of thermodynamically stable oil-water dispersions, i.e. microemulsions.

When γ is very small, the interfacial roughness is very large, and the ellipticity is dominated by the roughness contribution. One has, indeed, for the mean square amplitude of the surface roughness ζ:

$$\langle \zeta^2 \rangle = \int_0^\infty \frac{d^2q}{(2\pi)^2} \frac{kT}{\Delta\rho g + \gamma q^2 + \kappa q^4} , \qquad (5)$$

where q is the wavevector of the surface deformation modes, kT is the thermal energy, $\Delta\rho$ the density difference between the coexisting phases, g the gravity constant, and κ the bending elastic modulus

of the surfactant layer. The existence of κ allows to calculate the cut-off wave vector as

$$q_{max} = \sqrt{\gamma/\kappa} . \qquad (6)$$

When γ is measured independently, the determination of the ellipticity allows to measure the bending constant κ. This is a very interesting parameter, because it determines the maximum size in the microemulsion (i.e., the oil-water solubilization power) and the minimum oil-water interfacial tension [15].

We recently started a study of ternary water-alkane-non ionic surfactant mixtures [16]. We used very small surfactant concentrations in order to have equilibria between almost pure oil and pure water, the interface being nevertheless covered by a surfactant layer. Indeed, when a concentrated microemulsion is in equilibrium with an excess oil or water phase, the ellipticity of the reflected light contains contributions from the microemulsion microstructural elements close to the interface [17]. When three phases (oil, water and microemulsion) are formed, we remove most of the microemulsion

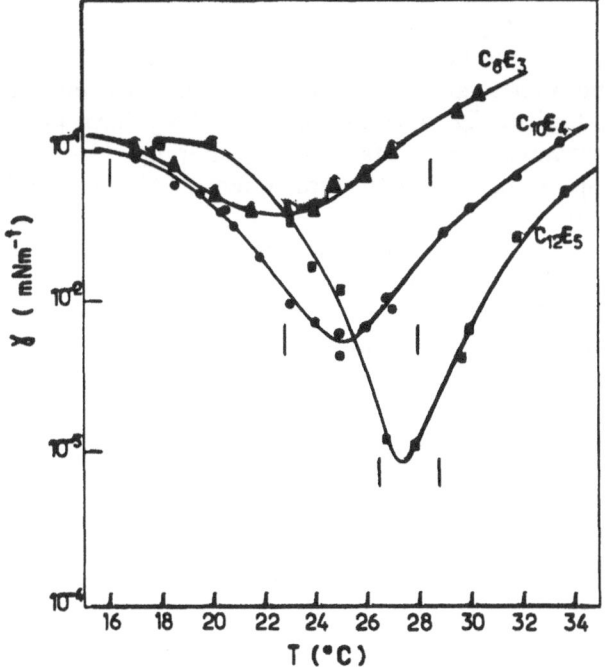

Fig. 4. Interfacial tension vs temperature for alkane-water-polyethylene glycol alkyl ethers systems: C_8E_3-decane, $C_{10}E_4$-octane, $C_{12}E_5$-hexane

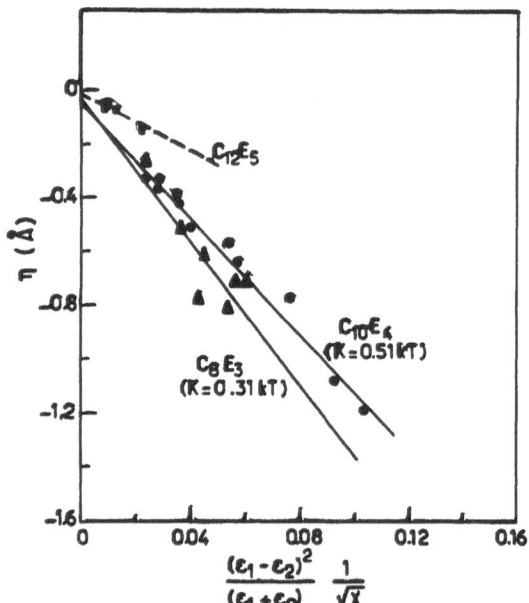

Fig. 5. Reduced ellipticity $\eta = \rho \dfrac{\lambda}{\pi} \dfrac{\varepsilon_1 - \varepsilon_2}{(\varepsilon_1 + \varepsilon_2)^{1/2}}$ vs reduced interfacial tension for the same systems as in Fig. 4. The value of κ is given in parenthesis (for $C_{12}E_5$ the measurements are not yet complete)

middle phase, just leaving a microemulsion drop in the cell boundary to ensure thermal equilibrium. The interfacial tensions of the studied systems are plotted in Fig. 4 vs temperature. They have been determined by surface light-scattering. The corresponding ellipticities are shown in Fig. 5 scaled with $1/\gamma^{1/2}$. The slopes, which are proportional to $\kappa^{-1/2}$, decrease with surfactant chain length. This shows that the bending elastic constant increases with chain length as predicted by recent simulations [18]. The maximum microemulsion size ξ can be calculated with these data and agree well with experiments. Since ξ increase exponentially with κ, and as surface tension scales like ξ^{-2}, there is a large difference between the minimum interfacial tensions (Fig. 4).

4. Surface waves methods

Interfacial dynamics can be probed by studying the propagation of surface waves. One can distinguish between three different kinds of waves at a liquid interface:

— capillary waves, whose propagation depends mainly on surface tension, but are also affected by dilational elasticity and viscosity because of their coupling with longitudinal waves;

— longitudinal waves, whose propagation depends mainly on dilational viscoelasticity, but are also affected by surface tension because of the above mentioned coupling;

— shear waves, entirely decoupled from the above two, and which depend on shear elasticity and viscosity.

We have developed optical methods in our laboratory to study both capillary waves: surface-light scattering [7], electrocapillary generated waves [20], and longitudinal waves, mechanically generated [21]. The combination of the three methods is particularly useful to investigate the frequency dependence of the dilational elasticity and viscosity. Indeed, the methods operate in very different frequency ranges: 5—50 kHz for the light scattering, 100—1000 Hz for the electrocapillary waves, and 0.1—1 Hz for the longitudinal waves.

The frequency dependence of the dilational viscoelasticity of the surface of surfactant aqueous solutions has been investigated by the first two methods [20, 21]. Figure 6 shows some of the data. The frequency dependence of the elastic modulus is particularly large close to the critical micellar con-

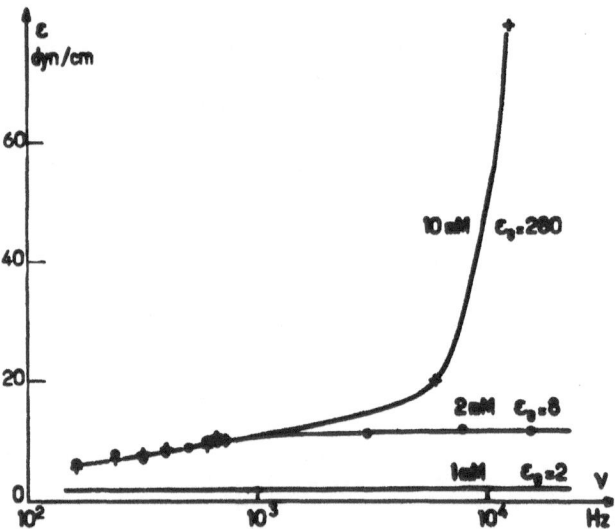

Fig. 6. Dilational elasticity of monolayers of dodecyl trimethyl ammonium bromide on water vs frequency for different surfactant bulk concentrations. The critical micellar concentration is 15 mM

centration, i.e., close to monolayer full coverage. The main part of this frequency dependence can be accounted for by a relaxation process associated to a desorption-adsorption process.

In order to investigate intralayer relaxation processes, we have recently started a study of myristic acid monolayers. Earlier measurements have shown that the high-frequency dilational modulus was up to 50% larger than the zero frequency modulus, as deduced from the surface pressure variation with surface concentration [22]. Preliminary measurements show that the moduli measured at low frequencies were close to the static values. The frequency variation of the surface dilational viscosity appears very large, even at low surface coverage: for $A = 60$ Å^2, η_s varies between 1 mN $\cdot$ s $\cdot$ m^{-1} at 1 Hz and 10^{-4} mN $\cdot$ s $\cdot$ m^{-1} at 10^4 Hz. Further work will hopefully clarify the origin of the relaxation processes involved.

5. Conclusion

A number of new techniques have recently been developed for the study of liquid interfaces that have allowed recent progress in the field. Among the techniques, optical methods are particularly interesting. In this paper, we have briefly reviewed the potentials of some of them for the study of surfactant monolayers. New information on these systems is clearly needed in view of the large number of potential applications, like Langmuir Blodgett films, and biological membranes. Very little is known about monolayer dynamics who plays an important role in numerous processes like coating, emulsion and foam stability, liquid-liquid extraction and others.

References

1. Mc Connell HM, Tann LK, Weiss RM (1984) Proc Natl Acad Sci — USA 81:3249; Weiss RM, Mc Connell HM (1984) Nature 310:47
2. Losche M, Sackmann E, Möhwald H (1983) Ber Bunsenges Phys Chem 87:848
3. Moore B, Knobler CM, Broseta D, Rondelez F (1986) J Chem Soc, Far Trans 2 82:1753
4. Huang JS, Webb WW (1969) J Chem Phys 50:3677
5. Meunier J, Langevin D (1982) J Phys Lett 43:L191
6. Azzam RMA, Bashara NM (1977) Ellipsometry and polarised light, Noster Holland
7. Langevin D, Meunier J, Chatenay D (1982) In: Mittal KL, Lindman B (eds) Surfactants in Solutions, Vol 3; p 1991, Plenum Press
8. Mann JA (1984) In: Matijevic E, Good RJ (eds) Surface and Colloid Science, Vol 13, Plenum
9. Lucassen J (1968) Trans Far Soc 64:2230
10. Miyano K, Abraham BM, Ting Li, Wasan DT (1983) J Coll Int Sci 92:297
11. Bercegol H, Gallet F, Langevin D, Meunier J (1989) J Phys France 50:2277
12. Peshan PS, Als-Nielsen J (1984) Phys Rev Lett 52:759
13. Bradley JE, Lee EM, Thomas RK, Willatt AJ, Penfold J, Ward RC, Gregory DP, Waschkowski W (1988) Langmuir 4:821
14. Zielinska BJA, Bedeaux D, Vlieger J (1981) Physica 107A:91
15. Binks BP, Meunier J, Abillon O, Langevin D (1989) Langmuir 5:415
16. Lee LT, Langevin D, Meunier J, Wong K, Cabane B (1990) Progr Colloid Polym Sci 81:209
17. Meunier J (1985) J Phys Lett 46:L-1055
18. Szleifer I, Kramer D, Ben-Shaul A, Roux D, Gelbart WM (1988) Phys Rev Lett 60:1966
19. Lucassen-Reynders EH, Lucassen J (1969) Adv Coll Int Sci 2:347
20. Stenvot C, Langevin D (1988) Langmuir 4:1179
21. Lemaire C, to be submitted for publication
22. Thominet V, Stenvot C, Langevin D (1988) J Coll Int Sci 126:54
23. Langevin D (1981) J Coll Int Sci 60:529

Author's address:

D. Langevin
Laboratoire de Physique Statistique
de l'Ecole Normal Supérieure
24 Rue Lhomond
75231 Paris Cedex 05, France

Progress in Colloid & Polymer Science   Progr Colloid Polym Sci 83:16—28 (1990)

Correlation between surface and interfacial tensions with micellar structures and properties of surfactant solutions

H. Hoffmann

Lehrstuhl für Physikalische Chemie I, Universität Bayreuth, Bayreuth, FRG

Abstract: It is shown that the surface activity of surfactants is expressed both in its critical micelle concentration (CMC), in the value of the surface tension of the surfactant solutions at the CMC (σ_s), in the value of the interfacial tension of a micellar solution against a hydrocarbon (σ_i), and in the area "a" which a surfactant molecule occupies at an interface. This area seems to be very similar for a micellar interface, a surface film, and an interfacial film. For zwitterionic single-chain surfactants the three parameters CMC, σ_s, and σ_i are all correlated with the absolute value of "a" and reach a minimum when "a" has such a value that is favorable for the formation of a planar interface, like in a lamellar phase. The interfacial tension σ_i is then very low, typically in the range of 10^{-2}—10^{-3} mN m^{-1}. The interfacial tension is thus a good measure for the curvature of an interface. It can be used to predict the type of micelles that are formed in surfactant solutions. Globular micelles are formed when $\sigma_i > 3$ mN m^{-1} and rodlike micelles are present when $3 > \sigma_i > 0.1$ mN m^{-1}. The rodlike micelles reach their highest stability when $\sigma_1 \simeq 0.3$ mN m^{-1}. — Micellar solutions with rodlike micelles in the semidilute entangled state reach their maximum viscosity under these conditions. Disklike micelles are formed when $\sigma_i < 0.1$ mN m^{-1}. The sequence of l.c. phases that are formed for higher concentrations can be predicted also from σ_i. — Finally, it is shown that other properties like the solubilization capacity and the bulk shear modulus of surfactant systems are also predetermined by σ_i.

Key words: Interfacial and surface tension; micellar structures and bulk properties; phase diagrams

Introduction

To a large extent during the last decade research in surfactant science has been focused on microemulsions [1]. The emphasis in many of these investigations was on phase diagrams, in particular on finding the conditions under which the surfactants can solubilize a maximum amount of oil in the aqueous phase [2]. This point is of great interest for many practical applications of surfactants, for example, in enhanced oil recovery [3]. This optimum condition can be obtained when parameters like the salinity, the cosurfactant concentration, the temperature, and the structure of the surfactant are varied or when the mixing ratio of surfactants is

changed [4]. This most favorable condition for the solubilization in surfactant solutions is reached when the interfacial tension σ_i of a micellar solution is adjusted against a hydrocarbon to its minimum value [5]. This value is usually in the range of 10^{-3} mN/m or even lower and it controls the radius r for microemulsion droplets that are formed in dilute solutions. The droplets have an interfacial energy of kT resulting in the simple equation $4\pi r^2 \cdot \sigma_i = kT$ [6].

For higher surfactant concentrations, a middle-phase microemulsion is formed that contains most of the surfactant. It is now generally agreed upon that the middle phase has a bicontinuous structure in which the water and the oil form interpenetrating networks, with the surfactant molecules being located at the interface [7]. On a local scale the shape of the structures will always fluctuate [8] because the network can easily be deformed by the thermal energy. There will be shape fluctuations in which the total area of the structure will remain constant and there will be fluctuations in shape in which the total area of a structure will vary. They can occur if the packing of the interfacial film is changing simultaneously. An increase or a decrease of the total surface area will be accompanied by a stretching or compression of the film. Fluids that consist of such deformable structures will have a completely different flow behavior than will a suspension of hard particles.

Therefore, we can expect that the macroscopic properties of micellar phases will change dramatically when the interfacial tension is minimized. Interfacial tensions of dilute surfactant solutions under nonoptimized conditions are up to 3 to 4 orders of magnitude higher. The absolute value of the interfacial tension is a very crucial parameter for the resulting macroscopic properties of binary and ternary surfactant systems. It will be shown in this paper that the value of the interfacial tension of the surfactant solution/hydrocarbon-interface can be used directly to predict what kind of micelles will exist in the aqueous solution, how much hydrocarbon can be solubilized per surfactant into the micelles, and what type of liquid crystalline phases will form at higher surfactant concentrations. This value can even be used to predict some bulk properties of these phases.

Surfactants can be of single-chain and double-chain type; the chains can be branched or unbranched. All these conditions will influence the packing of the surfactant at the interface and, therefore, have an effect on the interfacial tension. In order to avoid this complication, we used only single-chain surfactants with as similar headgroups as possible for the comparison. This paper begins with results of simple surface and interfacial measurements.

Results and discussion

Surface tension measurements

Surfactant molecules accumulate at the surface or at the interphase of an aqueous bulk phase against another bulk phase. By doing so, they modify the interfacial properties between the two phases, in particular the interfacial tension between the two phases. The surface activity of the surfactant as expressed, for instance, in mole of adsorbed surfactant per area depends on many parameters like chainlength, polarity of the head group, type of counter ions in case of ionic surfactants and so on. Generally, one finds Langmuir-type adsorption isotherms: the adsorption increases linearly with the concentration in the bulk at small concentration and reaches a saturation value at higher concentrations. At saturation the surfactant molecules form an interfacial film between two phases. We will be mainly concerned with the properties of this film and its effect on the interfacial tension [9].

When the surface activity of surfactants that all have the same hydrocarbon chain, but different polarities are compared, one finds that the surface activity increases with decreasing polarity. The difference in surface activity is reflected in three parameters, as is obvious in Fig. 1 where the results of surface tension measurements of surfactants with the same hydrocarbon chain are shown. Three zwitterionic systems of varying polarity were chosen for the comparison: the C_{14}dimethylaminoxide, the C_{14}dimethylphosphinoxide, and the C_{14}-dimethylammoniumpropanesulfonate. Results for a cationic surfactant were also included, but will not be discussed here. We note that the surfactant with the lowest polarity has the lowest CMC and also the lowest surface tension. On closer inspection of the data, one notes that even the slopes of the surface tension — ln$c$ curves at the CMC are somewhat different. For the zwitterionic-surfactants the system with the lowest CMC has the largest slope. In the Gibbs-adsorption isotherm the slope of the curve is a measure of the area per headgroup which the surfactant molecules requires at the interface. The steeper the slope, the smaller the area will be. Thus the different surface activities of the surfactants are reflected in their CMC, their surface tension at the CMC and in the slope of the curves at the CMC. The presented data are consistent with available theories on the adsorption process. The quantity which controls the surface active behavior is the area "a" that a surfactant molecule occupies at the interface. The smaller this value with respect to the cross-section of the hydrocarbon chain, the better the film can pack at the interface and the more water can be eliminated from the surface. Under optimized conditions the surface will have properties similar to a hydrocarbon surface. Many results from literature confirm this observed behavior.

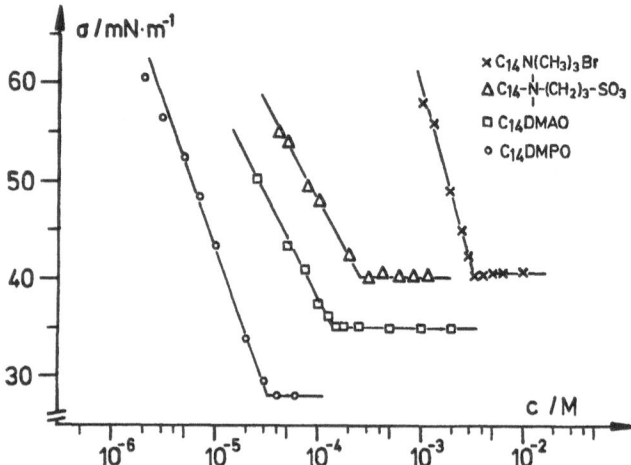

Fig. 1. The surface tension as a function of the concentration of different surfactants with the same chainlength at 25 °C

Surface tension measurements on alkylpolyglycolethers showed, when the alkylchain was kept constant and the degree of ethoxylation was varied, that, again as for the zwitterionic surfactants, all three parameters were modified [10]. The CMC's and the surface tensions at the CMC are increasing while the adsorbed molecules per interfacial area are decreasing with the number of EO-groups. This clearly shows that the hydrocarbon film from the surfactant molecules at the surface becomes less dense.

Results by Motomura et al. [11] on methyl substituted dodecylammonium chlorides show the same tendency. The surface activity of these compounds is decreasing with increasing methyl substitution, and this again is reflected in all three parameters. In this context it is also worthwhile to mention results for ionic surfactants with increasing excess salt concentration. Again, one notes a change of all three parameters [12]. A better shielding of the electric charge results in smaller "a" values and hence in denser films. These more qualitative remarks are the content of the importance of the packing parameter in the concepts that were originally proposed by Tanford and later refined by Israelachvili and Ninham [13, 14]. They showed, in particular, that the area "a" controls the packing parameter and this quantity controls the structure of the micelles.

Interfacial tension measurements

When the same surfactants are used in interfacial tension measurements we notice that the difference

for the different surface activities also shows up in these measurements. A comparison of surface and interfacial tension measurements is shown in Fig. 2 for tetradecyldimethylaminoxide. We note that the CMC of the surfactant is only slightly affected by the presence of decane. It is somewhat lower, but the difference is quite small with respect to the large changes that are possible for a surfactant when bulk properties are varied. We also note that the slopes at the CMC of the two systems are, within experimental accuracy, about the same. This then indicates that the packing of the surfactant film in both situations is probably very much the same.

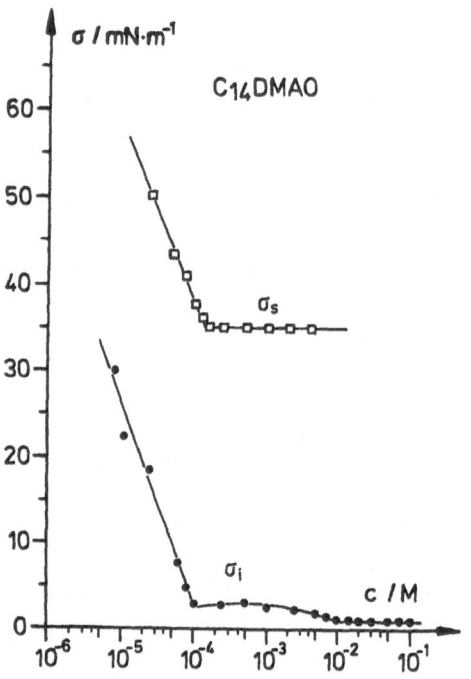

Fig. 2. Comparison of the surface tension σ_s and interfacial tension σ_1 of $C_{14}DMAO$ against decane as a function of the concentration

For C_{14}-aminoxide, we find an interfacial area of about 50 Å² from both the surface and the interfacial measurements. This is much more than the cross-section of a hydrocarbon chain in the liquid crystalline lamellar state, which is around 35 Å². We could, of course, imagine that the surfactant molecules at the interface are all in the same upright position and the space in between the surfactant molecules is taken up by hydrocarbon molecules. If this would be the case the interaction energy of the hydrocarbon chains of the surfactants would be completely different from the situation at the air/water interface. Since, however, the number of adsorbed molecules per area is the same in both cases, it is very likely that the interaction energy is the same also and, hence, the surfactant film is also the same. It could be argued that the behavior that is shown by the zwitterionic systems is accidental and the results of surface and interfacial behavior for other surfactants do not agree so well. This, however, seems not the case. Rosen has summarized surface and interfacial data and found rather good agreements for many surfactants [15]. The CMC-values by interfacial tension measurements are usually somewhat smaller than the values from the surface tension measurements, but the changes are small. We can conclude, therefore, that the presence of the hydrocarbon does not seem to affect the packing in the surfactant film; the hydrocarbon does not appear to penetrate the film. It should be mentioned, however, that Aveyard et al. concluded from their results that penetration of hydrocarbon does occur to some degree, if the films have a very low interfacial tension [16].

Since the discussed interfacial film is in a condensed state and the density of the film is likely to be the same as the density of a bulk hydrocarbon, it follows then that the thickness of the film should increase as the polarity of the surfactant is lowered. This is schematically shown in Fig. 3. This schematic drawing also suggests that the order parameter of the CH_2-groups in the chain should increase with increasing packing. This behavior has

Table 1. Values of the CMC/M, the headgroup area $a/Å^2$ of a surfactant molecule at the surface film, the interfacial tension σ_i of the surfactant solution against decane at the CMC, and the surface tension at the CMC

System	CMC/M	$a/Å^2$	σ_i/mN · m^{-1}	σ_s/mN · m^{-1}
C_{12} DMAO	$1.7 \cdot 10^{-3}$	74	4	32.6
C_{14} DMAO	$1.4 \cdot 10^{-4}$	51	1.1	30.9
C_{16} DMAO	$2.5 \cdot 10^{-/}$			30.9
C_{12} DMPO	$2.8 \cdot 10^{-4}$	42	0.42	28.1
C_{14} DMPO	$3.1 \cdot 10^{-5}$	38	0.11	27.2
C_{12} DMAPS	$3 \cdot 10^{--}$			38
C_{14} DMAPS	$2.5 \cdot 10^{-§}$	60	6	36

not yet been experimentally verified to our knowledge.

The evaluated parameters from interfacial tension measurements are summarized in Table 1. The values clearly show that the interfacial tension depends very strongly on the area parameter "a". A relative small change of this parameter can vary the value of the interfacial tension by orders of magnitude. It is interesting to note that the "a"-values for the zwitterionic systems vary by a factor of 2, while the interfacial tension vary about two orders of magnitude.

Interfacial tensions and micellar structures in dilute solutions

Micellar structures can be visualized as aggregates with hydrocarbon/water interfaces. The micellar interface will be curved, however, and for this reason the packing of the surfactant chains inside the micelle will be somewhat different than the packing for the surfactant molecules at a real hydrocarbon/water interface. This means that the configuration of the surfactant chain and, therefore, the number of kinks in the chain is different for a surfactant in the unlike environments even though the chemical potential will be the same [17]. The total energy for kinks per chain is, however, usually small in comparison to the total hydrophobic energy. We can expect, therefore, that the area that is occupied by a surfactant molecule in the micellar interface is very similar to that occupied area in a real interface. For some systems the area values for the micellar interfaces and hydrocarbon interfaces

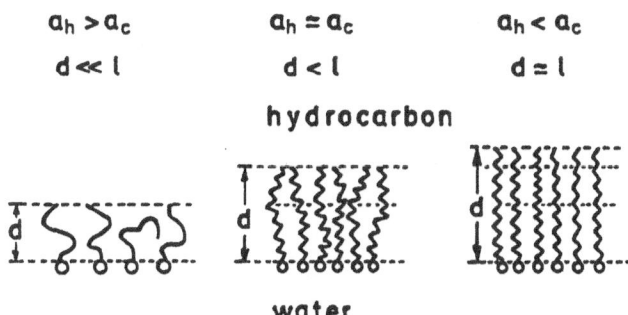

Fig. 3. Schematic drawing of surfactant films when the surfactants have the same chainlength, but different polarities. The area "a" that is occupied by the surfactant is decreasing with decreasing polarity

are available and these results show that within a few percent these areas are indeed the same [18].

The "a" value from surface or interfacial tension measurements can therefore be used to predict the type of micelles that exist in solutions, because their area determines the mean curvature of the micellar interface.

The curvature of a micellar interface is determined by the ratio a/a^0, where a^0 is the value for a surfactant molecule of an ideally packed lamellar layer. When a/a^0 will be larger than 1, the curvature is convex, and if it is smaller than 1 it will be concave and the surfactant can be used for W/O systems. The type of micelles that are formed in surfactant solutions can thus be estimated from the a/a^0 values. Since the "a" values correlate with the σ-values at the CMC, we can thus predict the type of micelles from interfacial tension measurements.

Interfacial tensions of around 2 mN/m or larger correspond to "a" values that are more than twice as large as a cross-section of a hydrocarbon. For this reason the curvature around the micelle must be highly curved and convex and results in spherical micelles. With decreasing interfacial tension and, hence, with lower "a" values rodlike micelles finally become possible in the solution. This is shown schematically in Fig. 4. According to the summarized results, rodlike micelles can be formed from surfactant solutions that have interfacial tensions against a hydrocarbon in the range between 1 and 0.1 mN/m.

When hydrocarbon is solubilized into a solution of rodlike micelles, the micelles undergo a rod-sphere transformation at a characteristic hydrocarbon/surfactant ratio [19]. This transformation is reached when the average curvature of the resulting microemulsion droplets is as large as the average curvature of the rodlike micelles. This characteristic hydrocarbon ratio depends again on the interfacial tension, but less strongly than does the saturation concentration of the hydrocarbon. For systems having interfacial tensions of more than 0.1 mN · m^{-1} the transformation is reached before saturation, while for $\sigma < 0.1$ mN · m^{-1} saturation is reached before transformation. The system under the last condition will, therefore, always be bicontinuous in the sense that the rodlike micelles with the solubilized hydrocarbon will form a continuous network in the solution. Systems that are at the minimum of the interfacial tension can form extremely stable lamellar phases with very large interlamellar spacings. The iridescent surfactant solutions are a result of this state [20].

Synergistic effects in mixed surfactants

We can expect to find synergistic effects when different surfactants are mixed. Two surfactants can have large area requirements for different reasons, namely due to steric effects or due to electric charge. In combination they can pack to a denser film more

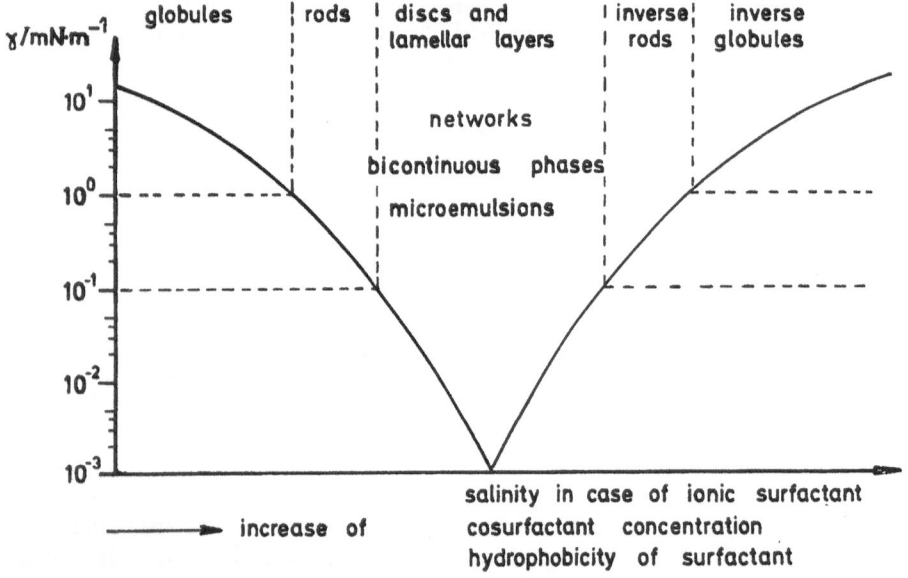

Fig. 4. Schematic drawing of the various micellar structures that exist in the surfactant solution against the interfacial tension that is measured against a hydrocarbon

so than they can individually. Interfacial results for mixtures are shown in Fig. 5. It is clear that the zwitterionic systems form ideal mixtures, while the combination aminoxide/SDS is highly synergistic. This is due to the cationic character of the aminoxide.

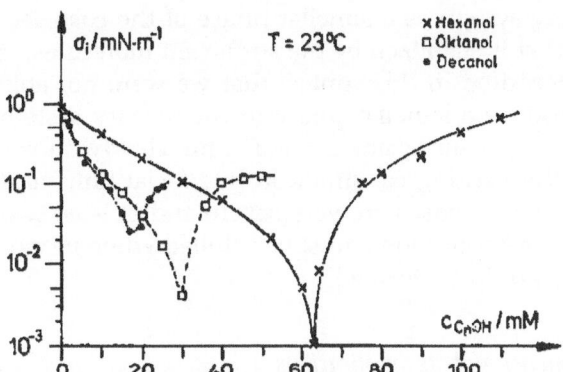

Fig. 6. The interfacial tension of a 30 mM C_{14} DMAO solution against the concentration of *n*-alcohols with different chain length. Note that the position of the minimum is shifted to smaller concentrations with increasing chain length of the cosurfactant while the minimum of the interfacial tension is increasing in value

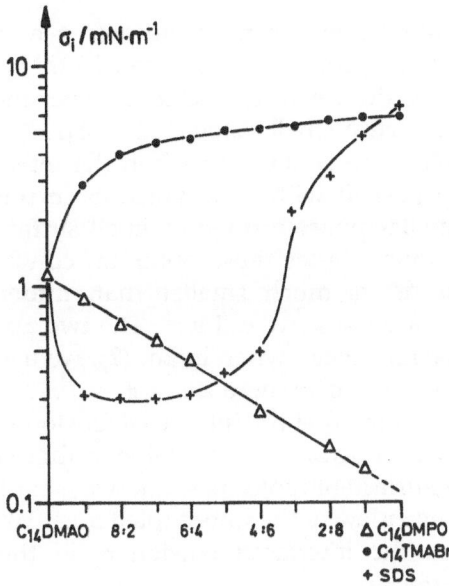

Fig. 5. Interfacial tensions σ_1 of mixtures of 10 mM aqueous surfactant solutions against decane at 25°C

An often used method to lower the interfacial tension of a surfactant solution is to add cosurfactant to the system. As cosurfactants, short or intermediate chain length alcohols are usually used. Results of such data are shown in Fig. 6. The experiments show that even undecanol can still produce a minimum in the interfacial tension. However, the absolute value is not as low as the one for short chain alcohols. It was emphasized before that the minimum corresponds to a planar film of tightly packed surfactant molecules. The different values at the minimum might be related to the different bending constants of the resulting films [18]. Low chain cosurfactant probably results in more flexible films. The results of Fig. 6 also make it clear that the more cosurfactant required to produce the minimum, the shorter is the chain length of the cosurfactant. Some of this effect is due to the increased solubility of the cosurfactant in the bulk water, but this is not the only reason. Some of the

effect must be due to the packing of the hydrocarbon chains in the film.

In the presence of the longer chain cosurfactants only about one cosurfactant molecule per three surfactant molecules is required to form the minimum in the interfacial tension. At approximately this cosurfactant/surfactant ratio lamellar phases can be formed in the ternary system when no hydrocarbon is present. Qualitatively this is sensible because the headgroup area for an alcohol on an interface is only about 25 Å² and the minimum of the interfacial tension or the formation of the lamellar phase occurs when the average headgroup area is around 35 Å². The system cannot form a lamellar phase in the ternary system for small cosurfactants. For a combination of cosurfactant/surfactant with unequal chain length the ideal packing at the headgroup would still lead to strong curvature of the interface. However, when some excess hydrocarbon is present in the system, then the cosurfactant/surfactant ratio can be very similar for short-chain cosurfactants as for long-chain cosurfactants. A combination of a short-chain hydrocarbon in combination with a hydrocarbon molecule can act like a long-chain cosurfactant. Such situations were recognized by Shah on other multicomponent systems [13]. It is, however, also possible to form a lamellar phase in the ternary system hexanol/C_{14}-DMAO/H_2O but the composition occurs at a very much higher cosurfactant/surfactant ratio than in the presence of hydrocarbon. It is therefore ap-

ternary system as a lamellar phase of the cosurfactant that is stabilized by the surfactant molecules. It is interesting in this context that we were not able to produce a lamellar phase in the ternary system with the cosurfactant octanol. This clearly shows that the packing constraint in interfacial films and in lamellar phases are very severe and at least two boundary conditions must be fulfilled when monolayers are to be formed.

The model and its implications

Experimentally, we observe that the interfacial tension values correlate with the "a" values. A theoretical tratment of this problem is very complex and has to take into account the interaction between the chains of the hydrocarbon and their conformation [17]. Qualitatively, the data can easily be explained. The better the coverage of the aqueous solution by the surfactant film, the more the surface will resemble a hydrocarbon surface. The most favorable situation is reached at the minimum of the interfacial tension.

A qualitative model for the data can be based on very few equations that are generally agreed upon. The surface tension and interfacial tension results will be discussed on the basis of the Gibbs adsorption isotherm

$$\Gamma = -\frac{1}{RT} \frac{d\sigma}{d\ln c} . \tag{1}$$

This equation gives the mol of surfactant molecules per area and, hence, the area "a" which a surfactant molecule occupies.

The energy that is gained when a molecule is brought from the bulk to the interface or the micelle is usually divided into different contributions, namely the bulk term g that is proportional to the number of CH_2 groups, a polar term which is governed by the interfacial tension and the area "a", and an electrostatic term in the case of ionic surfactants. Sometimes in more refined theories additional terms for the packing and the number of kinks are included. We will mainly use results for zwitterionic terms and will need, therefore, only the first two terms:

$$\Delta\mu = g + \sigma \cdot a . \tag{2}$$

Usually, in theoretical treatments for micelle formation, the interfacial tension σ of a hydrocarbon/water interface is used together with the total area "a" in Eq. (2).

Previously, we have proposed to directly use the value σ which can be measured experimentally at a planar interface at the CMC of the surfactant [18]. This value might have to be modified in order to make corrections, due to the curvature of the interface in order to obtain a more quantitative description.

The interfacial tension σ_i can, in the presence of surface active compounds, vary between 50 mN $\cdot$ m^{-1}, the value of the clean hydrocarbon water interface σ^0, and a very small value that is close to zero. This small value is obtained when the interface is densely packed with surfactants, as in the case of the lamellar phase formed at small surfactant concentrations. Under these optimum conditions the area "a" is much smaller than under nonoptimized conditions; we call it a^0. The two approaches can be reconciled when in Eq. (2), we use $\Delta a \cdot \sigma^0$ instead of $a \cdot \sigma^0$, where $\Delta a = a - a^0$.

We can then assume that the interfacial tension is determined by the fraction of the total area that is not covered by surfactant molecules, and we can fit the data by a power law of the unoccupied area. We then obtain for the interfacial tension σ_i in the presence of surfactant

$$\sigma_i = \left(\frac{a - a^0}{a}\right) \cdot \sigma^0 . \tag{3}$$

This simple equation fulfills our two boundary conditions, namely $\sigma \to 0$ for $a \to a^0$ and $\sigma \to \sigma^0$ for $a \gg a^0$.

Similarly, the surface tension of a surfactant solution σ_s can be written as

$$\sigma_s = \sigma_{HC} \left(\frac{a^0}{a}\right)^x + \left(\frac{a - a^0}{a}\right)^x \cdot \sigma H_2O , \tag{4}$$

where σ_{HC} is the surface tension of a hydrocarbon and σ_{H20} is the surface tension of water.

For $a \gg a_0$ the surface tension will be σH_2O and for complete coverage σ will be σ_{HC}.

Interfacial tensions and macroscopic properties

The system C_{14} DMAO is a surfactant system that, depending on the concentration and temperature, can both form globular and rodlike micelles in the aqueous solutions. Its interfacial tension is around 1 mN/m. The analogue phosphinoxide at the same concentration and temperature has an interfacial tension of 0.1. Solutions of this system at room temperature separate into concentrated and dilute phase [21]. Mixtures of the aminoxide and phosphinoxide form stable viscoelastic solutions in the whole mixing range. The interfacial tension varies linearly with the mixing ratio. It is of interest to investigate the macroscopic properties of these solutions as a function of the mixing ratio. Results of rheological measurements are shown in Fig. 7 where the longest structural relaxation time is plotted against the mixing ratio for a 50 mM solution. While the modulus G^0 of these solutions remains more or less constant, both the zero-shear viscosities and the structural relaxation times pass over a miximum. The maximum occurs at an interfacial tension of around 0.3 mN/m. For these conditions the system froms the most stable rodlike micelles. It is conceivable that at this state the rodlike micelles posses their highest intrinsic rigidity. From a theoretical consideration this point is probably reached when the average headgroup area per surfactant has reached the value that allows a radius for the rods and which corresponds to the completely all-trans conformation of a hydrocarbon chain. If the headgroup area is larger than this value, the rods will be thinner and hence more flexible. If the headgroup area is smaller than the optimum value, fluctuations of the cross-section of the rods will become possible, which again probably increases the flexibility of the rods and shortens the structural relaxation time. Higher flexibility would also conceivably lead to larger concentration fluctuations and, finally, to phase separations. It is likely, however, that even in the coacervate state in which two macroscopically separated phases coexist, rodlike micelles will be present in both solutions.

The dilute solution will contain rods that, in the thermodynamic sense, are in vapor state, while in the more dense state the rods are in the condensed state. The observed behavior for the aminoxide/phosphinoxide mixtures seems to be a very general one. Whenever one thermodynamic variable in a surfactant solution is changed in such a way as to decrease the average headgroup area the system will be transformed from globular to rodlike micelles. The rodlike micelles will reach their optimum size and, finally, the conditions for disk-like or lamellar phases will be met. In the absence of

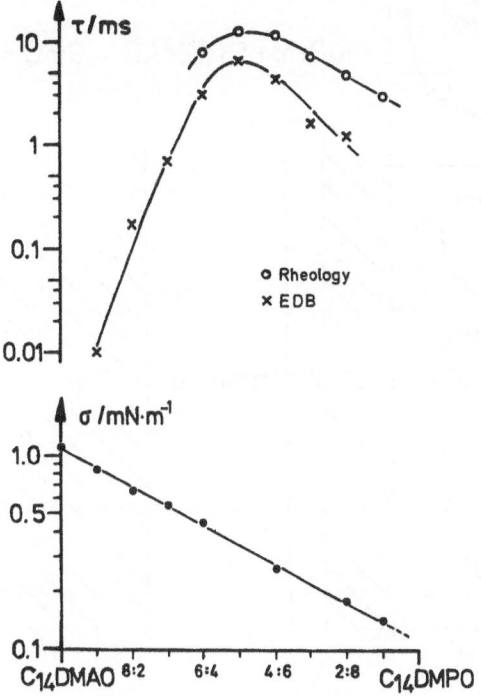

Fig. 7. Comparison of the largest structural relaxation time τ of viscoelastic surfactant solution with the interfacial tension in mixtures of C_{14} DMAO and C_{14} DMPO

strong repulsive interactions between the micellar structures, like in the discussed situation, the system will pass through a two-phase region. Results demonstrating this general behavior are shown in Fig. 8 where for a 100 mM C_{14} DMAO various parameters are plotted against the decanol concentration that is added to the aminoxide. We again observe a maximum in the viscosity before the system undergoes phase separation. The maximum in the viscosity again correlates with the maximum in the structural relaxation time and occurs at the interfacial tension of around 0.3 mN/m. The particular chemistry of the system or how this state is reached does not seem to be of importance. The absolute value of the relaxation time depends, however, on the way the interface will be formed.

The maximum in the viscosity is even visible when a shorter chainlength cosurfactant is used for the experiment, such as hexanol. Figure 9 shows some results of the sequence of phases and macroscopic rheological properties of the phases that occur when hexanol is added to a 100 mM C_{14}DMAO-solution. After the two-phase region, an optically isotropic phase is encountered; we called

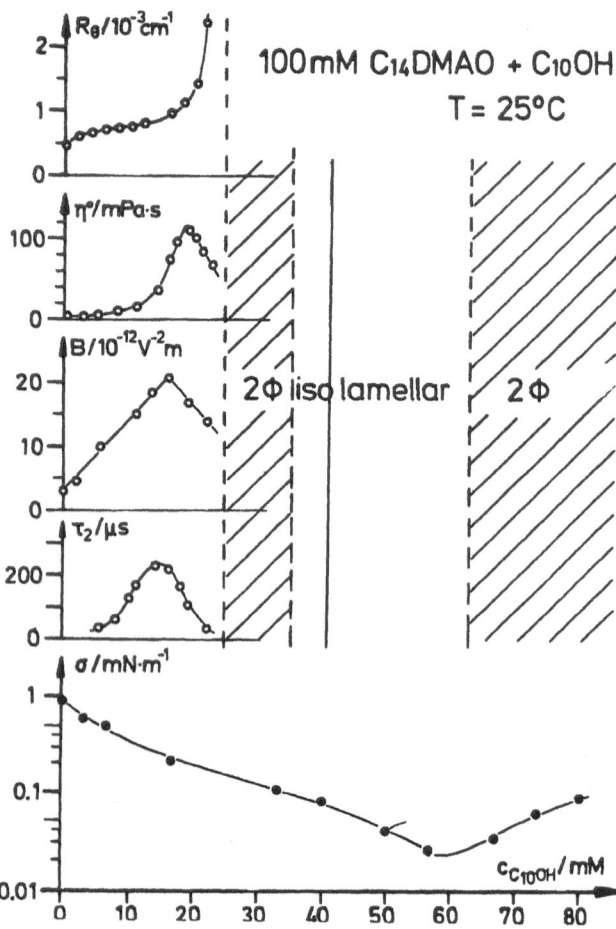

Fig. 8. Comparison of different properties in a 100 mM solution of C_{14}DMAO with the concentration of decanol against the interfacial tension. Note that the maximum of the zeroshear viscosity, the Kerr-constant and the structural relaxation time occur at about the same composition of the solution, which has an interfacial tension of about 0.3 mN m^{-1}

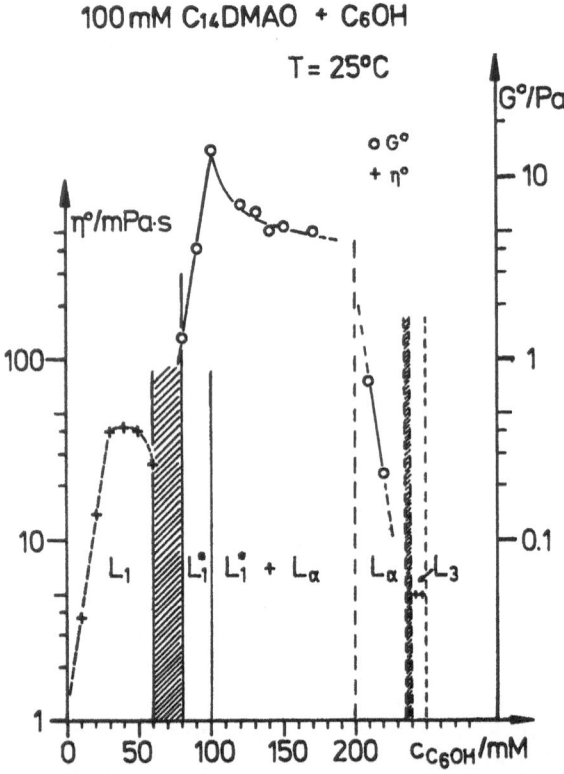

Fig. 9. Diagram of the different phases that occur in a 100 mM C_{14}DMAO solution with increasing concentrations of hexanol. Results of rheological measurements of the various phases are plotted against the hexanol concentration. Note the maximum of η^0 in the L_1-phase and the low value of η^0 in the L_3-phase. This is in contrast to the yield stress value of the L_a-phase. Note also the break of the shear modulus G^0 at the beginning of the two-phase region $L_1^x/L_1^* + L_a$

this phase a condensed phase of the rodlike micelles. This phase has remarkable rheological properties. In the phase diagram it is marked with an asterisk in order to distinguish it from the normal L_1-phase. In spite of the fact that 100 mM hexanol are present in comparison to the 50 mM that are present for the L_1-phase, the structural relaxation time is extremely long. The oscillating rheological measurements given in Fig. 10 indicate that the structural relaxation time is larger than 10 s. Actually, the data look as if the system has a yield stress. This is probably not the case and the systems seem to have a finite relaxation time, which, however, is probably in the range of minutes.

The modulus G^0 in the L_1^*-phase follows the same power law behavior as in the L_1-phase. In the L_1-phase the elasticity in the system and, hence, the shear modulus is due to a temporary network that behaves like an entanglement network. The elasticity is an entropy-elasticity and the network can be deformed or stretched considerably before the modulus breaks down. In this respect the system in the L_1^*-phase behaves differently. This is shown in Fig. 11 where the moduli of a L_1^* is plotted against the deformation. Even for deformation as small as a few percent the modulus begins to decrease. For comparison similar measurements are shown for typical viscoelastic solution in a L_1-phase. Under these conditions the network can be stretched much further before the modulus

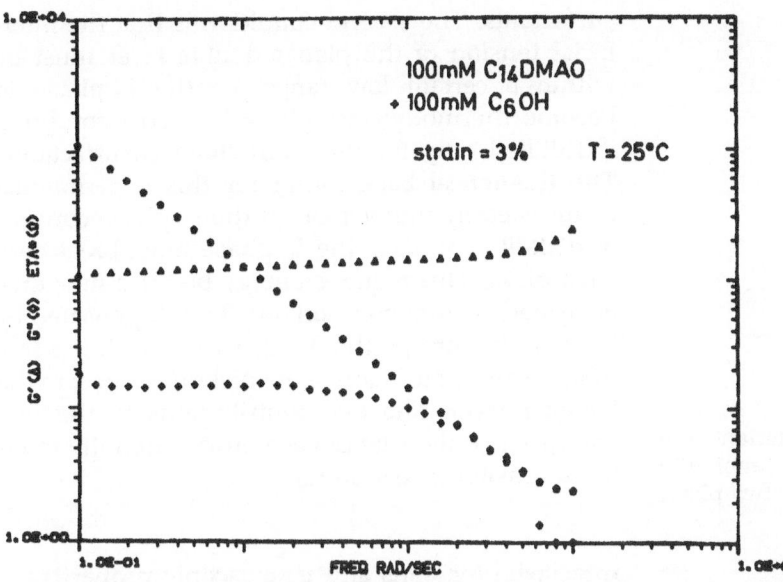

100mM C$_{14}$DMAO
+ 100mM C$_6$OH

strain = 3%   T = 25°C

Fig. 10. Rheological data in a solution of 100 mM C$_{14}$DMAO + 100 mM hexanol (L_1^*-phase) note that the storage modulus is practically independent of the oscillating frequency, while the complex viscosity η^* is decreasing with frequency. The structural relaxation time is larger than 10 s. The amplitude of the deformation was 3%

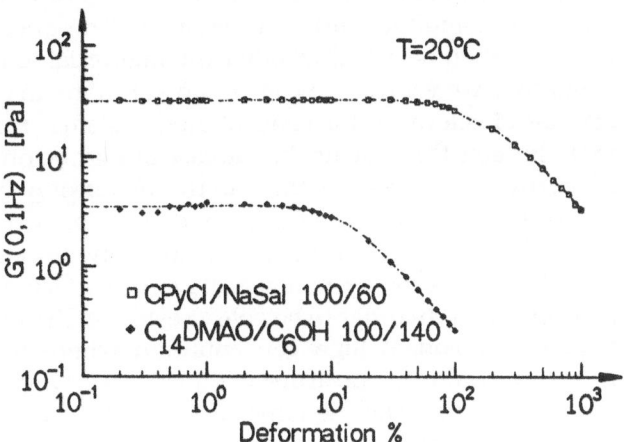

Fig. 11. A comparison of the storage modulus G^0 for two different viscoelastic solutions as a function of the deformation

decide whether the system is a single phase or, indeed, a two-phase system. One of the reasons for the difficulty in determining phase separation seems to come from the yield stress which these systems possess. It is likely that the system phase separates on a microscopic level and that the L_a-phase forms a continuous network throughout the macroscopic volume. The network of the L_a-phase has a yield stress and prevents sedimentation in the system and hence macroscopic phase separation. In order to help to decide whether the systems are really two-phase systems, rheological measurements, in particular, the measurements of the modulus can be very helpful. The values of the moduli increase in the L_1^*-phase with the hexanol concentration and then become smaller again in the 2-φ region, and level off against the module of the L_a-phase. Both the L_a-phase and the biphasic system in front of the L_a-phase have a yield stress. Results of the yield stress against the concentration are shown in Fig. 12. The yield stress values are large enough to prevent particles with densities that are considerably larger than the density of H$_2$O and dimensions in the μm-range from sedimentation. The rheological properties in the very narrow phase that follows the L_a-phase are completely different. This phase is usually called the L_3 phase and according to recent publications it is supposed to consist of sheetlike structures which form a continuous network in the system [22]. On the basis of the rheological data this is very unlikely. If the topology

decreases. This is a strong indication that the underlying mechanism of the elasticity in the L_1^*-phase is different from the situation in the L_1-phase. The situation is more reminiscent of a situation in which contacts are broken in a network which is fixed in space.

In Fig. 9 the L_1^*-phase is followed by a two-phase region in which the L_1^*-phase is in equilibrium with the birefringent L_a-phase. Systems in this composition range usually do not phase separate macroscopically and it is, therefore, difficult to

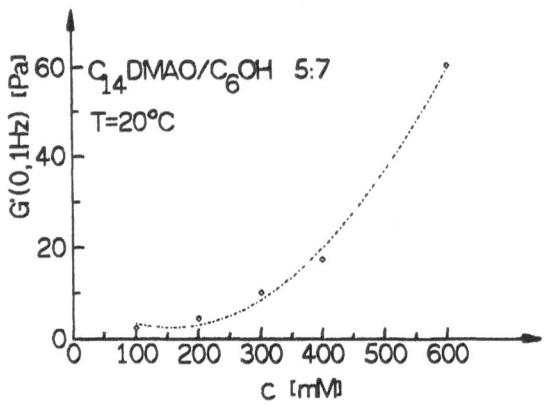

Fig. 12. Yield stress values G' against the surfactant concentration C in solutions of $C_{14}DMAO$ and hexanol with a mixing ratio of 5:7. The system is in the two-phase ($L_a + L_1^x$) region

of this phase would be very similar to the L_a-phase and would consist of a molten L_a-phase, we would expect that the systems should have a shear modulus of similar values as in the L_a-phase. Instead, the rheological data reveal no shear modulus and the longest structural relaxation time of the system is in the μs-region. It is, therefore, likely that the L_3-phase does not have a bicontinuous structure, but consists of discrete disklike micelles that have about the same main diameter as the interlamellar spacing of the L_a-phase [23]. Such a structure was originally proposed for the L_3-phase by Miller et al. [24]. Later, it was shown that SANS-measurements can indeed be fitted by assuming such structures. This structure, however, was disregarded on the basis of theoretical arguments [25].

The L_3-phase is directly adjacent to the lamellar phase over the whole composition range [25]. In this connection it is noteworthy that the L_a-phase can also be formed by $C_{14}DMAO$ and the homologues cosurfactants with different chainlengths. With increasing chainlength of the cosurfactant the L_a-phase shifts to a lower cosurfactant/surfactant ratio. The L_3-phase, however, is only observed for pentanol and hexanol as cosurfactant. With respect to Fig. 6, it seems, therefore, that at least two conditions must be met in order that the L_3-phase can be formed. The first condition is that the curvature of the interface on a micellar system has to be adjusted so that it is close to planar. This can be done with all cosurfactants or with mixtures of different

surfactants. The second condition is that the interfacial tension of the planar double layer must be within a certain low range for the L_3-phase to become thermodynamically stable. This condition is fulfilled only for the short-chain cosurfactants. The theoretical background for this experimental result is easily understood. If the L_3-phase consists of disk-like micelles, the L_a-phase must break into such disks. This requires energy because new area is formed. This energy can only be compensated by the gain in entropy that is associated with the free rotation and translation of the disks in respect to the lamellar layer. The two contributions to the free energy can only balance each other when the interfacial tension is low enough.

Interfacial tensions and macroscopic properties in the concentrated systems

Most of what has been said so far was relevant in dilute or semidilute surfactant systems. The question may then be asked whether the interfacial tensions that were measured in dilute solutions may also be of relevance for concentrated systems. For ionic systems this may not be the case or may be only partly true, because the interfacial tension is determined by the packing, and the packing depends on the ionic strength. The ionic strength is changing, however, with increasing ionic surfactant concentration. In order to be able to estimate the interfacial tension at high concentration where we probably could not measure it directly, we would have to first estimate the ionic strength. If we then measure the interfacial tension at low surfactant concentration, but higher ionic strength the value might be close to reality. For zwitterionic systems this procedure is not necessary and we probably can directly use the value that is measured at low concentration. This interfacial tension is relevant for the solubilization of hydrocarbon into the micellar structures. It also determines the moduli in the bulk properties.

All liquid crystalline phases can, in a way, be looked at as systems consisting of interphases that have different topologies [26]. Qualitatively, any deformation that is applied to a macroscopic sample results in a stretching of this interface that is associated with a restoring force. If the interfacial tension is high the restoring forces must be also high. This qualitative picture shows that the moduli of phases should be proportional to the interfacial tension of the interfaces. This simple model has actually been used to describe the shear moduli of concentrated emulsions [27]. Under highly concentrated conditions of O/W-emulsions the aqueous surfactant solution is only present in 1%. Under these conditions it forms a polyhedral foam structure that has a rheological yield stress [28]. It was shown by Princen [27] that both the yield stress and the shear modulus of these systems can quantitatively be explained on the basis of this simple model with the interfacial tension.

Concentrated binary and ternary surfactant systems usually have solid-like properties, even though all the components are in the liquid state. These properties come from the fact that the systems have a yield stress, and this is due to the fact that the micellar structures, whatever their shape, are so densely packed that they cannot pass each other when a small shear stress is applied. The particles are restricted to a certain volume element. The situation is very similar to that in dense dispersions [29]. When the interfacial tension of the structures is low enough the structure can easily deform and the particles can pass each other. Qualitatively, we can, therefore, say that only systems with a high interfacial tension can have a yield stress. This model is also suitable for the calculation of yield stress values. Any arrangements of a cubic or a hexagonal array of particles are characterized by an overlap of the particles. In a shear flow the particles have to be deformed to such a state that they can slide past each other. The required energy per unit volume for this deformation would be the yield stress. With a model that consists of well defined structures with an interface, such a calculation can be carried out.

In this respect it is worth mentioning that the shear modulus of cubic phases of ternary systems consisting of surfactant, hydrocarbon and water could quantitatively be explained by the interfacial tension and a radius of the globules [30]. The radius of the globules is, however, also determined by the interfacial tension.

Conclusions

It is shown that the value of the interfacial tension σ of a surfactant solution at the CMC for the surfactant system, measured against a hydrocarbon solvent, is a most important parameter. Its value can be used to predict macroscopic properties of the surfactant system. The value can vary anywhere from 10 mN/m to about 0.01 mN/m in the two-phase system. Its absolute value is controlled by the packing density of the surfactant molecules in the interfacial layer. The area per headgroup of surfactant can directly be obtained from the interfacial tension.

For zwitterionic surfactants this area is mainly determined by the polarity of the surfactant. The value is about the same for the surfactant at the O/W interface or at the hydrocarbon interface. Within experimental error it is even the same at the micellar interface.

The absolute value of the interfacial tension directly controls the amount of hydrocarbon that can be solubilized per surfactant in the micellar solution. The molar ratio of solubilized hydrocarbon per surfactant increases from less than 0.1 for $\sigma > 5$ mN/m to about 2 for $\sigma = 0.1$ mN/m. Since the solubilization ratio also determines the size of the swollen micelles, it is possible to predict the maximum size of the aggregates in a micellar solution that is in contact with a hydrocarbon. The value of σ further determines the amount of hydrocarbon necessary to transform rodlike micelles in the micellar solution into globular aggregates. The transformation of the rods into globules is normally accomplished before the solution is saturated with the hydrocarbon.

However, it is shown that when $\sigma < 0.05$ mN/m, saturation of the micellar phase is reached before the transformation has taken place. In this situation bicontinuous phases are formed. At the other extreme, we can have a situation in which the hydrocarbon is not soluble enough to transform the rods into globules. This condition is found in micellar solutions in contact with a hydrocarbon solvent, where the chainlength of the solvent is larger than the chainlength of the surfactant.

The shape of the micelles in the micellar solution determines the type of the liquid crystalline phases. Since the shape of the micelles can be recognized by the value of the interfacial tension, it is also possible to predict the type of the first l.c. phase that will be formed in the binary system. Systems with values of $\sigma > 1$ mN/m at the appropriate volume fraction will form a cubic phase. Systems with a value between 0.2 and 1 mN/m will form a hexagonal phase and systems with $\sigma < 0.15$ will first form a lamellar phase.

Within the existence region of the rodlike micelles the properties of solutions vary systematically with the interfacial tension as a parameter. The zero shear viscosity and the structural relaxation time, as determined from rheological or electric birefringence measurements, pass over a maximum for a constant surfactant concentration. The longest relaxation times and, therefore, the most stable rods are reached with systems where an interfacial tension is around 0.3 mN/m. It is believed that at the maximum of the structural relaxation time the rods have their largest persistence length. For smaller interfacial tensions the area per headgroup at the micellar interface becomes less than the value required for the largest spherical cross-section. It, thus, becomes easier to deform the cross-section, and the persistence length breaks down.

The most stable lamellar phases can be reached in surfactant solutions that show a minimum in the interfacial tension against a hydrocarbon. This shows that the interaction in the surfactant systems is such that the surfactant layer has the tendency to form a planar film. When surfactant solutions are adjusted to its minimum value by the addition of different chainlength alcohols, the absolute values at the minimum depend on the chain length of the alcohol. The interfacial tension values increase with increasing chain length of the alcohol, between 5 and 12.

The lamellar phases, which in the ternary system have a surfactant/cosurfactant ratio that corresponds to the minimum of the interfacial tension, can be swollen by water to very large interlamellar distances. With some systems iridescent colors can be observed under favorable conditions.

References

1. Barakat Y, Fortney LN, Schechter RS, Wade WH, Yiv SH (1983) J Colloid Interf Sci 92:561
2. Kahlweit M, Strey R (1985) Angew Chem 97:655
3. Shah DO (1981) In: "Surface Phenomena in Enhanced Oil Recovery", Plenum Press, New York
4. Baviere M, Schechter R, Wade W (1981) J Coll a Interf Sci 81
5. Chan KS, Shah DO (1980) J Dispersion Science and Technology 1:55
6. Pouchelon A, Chatenay D, Meunier J, Langevin D (1981) J Colloid Interf Sci 82
7. Jahn W, Strey R (1988) J Phys Chem 92:2294
8. Farago B, Richter D, Huang JS (1989) Physica B 156:452
9. Miller CA (1985) In: "Surfactant Science Series", Vol. 17, Marcel Dekker, New York
10. Schick MJ (1962) J Coll Sci 17:801
11. Aratono M, Okamoto T, Ikeda N, Motomura K (1988) Bull Chem Soc Jpn 61:2773
12. Verhoeckx GJ, De Bruyn PL, Overbeek JKT (1987) J Colloid Interf Sci 119:409
13. Tanford C (1980) In: "The Hydrophobic Effect", J. Wiley & Sons, New York
14. Israelachvili IN, Mitchell DJ, Ninham BW (1976) J Chem Soc Faraday Trans II, 72:1525
15. Murphy DS, Rosen MJ (1988) J Phys Chem 92:2870
16. Aveyard R, Binks BP, Mead J (1988) J Chem Soc, Faraday Trans 1, 84:675
17. Cantor RS, Mellroy P (1989) J Chem Phys 90(8):4423
18. Oetter G, Hoffmann H (1988) J Dispersion Sci and Techn 9:459
19. Hoffmann H, Ulbricht W (1989) J Coll a Interf Sci 129:388
20. Thunig C, Hoffmann H, Platz G (1989) Progr Colloid Polym Sci 79:297
21. Pospischil KH (1986) Langmuir 2:170
22. Anderson D, Wennerström H, Olsson U (1989) J Phys Chem 93:4243
23. Miller CA, Gradzielski M, Hoffmann H, Krämer U, Thunig C, J Phys Chem, in press
24. Miller CA, Ghosh O (1986) Langmuir 2:321
25. Larche FC, Appell J, Porte G, Bassereau P, Marignan J (1986) Phys Rev Lett 56:1700
26. Charvolin J, Sadoc JF (1988) 92:5787
27. Princen HM, Kiss AD (1989) J of Coll a Interf Sci 128, March 1
28. Rehage H, Platz G, Ebert G (1988) Ber Bunsenges Phys Chem 92:1158
29. van der Werff JC, de Kruif CG (1989) J of Rheology 33:421
30. Gradzielski M, Hoffmann H, Oetter G (1990) Colloid Polym Sci 268:167

Author's address:

H. Hoffmann
Lehrstuhl für Physikalische Chemie I
Universität Bayreuth
Universitätsstraße 30
8580 Bayreuth, FRG

Progress in Colloid & Polymer Science

Progr Colloid Polym Sci 83:29—35 (1990)

Predicting conditions for intermediate phase formation in surfactant systems

J. C. Lim and C. A. Miller

Department of Chemical Engineering, Rice University, Houston, Texas, USA

Abstract: For a pure surfactant below its cloud point, only the location of the limiting tie-line between L_1 and L_2 regions is needed to predict with diffusion path theory whether an intermediate phase such as a liquid crystal or microemulsion forms when an aqueous solution of the surfactant is brought into contact with a pure oil. A critical surfactant concentration should exist above which intermediate phase formation occurs. Experiments for anionic and nonionic surfactants with pure polar oils are consistent with this prediction. In view of results presented previously, removal of pure oils from polyester/cotton fabric during washing processes should improve significantly when intermediate phases develop. If the surfactant is above its cloud point, at least one intermediate phase forms. In this case the criterion for improved detergency is that water, which solubilizes little oil, not be the only intermediate phase. An approximate method for predicting when such behavior will occur is presented. The results seem generally consistent with available experimental observations.

Key words: _D_etergency; _s_urfactants; _d_iffusion _p_ath

Introduction

Previous papers from this laboratory described the dynamic behavior observed by videomicroscopy when pure hydrocarbons were carefully brought into contact with mixtures of pure nonionic surfactants and water in the absence of external agitation [1, 2]. Prominent among the phenomena seen in many of the experiments was growth near the initial surface of contact of intermediate phases which were not present initially, e.g., microemulsions, liquid crystals, and even water. Recently, results of a similar study were presented where the pure triglyceride triolein was used instead of pure hydrocarbons [3]. Here, too, intermediate phases were frequently observed. For both types of oils, the appearance of different intermediate phases under different conditions could be understood in terms of diffusion-path analysis. As discussed below, diffusion paths, used previously to predict the spontaneous formation of emulsions in oil-water-alcohol systems [4, 5], are basically plots on the equilibrium phase diagram of solutions of the one-dimensional

diffusion equations for the case of two semi-infinite phases brought into contact.

These observations were shown to have relevance for the removal of oily soils from synthetic fabrics by nonionic surfactants [2, 3]. The dominant mechanism of detergency for these conditions is not the well known rollback mechanism in which adsorption of surfactant on the fibers increases the contact angle measured through the oil. The result is contraction of oil films to form drops which are broken off and dispersed by the imposed agitation in the washing bath. Instead, soil removal in these systems is by a solubilization-emulsification mechanism, the solubilization frequently being into intermediate phases that are subsequently emulsified. For hydrocarbon soils the most favorable conditions for detergency occur near the system Phase Inversion Temperature (PIT) where solubilization is substantial and low interfacial tensions facilitate emulsification [2].

As is well known, phase diagrams in surfactant systems are rather complex. Thus, considerable effort is required to generate an entire diagram for use in the diffusion path analysis. In this paper, we examine the diffusion path approach to see if predictions about when intermediate phases develop can be made without having full details of the phase diagram. We show that the surfactant concentration in an aqueous solution required to initiate intermediate phase formation following contact with an oil can often be predicted with no information other than the limiting tie-line of the coexistence region between isotropic aqueous (L_1) and oleic (L_2) phases. This conclusion holds for pure hydrocarbon, triglyceride, and polar oils, and we employ videomicroscopy to demonstrate its validity for systems where the oil is a long-chain alcohol. Even if the surfactant is above its cloud point, we show that it is still possible in many cases to predict whether formation of an intermediate phase other than water can be expected following contact of the surfactant-rich phase and oil.

Diffusion path analysis for a surfactant below its cloud point

Consider an aqueous micellar solution L_1 having some uniform composition below the cloud point of the surfactant. When it is brought into contact with an oil phase L_2 which is also of uniform composition, diffusion within each phase and mass transfer across the interface occur. If density is uniform, convection is negligible, and diffusion of each species is proportional only to its own concentration gradient, then the governing equations take the form

$$(\partial \omega_i / \partial t) = D_i (\partial^2 \omega_i / \partial x^2) , \qquad (1)$$

where ω_i is the local mass fraction of species i and D_i its (uniform) diffusion coefficient. In a ternary system Eq. (1) need be written for only two species, since the third mass fraction is readily obtained from the requirement $\Sigma \omega_i = 1$.

We seek a solution where the ω_i's in each phase approach their initial values far from the interface. In the absence of intermediate phases the boundary conditions at the interface are simply conservation of mass for each species and the existence of local equilibrium. For these conditions it is well known [6] that the solution of Eq. (1) is given by

$$\omega_i = a_{1i} + a_{2i} \operatorname{erf} \eta_i = a_{1i} + a_{2i} \operatorname{erf}(x / \sqrt{4 D_i t}) , \qquad (2)$$

where a_{1i} and a_{2i} are constants to be determined from the boundary conditions. Moreover, the set of compositions in each phase and the interfacial compositions are independent of time. Thus, they can be plotted on the equilibrium phase diagram, giving a diffusion path such as ABCD of Fig. 1 (see [4] for further details).

We have reported previously that detergency is poor in nonionic surfactant systems at high temperature for both hydrocarbon and triglyceride oils [2, 3]. Under these conditions the surfactant is lipophilic and diffuses along with some water from the initial surfactant-water mixture into the oil, forming a water-in-oil microemulsion. No intermediate phases that solubilize appreciable oil are formed; indeed, for a surfactant below its cloud point, no intermediate phases are formed at all. The objective of the analysis given below is to determine the limiting condition when such behavior no longer occurs and intermediate phases begin to form as surfactant concentration is increased or the surfactant made more hydrophilic.

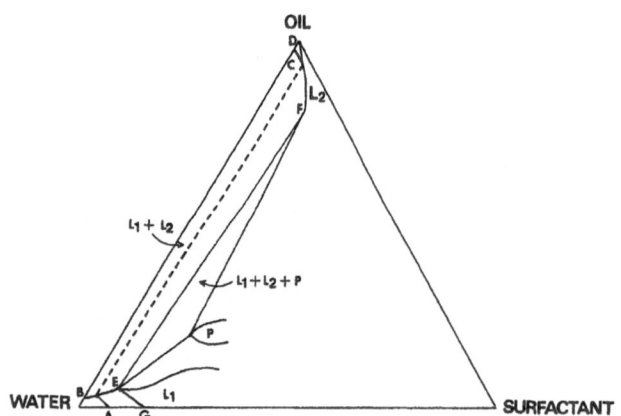

Fig. 1. Schematic diffusion paths for contact of surfactant solution with oil. For surfactant concentrations above ω_{sG}, formation of the intermediate phase P is expected

The concentration distribution of oil in an L_2 phase which initially is pure oil and occupies the region $x > 0$ is readily obtained from Eq. (2):

$$\omega_0 = a_{10} + (1 - a_{10}) \operatorname{erf} \eta_0 . \qquad (3)$$

The solubility of oil in the L_1 phase at point B is often quite low for the conditions of interest here, as described in the preceding paragraph. As a result, the flux of oil into the L_1 phase is low as well. If both these quantities can be neglected in the oil mass balance at the interface, it simplifies to

$$a_{10} = [(\omega_{0c} - \mathrm{erf}\,\eta_{0c})/\mathrm{erfc}\,\eta_{0c}]$$

$$= [\exp(-\eta_{0c}^2) + \eta_{0c}\,\mathrm{erf}(\eta_{0c})\sqrt{\pi}]/[\exp(-\eta_{0c}^2)$$

$$- \eta_{0c}\,\mathrm{erfc}(\eta_{0c})\sqrt{\pi}] \tag{4}$$

with ω_{0c} and η_{0c} the values of ω_0 and η_0 at point C of Fig. 1. Thus, for any tie-line BC, η_{0c} can be calculated.

The surfactant concentrations in the L_2 and L_1 phases are given by

$$\omega_s = a_{1s}(1 - \mathrm{erf}\,\eta_s) \tag{5}$$

$$\omega_s = a_{3s} - (\omega_{sA} - a_{3s})\,\mathrm{erf}\,\eta_s' , \tag{6}$$

where $\eta_s' = (x/\sqrt{4D_s't})$ in the L_1 phase, which initially occupies the region $x < 0$. Since $\eta_{sc} = \eta_{0c}(D_0/D_s)^{1/2}$, a_{1s} may be found in terms of ω_{sc} by applying Eq. (5) at the interface. In a similar manner Eq. (6) may be used to eliminate a_{3s}. Then the surfactant mass balance at the interface, when solved for ω_{sA}, has the form

$$\omega_{sA} = \omega_{sB}F_1(\omega_{0c}) + \omega_{sc}F_2(\omega_{0c}) , \tag{7}$$

with F_1 and F_2 given by

$$F_1(\omega_{0c}) = \{1 + [\eta_{0c}\sqrt{\pi}\,(\exp\eta_{0c}^2)$$

$$\times\,(1 + \mathrm{erf}\,\eta_{0c})D_0/D_s']\} \tag{8}$$

$$F_2(\omega_{0c}) = \{[(1 + \mathrm{erf}\,\eta_{0c})D_s/(1 - \mathrm{erf}\,\eta_{0c})D_s']$$

$$- [\eta_{0c}\sqrt{\pi}\,(\exp\eta_{0c}^2)(1 + \mathrm{erf}\,\eta_{0c})D_0/D_s']\} . \tag{9}$$

These functions are shown in Fig. 2 for the case $D_0 = D_s = D_s'$.

If the initial surfactant concentration ω_{sA} in the L_1 phase increases, the tie-line joining the interfacial compositions shifts to higher surfactant concentrations. Eventually, for some point G with concentration ω_{sG}, the limiting tie-line EF of the L_1–L_2 region is reached. If the initial surfactant concentration increases still further, a different diffusion path involving formation of the intermediate phase P applies. As indicated above, our previous studies sug-

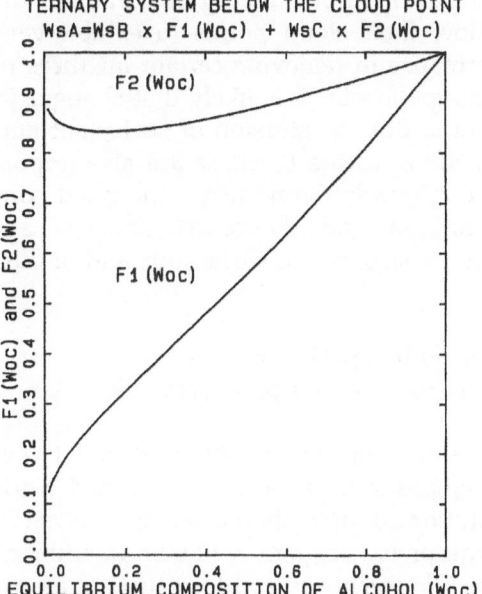

Fig. 2. The functions F_1 and F_2 as defined by Eqs. (8) and (9) for the case $D_0 = D_s = D_s'$

gest that a marked improvement in detergency should be seen when such an intermediate phase forms, provided that it solubilizes a substantial amount of oil.

According to Eq. (7), the higher the surfactant solubility in the L_2 phase at the limiting tie-line, the higher the initial surfactant concentration ω_{sG} required to produce intermediate phase formation. That is, higher surfactant concentrations are needed for more lipophilic surfactants as well as for shorter chain hydrocarbons, triglycerides, and more polar oils. Since surfactant concentration is frequently of the order of 0.1 wt% or even lower during household washing, this analysis suggests that surfactants below the cloud point are likely to be ineffective in removing pure oils in which surfactant solubility is appreciable.

It should be emphasized that this last conclusion and indeed the entire present analysis apply only for the case of a ternary system where both the surfactant and the oil are pure components. If the oil is a mixture and its volume is small compared to that of the surfactant solution L_1, preferential solubilization of some components into the L_1 phase could cause oil composition to change with time. Raney and Benson [7] have suggested that the effect on dynamic behavior of such changes may be responsible for the known ability of nonionic surfactants below their cloud points to exhibit very good performance in removing certain mixtures of polar and nonpolar oils. It is likely that changes in oil composition due to diffusion of surfactant and water from the L_1 to the L_2 phase are also important. We are currently developing a modified contacting technique and theoretical analysis applicable to the case of mixed surfactants and/or oils.

Comparison with experiment for anionic surfactant and pure polar oil

Kielman and van Steen [8] performed experiments where n-decanol was contacted with aqueous solutions of potassium octanoate, a system for which the phase diagram is known at ambient temperature [9]. They observed formation of an intermediate liquid crystalline phase for surfactant concentrations above about 3 wt%, and noted that this critical concentration must exceed ω_{sE}, which is 2.4 wt%. This concentration has been called the "limiting association concentration" or LAC by Ekwall [9]. While their conclusion that ω_{sG} must exceed the LAC is certainly correct, a better estimate of ω_{sG} can be obtained by the diffusion path analysis described above. Our calculations using Eq. (7) yield a value for ω_{sG} of about 3.8 wt%. It may be that their observations of some liquid crystal at slightly lower surfactant concentrations was a result of mixing during the initial contact of the phases (see below).

Kielman and van Steen [8] also conducted detergency experiments for this system and found that removal of n-decanol from polyester fabric increased from less than 10% when no liquid crystal formed to nearly 100% when it did form.

Because the development of intermediate phases can be seen more clearly at higher magnifications and the critical concentration ω_{sG} more accurately determined, we performed contacting experiments with our videomicroscopy system, which employs a vertical-stage microscope. The experimental technique was the same as used in our previous work [1—3]. Because sodium octanoate was available in our laboratory, we used it instead of potassium octanoate. The phase diagrams for the two systems are very similar, the LAC ω_{sE} for sodium octanoate being about 2.1 wt% [9].

No liquid crystal was seen for initial surfactant concentrations below 3.25 wt%. Between about 3.25

and 3.98 wt% some liquid crystal was observed, but it appeared to be transient in nature, having been formed during the initial contacting process in which a small amount of mixing inevitably occurs. Above 4.10 wt% the liquid crystal continued to form at times well beyond the initial stages of contact. As Fig. 3 shows, an interface was seen between L_1 and L_2 regions, both of which contained dispersed liquid crystal L_a. That is, the diffusion path apparently passed through the $(L_1 + L_a)$ region, crossed the $(L_1 + L_a + L_2)$ three-phase triangle at the interface, and then passed through the $(L_2 + L_a)$ region. Such diffusion paths crossing a three-phase triangle in surfactant systems have been discussed by Raney and Miller [10].

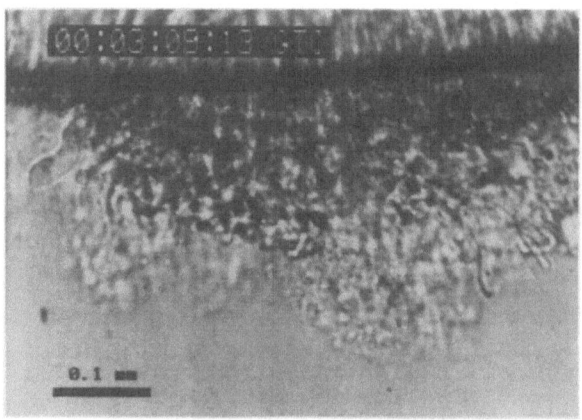

Fig. 3. Videoframe showing liquid crystal formation when a 4.10 wt% solution of sodium octanoate is contacted with n-decanol

Using the analysis developed in the previous section and the limiting tie-line from the equilibrium phase diagram [9], which has $\omega_{sF} = 2$ wt% and $\omega_{0F} = 93$ wt%, we calculated ω_{sG} to be about 3.9 wt%, which is good agreement with the experimental results just described. Using the full diffusion path solution, we confirmed that the neglect of n-decanol transfer into the L_1 phase, used in deriving Eq. (7), is valid in this system.

Results for nonionic surfactants below the cloud point with polar soils

Nonionic surfactants typically have critical micelle concentrations (CMC) much lower than those of ionic surfactants with comparable hydrocarbon chain lengths. Since the LAC (point E of Figure 1) cannot exceed the CMC, the L_1 end of the limiting tie-line must be very near the water corner of the ternary diagram in nonionic surfactant systems. Under these conditions the first term of Eq. (7) is negligible in comparison to the second.

Contacting experiments similar to those described above were performed for the $C_{12}E_8$-n-decanol-water system at 25°C. The limiting tie-line in this system has $\omega_{sF} = 17$ wt% and $\omega_{oF} = 72$ wt% [11]. The prdicted critical surfactant concentration ω_{sG} is about 16 wt%. Both liquid crystal and the isotropic D' phase were seen as intermediate phases for surfactant concentrations above about 13 wt%, in reasonable agreement with the prediction. Note that this concentration is orders of magnitude above the LAC since the CMC is only 0.005 wt%. That is, use of the LAC to estimate the critical surfactant concentration is inadequate and the diffusion path approach must be employed.

We have also studied the system $C_{12}E_5$-oleyl alcohol-water at 30°C, which is just below the cloud point of the surfactant. We found that the alcohol-rich end of the limiting tie-line has surfactant and alcohol mass fractions ω_{sF} and ω_{0F} of about 19 wt% and 11 wt%, respectively. According to Eq. (7) and Fig. 2, an intermediate phase should be seen when initial surfactant mass fraction ω_{sG} exceeds about 16 wt%. This value is again orders of magnitude above the LAC.

Our contacting experiments are in fair agreement with this prediction. Liquid crystal was observed for initial surfactant concentrations exceeding about 21 wt% although it did not appear immediately at this and even somewhat higher concentrations as would be expected from the diffusion path analysis. A possible explanation of this behavior can be given based on the fact that alcohol concentration at the limiting tie-line is much lower in this system than in those discussed above. As a result, the predicted compositions for the L_2 phase are far inside the $(L_1 + L_2)$ region. Thus, extensive spontaneous emulsification is predicted in the L_2 phase [4, 5] and was, in fact, observed. The emulsification may well have been so extensive that neglect of the disperse phase in formulating the governing equations was a poor approximation. The analysis also assumes uniform diffusion coefficients in each phase, which becomes less realistic when the composition variations within the phases are large, as in these experiments.

Diffusion path analysis for a surfactant above its cloud point

At temperatures above the cloud point, dilute nonionic surfactant-water mixtures separate into a surfactant-rich phase and a surfactant-lean phase. For sufficiently high temperatures the latter is just water containing some molecularly dissolved surfactant. The former is an isotropic liquid for temperatures somewhat above the cloud point, a lamellar liquid crystal at higher temperatures, and another isotropic liquid known as the L_3 phase at still higher temperatures [12]. During the washing process intermediate phase formation, if any, occurs when drops or particles of the surfactant-rich phase are brought into contact with oily soil as a result of agitation in the washing bath.

One obvious possibility is that the only intermediate phase produced by such contact is water containing molecularly dissolved surfactant. In this case the diffusion path is as shown by QABCD in Fig. 4. Of course, such a diffusion path is undesirable for purposes of detergency because the intermediate phase solubilizes negligible oil. Accordingly, the objective here is to determine the conditions when such a diffusion path is not possible. Presumably, other intermediate phases having significant solubilization capabilities would then form.

Extension of the diffusion path approach to incorporate formation of one or more intermediate phases is straightforward [13]. However, more information on equilibrium phase behavior is re-

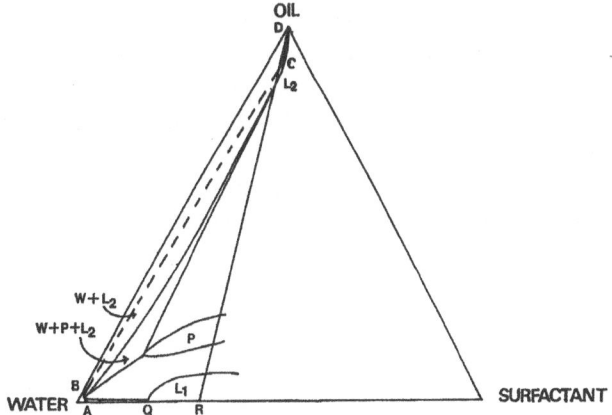

Fig. 4. Schematic diffusion path for surfactant above its cloud point showing water as the only intermediate phase formed

quired for a complete solution than in the previous case where only one limiting tie-line was needed. An approximate solution is nevertheless possible when water is the only intermediate phase formed if it is assumed that virtually no oil leaves the L_2 phase and that the low solubility of surfactant in the water phase causes the rate of diffusion of surfactant from the surfactant-rich phase to be low. The first assumption is the same as before, and Eq. (4) applies. The second assumption implies that surfactant concentration is nearly uniform in the surfactant-rich phase. A surfactant mass balance for the intermediate water phase, neglecting accumulation because of the low surfactant solubility, then leads to

$$\omega_{sQ} = \omega_{sc}[\eta_{sc} \, \mathrm{erf}(\eta_{sc}) \, \sqrt{\pi} - \exp(-\eta_{sc}^2)]/$$

$$[\eta_{sA} \, \mathrm{erf} \, \eta_{sc} \, \sqrt{\pi}] \, , \qquad (10)$$

where η_{sc} refers to the interface between water and oil phases and η_{sA} to the interface between water and surfactant-rich phase. We note that η_{sc} and η_{0c} are related by the expression given following Eq. (6).

Since we have assumed that negligible oil leaves the L_2 phase, both interfaces move toward the surfactant-rich phase, so that both η_{sc} and η_{sA} are negative. However, the interface between water and L_2 must move more slowly than the other and thus not overtake it. After some manipulation which involves invoking Eqs. (3) and (4) it can be shown that this constraint requires that ω_{sQ} satisfy

$$\omega_{sQ} < \omega_{sc}/(1 - \omega_{0c}) \, . \qquad (11)$$

However, this inequality is equivalent to an overall mass balance constraint, which dictates that Q lies to the left of point R where the line joining D and C meets the base of the triangular diagram of Fig. 4. In the limiting case that (11) becomes an equality, the intermediate water phase has zero thickness and an equivalent diffusion path is the straight line RFD, where F is as in Fig. 1, with no intermediate phase.

Thus, for the special case where C lies on the limiting tie-line between water and L_2 phases, a solution with water as the only intermediate phase is possible only if surfactant concentration in the initial surfactant-rich phase is low enough to satisfy the inequality (11). Clearly, good detergency (where such a solution is impossible) is favored by more concentrated surfactant-rich phases. We note that

drops of such a phase can form above the cloud point, even when overall surfactant concentration is low. Indeed, in the $C_{12}E_5$-water-oleyl alcohol system discussed above, we have observed formation of an intermediate liquid crystalline phase at temperatures above the cloud point of about 31 °C when the pure alcohol was contacted with dilute surfactant-water mixtures. In contrast, very high surfactant concentrations were required to initiate liquid crystal formation below the cloud point in this system as discussed above. In a similar manner, formation of intermediate microemulsion phases has been observed in nonionic surfactant-water-hydrocarbon systems near the PIT for systems where the PIT is well above the cloud point [1, 2].

The effect of an increase in ω_{sc} on the value of ω_{sQ} needed for intermediate phase formation depends on water solubilization in the L_2 phase. It can be shown that the limiting value of ω_{sQ} which makes (11) an equality increases as the solubilization parameter (ω_{wc}/ω_{sc}) decreases, being 0.25 and 0.50 when this parameter is 3 and 1, respectively. That is, in more lipophilic systems with lower water solubilization, a solution with only water as an intermediate phase is likely. Indeed, we have frequently observed such behavior and the associated poor detergency at high temperatures where nonionic surfactants are lipophilic [1—3].

Conclusion

When an aqueous surfactant solution below the cloud point is brought into contact with a pure oil in which the surfactant has an appreciable solubility, diffusion path theory can be used to predict the minimum surfactant concentration needed to produce formation of an intermediate phase. The only information on phase behavior required for this calculation is the location of the limiting tie-line of the L_1—L_2 region. Based on available experimental results with hydrocarbons, triglycerides, and long-chain alcohols [2, 3, 8], formation of such an intermediate phase brings about a substantial improvement in detergency for synthetic fabrics, provided that the phase is capable of solubilizing significant quantities of the oil. Generally speaking, higher surfactant solubilities in the oil lead to higher initial surfactant concentrations required for intermediate phase formation. The required surfactant concentration can be far above the LAC, especially in nonionic surfactant systems.

For a nonionic surfactant above its cloud point, one can, knowing the location of the limiting tie-line between water and L_2 phases, estimate the minimum surfactant concentration in the initial surfactant-rich phase for which an intermediate phase other than water must be formed. Here too high surfactant solubilities in the oil require higher minimum surfactant concentrations. That is, formation of water as the only surfactant-rich phase and the associated poor soil removal are more likely for lipophilic surfactants, a prediction that is in agreement with our experimental observations.

Acknowledgements

This work was supported by grants from NSF and Shell Development Company. Discussions with Shell personnel were helpful.

Appendix

Since this paper was submitted, the authors have developed a theory and conducted experiments for the case of a small drop of oil brought into contact with a large volume of surfactant solution. Both theory and experiment show that even when the initial surfactant concentration is below the minimum value ω_{sG} needed to produce *immediate* formation of an intermediate phase (see Fig. 1), such a phase can begin to grow at a later time. Details of this work have been submitted for publication.

References

1. Bention WJ, Raney KH, Miller CA (1986) J Colloid Interface Sci 110:363
2. Raney KH, Benton WJ, Miller CA (1987) J Colloid Interface Sci 117:282
3. Mori F, Lim JC, Raney OG, Elsik CM, Miller CA (1989) Colloids and Surfaces 40:323
4. Ruschak KJ, Miller CA (1972) Ind Eng Chem Fundam 11:534
5. Miller CA (1988) Colloids and Surfaces 29:89
6. Bird RB, Stewart WE, Lightfoot EN (1960) Transport Phenomena, New York, Wiley
7. Raney KH, Benson HL, The Effect of Polar Soil Components on the PIT and Optimum Detergency Conditions, J Am Oil Chem Soc, in press
8. Kielman HS, van Steen PJF (1979) Influence of Mesomorphic Phase Formation on Detergency, in Surface Active Agents, Proceedings of S.C.I. Conference, Nottingham, England
9. Ekwall P (1975) Liq Crystals 1:1
10. Raney KH, Miller CA (1987) AIChEJ 33:1791
11. Kunieda H, Miyajima A (1989) J Colloid Interface Sci 129:554
12. Mitchell DJ, Tiddy GJT, Warring L, Bostock T, McDonald MP (1983) J Chem Soc Faraday Trans 1 79:975
13. Danckwerts PV (1950) Trans Faraday Soc 46:701

Authors' address:

C. A. Miller
Department of Chemical Engineering
Rice University
P.O. Box 1892
Houston, Texas 77251
713) 527-4904

Progress in Colloid & Polymer Science   Progr Colloid Polym Sci 83:36—45 (1990)

Concentrated sterically stabilized dispersions and their flocculation by depletion layers

Th. F. Tadros

ICI Agrochemicals, Jealotts Hill Research Station, Bracknell, UK

Abstract: Model systems of polystyrene latex dispersions (R = 175 nm) containing grafted polyethylene oxide chains (M_w = 2000) were prepared using dispersion polymerisation. Their properties were investigated using steady-state shear stress-shear rate and oscillatory measurements. The viscosity-volume fraction φ results were fitted to the hard-sphere model of Dougherty and Krieger. This showed a reduction in adsorbed layer thickness with increase in φ. The viscoelastic results showed a gradual increase of the elastic response with increase in φ. At a critical φ value, G' (storage modulus) becomes greater than G'' (loss modulus) indicating the onset of interaction between the steric layers. $G' - \varphi$ curves were converted to $G' - h$ (interparticle distance) curve to illustrate the analogy with the directly measured force (F) distance curves (using mica sheets with adsorbed graft copolymer). The $F - h$ curves could be also converted to $G' - h$ curves and a comparison could be made with the curves obtained from rheological measurements. Although the values of G' obtained from the $F - h$ curves were different from these obtained from rheology, the trend was the same; this illustrates the value of rheological measurements for studying steric interaction. — The above concentrated sterically stabilised dispersions were then flocculated by adding a "free" (non-adsorbing) polymer (polyethylene oxide or hydroxyethyl cellulose). The flocculation was followed using rheological measurements. As a result of the presence of depletion layers, flocculation became significant above a critical volume fraction of free polymer φ_p^+. The results were analysed using the available theories on depletion flocculation and a comparison was made between the free energy of depletion and the energy of separation of particles calculated from rheology.

Key words: Concentrated dispersions; depletion flocculation; polystyrene latex; oscillatory measurements

Introduction

Concentrated sterically stabilised dispersions are encountered in many industrial applications of which we mention paints, dyestuffs, printing inks, cosmetics, agrochemicals, ceramics, paper coatings, etc. It is now fairly well established that the stabilising mechanism in these dispersions arises from two main effects: an unfavourable mixing of the stabilising chains when these are present in good solvent conditions (usually referred to as mixing or osmotic effect) and a reduction in the configurational entropy of the chains when the latter undergo significant interpentration and/or compression on close approach (usually referred to as elastic or volume restriction effect). Several theories have been developed to quantify steric repulsion and this has been summarised in a recent monograph (1). Such theories have been tested experimentally using dilute dispersions whereby the conditions of stability/instability have been clearly defined. Although these theories can be usefully applied for the design of concentrated sterically stabilised dispersions, they cannot predict the properties of the systems, in particular their flow behaviour during application. For that reason we have initiated systematic studies of the properties of concentrated sterically stabilised dispersions using rheological techniques [2]. For that purpose model systems of polystyrene latex dispersions containing grafted or physically adsorbed polymer layers have been designed and their viscoelastic properties measured as a function of various variables such as volume fraction, and chain solvency. The results obtained from these measurements are complimentary to the more quantitative studies that are obtained by small-angle neutron scattering (SANS) [3]. In the latter case, information on the microstructure of the system can be obtained and this can be correlated with the bulk properties of the concentrated dispersion.

Another important phenomenon with sterically stabilised dispersions is their behaviour on addition of water-soluble polymers that do not adsorb on the particle surface (usually referred to as "free" polymer). At a critical volume fraction of the added "free" polymer, flocculation occurs as a result of the interaction of the polymer free zones near the particle surface (usually referred to as depletion layers). These depletion zones arise from the fact that any reduction in the configurational entropy of the chains near the surface is not compensated by an adsorption energy [4]. The thickness of the depletion layer Δ is comparable to the radius of gyration of the free polymer chains, at least at moderate polymer volume fraction. When two particles with their depletion layers approach each other to a surface-to-surface distance h that is smaller than 2δ, attraction occurs and this results in a depletion free energy of attraction G_{dep} that is proportional to the osmotic pressure of the free polymer solution [5]. Several theories [4—8] are now available to calculate G_{dep}, and these need to be verified experimentally.

In this paper, we will give a summary of some recent results obtained using a model concentrated dispersion of polystyrene latex containing grafted poly(ethylene oxide) chains. Their properties were investigated by steady state shear stress (τ) — shear rate $\dot{\gamma}$ and oscillatory measurements. From the $\tau - \dot{\gamma}$ results, the viscosity (η)-volume fraction (φ) curve was established and the data were fitted to the hard sphere model of Dougherty and Krieger [9,

10] in order to obtain the adsorbed layer thickness δ as a function of φ. From the oscillatory measurements, the complex modulus G^*, storage modulus G' and loss modulus G'' were obtained as a function of frequency (ω) and φ. $G' - \varphi$ curves were converted to a $G' - h$ curves to illustrate the analogy with the directly measured force (F) distance curves [11]. The latter was obtained using macroscopic mica sheets to which a polymer that is similar to that grafted on polymer latex particles has been physically adsorbed.

The above sterically stabilised dispersions was also used as a model to study depletion flocculation obtained by addition of poly(ethylene oxide) (PEO) or hydroxy ethyl cellulose (HEC) with various molecular weights. With such concentrated dispersions, the flocculation was investigated using rheological technique. From such measurements, the energy required to separate the particles in a flocculated structure was obtained and compared to that calculated using available theories.

Experimental

Preparation of latex dispersions

Concentrated polystyrene latex dispersion with grafted poly(ethylene oxide) was prepared using the aquersemer process described before [12, 13]. Basically, styrene is polymerised in alcohol-water mixture in the presence of methoxy poly(ethylene oxide) methacrylate. The PEO molecular weight of the poly(ethylene oxide) was 2000. The final solids content of the latex was 46.9% w/v and its z-average particle diameter was 350 nm. The z-average diameter was determined using a Malvern photon correlation spectroscopy apparatus (Malvern Instruments, Malvern, U.K.). The results showed that the latex has a narrow particle size distribution.

Latex dispersions with various volume fractions were prepared by centrifugation of the stock latex suspension and dilution with the supernatant liquid.

Rheological measurements

Steady state shear stress (τ) — shear rate ($\dot{\gamma}$) measurements: These were obtained using a Haake Rotovisco (Model RV100 and RV2) fitted with an M50/M150 or M500 head, concentric cylinder platens, a programmer for continuous increase of the shear rate and a chart recorder. From the $\tau - \dot{\gamma}$ curves, the yield value was obtained using two procedures. In the first procedure a Bingham model was applied, where

$$\tau = \tau_\beta + \eta_{pl}\dot{\gamma} , \tag{1}$$

from which τ_β was obtained by extrapolation of the linear portion of the shear stress — shear rate are to $\dot{\gamma} = 0$.

In the second procedure the yield value τ_{2c} was obtained using Casson's equation [14]:

$$\tau^{1/2} \; 3 \; \tau_c^{1/2} \; + \; \dot{\gamma}^{1/2} \eta_c^{1/2} \; . \tag{2}$$

Plots of $\tau^{1/2}$ vs $\dot{\gamma}^{1/2}$ are linear and by least squares analysis of the data, τ_c and η_c can be obtained.

Shear modulus measurements

The shear modulus G_∞ was measured by a pulse shearometer (Rank Bros, Brottisham, Cambridge, U.K.) based on the model originally described by Van Olphen [15] and later developed by Goodwin et al. [16]. In this instrument the dispersion is placed between two parallel perspex plates which are connected to piezoelectric crystals. A shear wave is generated at the bottom plate using a pulse generator which causes a small rotation (10^{-4} rad) of the plate, and the shear-wave velocity is measured from a plot of plate separation and the time of propagation of the shear wave. The instrument operates at small amplitudes and high frequency (~ 200 Hz) and, therefore, the modulus obtained is the high frequency modulus, G_∞. The data are analysed using an Acorn Atom computer supplied with the instrument. The shear modulus G_∞ is calculated from the equation,

$$G_\infty = u^2 \rho \tag{3}$$

where ρ is the density of the dispersion.

Oscillatory measurements: A model VOR Bohlin rheometer (Bohlin Reolgi, Lund, Sweden) interfaced with a Facit DTC2 or IBM microcomputer was used for such measurements. The instrument can operate in the frequency range $10^{-3} - 20$ Hz and has interchangeable torsion bars covering a wide range of sensitivities. A coaxial cylinder (C25) with a moving cup of radius 27.5 mm and a fixed bob with radius 25.0 mm was used. The Bohlin rheometer performs oscillation tests by turning the cup back and forth in a sinusoidal manner. The shear stress in the sample is measured by measuring the deflection in the bob which is connected to interchangeable torsion bars. The phase angle shift is automatically computed from the time displacement between the sine waves of stress and strain (Δt), i.e., $\delta = \nu \Delta t$, where ν is the frequency in radians ($\nu = 2 \pi \omega$, where ω is the frequency in Hertz). The complex modulus G^*, storage modulus G' and loss modulus G'' are calculated from the stress and strain amplitude (τ_0 and γ_0, respectively) and the phase angle shift δ,

$$G^* = \tau_0/\gamma_0 \tag{4}$$

$$G' = G^* \cos \delta \tag{5}$$

$$G'' = G^* \sin \delta \tag{6}$$

$$G^* = G' + iG'' \, , \tag{7}$$

where i is equal to $(-1)^{1/2}$.

In oscillatory measurements, one initially fixes the frequency and measures the rheological parameters as a function of strain amplitude. This enables one to obtain the linear viscoelastic region whereby G^*, G' and G'' are independent of applied strain, at any given frequency. Once the linear region is indicated, then measurements are made as a function of frequency at a fixed amplitude. Many systems only show a linear region at very low amplitudes, and therefore the measurement should be carried out at such low deformation.

Results and discussion

Relative viscosity η_r — volume fraction φ relationship for sterically stabilised dispersions

$\eta_r - \varphi$ relationship followed the trend obtained for hard sphere dispersions and this may be fitted using the Dougherty-Krieger equation [9, 10], i.e.

$$\eta_r = [1 - (\varphi/\varphi_{sp})]^{-[\eta]\varphi_{sp}} \, , \tag{8}$$

where $[\eta]$ is the instrinsic viscosity that is equal to 2.5 for hard spheres, φ is the volume fraction which must be replaced by the effective volume fraction φ_{eff} to take into account the presence of the grafted polymer chains, i.e.

$$\varphi_{eff} = \varphi \left(1 + \frac{\delta}{R} \right)^3 \, , \tag{9}$$

where δ is the adsorbed layer thickness, φ_{sp} is the maximum packing fraction which is 0.64 for random packing and 0.74 for hexagonal packing.

To fit the experimental results to Eq. (8), one needs to know φ_{eff} and φ_{sp}. The latter could be obtained by an empirical method whereby a plot $1/\eta_r^{1/2}$ vs φ gave a straight line that could be extrapolated to $1/\eta_r^{1/2} = 0$ to obtain φ_{sp} (note that when $\varphi \to \varphi_{sp}$, $\eta_r \to \infty$). However, the problem of calculating φ_{eff} is not tirivial, since one would expect δ to change with φ, as a result of interpenetration and/or compression of the chains at high φ values. For that reason, δ was used as an adjustable parameter to fit the data to Eq. (8) and this clearly showed that δ does indeed decrease with increase in φ. The calculations are summarised in Table 1. The value of δ at the lowest φ value, namely 20.5 nm, is a reasonable value for a terminally an-

Table 1. Adsorbed layer thickness values used to fit the $\eta_r - \varphi_{eff}$ curve to the Dougherty-Krieger equation [9, 10]

η_r	φ	δ/nm	φ_{eff}
6.5	0.33	20.5	0.460
9.0	0.36	20.0	0.500
11.0	0.38	19.3	0.520
15.0	0.41	18.0	0.550
19.0	0.43	17.0	0.567
21.0	0.44	16.3	0.575
26.0	0.46	15.1	0.590
35.0	0.48	15.0	0.615
53.0	0.50	15.0	0.640
78.0	0.52	13.3	0.648
110.0	0.54	11.1	0.650
220.0	0.57	9.2	0.665

chored PEO chain with molecular weight M_w = 2000. Neutron-scattering measurements [17] using deuterated polystyrene latex dispersions with terminally anchored PEO chains with M_w = 4800 showed highly extended PEO tails reaching more than 30 nm. Thus, the value of 20.5 nm for M_w = 2000 is very reasonable. As mentioned above, the reduction in φ values, especially at high φ values is accounted for in terms of compression of the PEO chains as the particles approach each other very closely in a concentrated dispersion. This compression is reflected in the elastic properties of the dispersion as discussed below. Similar effects were obtained for sterically stabilised non-aqueous dispersions [18].

Viscoelastic properties of concentrated sterically stabilised dispersions

The variation of G^*, G' and G'' with frequency showed an interesting transition as the volume fraction of the dispersion was increased. At $\varphi < 0.5$, G'', i.e., the dispersion showed a predominantly viscous response within the frequency range 10^{-2} — 1 Hz. This is not surprising since at $\varphi < 0.5$, the surface-to-surface distance of separation between the particles h is less than 2δ and, hence, interaction between the terminally anchored PEO chains is weak. These chains are only slightly compressed (δ decreases from 20.5 to 15.0 nm as φ increases from 0.33 to 0.48). Such compression probably occurs with the larger PEO chains (note that the grafted

PEO chains are polydisperse) and such small effect does not lead to an elastic response within the frequency range studied. However, when $\varphi > 0.5$, $G' > G''$ and the magnitude of the moduli starts to rise sharply with further increase in φ. For example, when φ is increased from 0.5 to 0.575, G' at $\omega = 1$ Hz increases by an order of magnitude, and on further increase to 0.585 it increases by another two orders of magnitude. Moreover, at such high φ values, the moduli tend to show little dependence on frequency within the range studied.

A summary of the results to illustrate the above effects is shown in Fig. 1 in which plots of G^*, G' and G'' vs φ are shown at $\omega = 1$ Hz. It can be seen that at $\varphi < 0.5$, $G'' > G'$, whereas at $\varphi > 0.5$, $G' > G''$ and at $\varphi > 0.53$, $G' \sim G^*$ and it increase very rapidly with increase in φ. Within this high φ value considerable compression of the PEO chains occur and δ starts to fall rapidly (see Table 1). Under these conditions, strong steric interaction occurs as a result of interpenetration of the peripheries of the longer chain PEO molecules and further compression of the whole grafted polymer layers. Under these conditions the concentrated sterically stabilised dispersion behaves as a near elastic network. This is confirmed by the lack of dependence of the moduli on the applied frequency. Indeed at $\varphi = 0.585$, $G^* \sim G' = 4.8 \times 10^3 \, \text{Nm}^{-2}$ and $\eta' = 8.82 \times 10^3$ Pas, whereas at $\varphi = 0.65$, $G^* \sim G' = 1.12 \times 10^5$ and $\eta' = 1.6 \times 10^5$ Pas, i.e. the latex behaves as a near elastic solid.

The above viscoelastic measurements clearly show their value in studying steric interaction in concentrated dispersions. It is possible to convert the moduli — φ curves into $G' - h$ curves and compare this with the $F - h$ curves obtained from direct force measurements. This is illustrated in Fig. 2 in which G' is plotted vs h. For a monodisperse ordered array of latex dispersion, h is simply given by the expression,

$$h = 2R[(\varphi_m/\varphi)^{1/3} - 1], \qquad (10)$$

where φ_m is a constant that is characteristic of the type of array, e.g., 0.74 for hexagonal or face-centered cubic arrays, 0.68 for body-centred cubic. For the present calculation of h, a value of $\varphi_m = 0.68$ was used.

In Fig. 2, G_{th}^0 vs h obtained from the $F - h$ curves is also shown. These calculations were recently obtained by Costello [19] and will be published [20]. Although the values of G' and G_{th}^0 vary con-

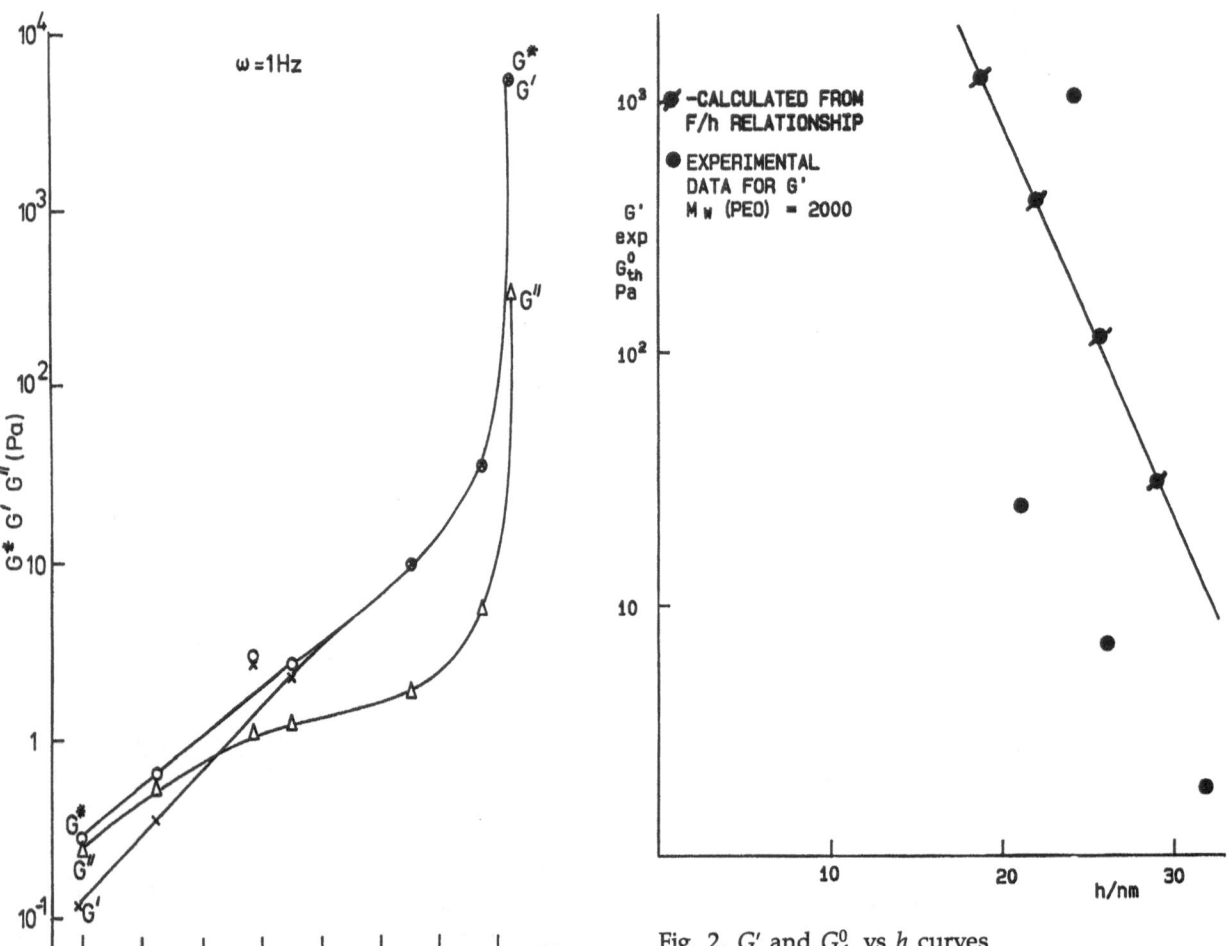

Fig. 1. G^*, G' and G'' ($\omega = 1$ Hz) vs φ for polystyrene latex dispersions containing grafted PEO chains

Fig. 2. G' and G_{th}^0 vs h curves

siderably, the general trend is the same, namely a rapid increase in elastic modulus with decrease in h, when the latter is less than 30 nm. As mentioned before, this h value is less than 2δ and hence interpenetration and/or compression of the PEO chains occurs, resulting in such elastic response.

Similar results were obtained for physically adsorbed polymers on polystyrene latex [21] and this is illustrated in Fig. 3. In this case a polystyrene latex dispersion ($R = 700$ nm) was sterically stabilised using polyvinyl alcohol (PVA) with $M_w = 45\,000$. The PVA chains form long dangling tails giving a hydrodynamic polymer layer thickness of 46 nm [22]. It can be seen from Fig. 3 that at $\varphi > 0.53$ both G^* and G' increase rapidly with further increase in

φ, whereas G'' remains fairly low. At $\varphi > 0.53$, h becomes smaller than 2δ (assuming $\varphi_m = 0.64$, i.e. random packing of the particles that is more probable for a system with long dangling tails) and any increase in φ results in a predominantly elastic response as a result of the steric interaction between the polymer tails.

Influence of addition of "free" (non-adsorbing) polymer (depletion flocculation)

Two systems were investigated, namely effect of addition of PEO (with $M_w = 20\,000$, $35\,000$ and $90\,000$) [13] and HEC (with $M_w = 70\,000$, $124\,000$ and $223\,000$) [23]. In both cases, polystyrene latex dispersion with terminally anchored PEO chains (i.e., the system described in the above section) was used and the volume fraction of the latex was kept constant at 0.3. The volume fraction of the free

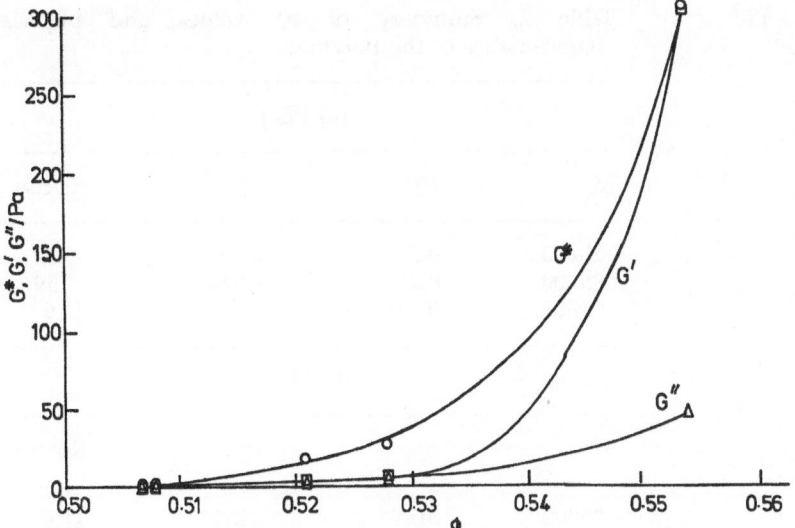

Fig. 3. G^*, G' and G'' (ω = 1 Hz) vs φ for polystyrene latex dispersions containing physically adsorbed PVA chains

polymer φ_p in bulk solution was gradually increased and rheological measurements carried at each φ_p value.

Steady state shear stress — shear rate measurements showed that below a critical φ_p value, the latex dispersion was Newtonian (this is to be expected since φ = 0.3), whereas above this critical value, non-Newtonian (pseudoplastic) flow was observed. The φ_p value at which such transition occurred was denoted by φ_p^+, i.e. the polymer volume fraction at which depletion flocculation starts. Above φ_p^+, the dispersions also showed a measurable shear modulus G_∞, which increased with further increase in φ_p.

As an illustration, Fig. 4 shows a plot of G_∞, τ_β and τ_c as a function of φ_p for PEO, M_w = 20000. At the $\varphi_p > 0.02$, all rheological parameters increase with further increase in φ_p. Similar trends were observed for all other systems studied. The influence of increasing molecular weight of the added free polymer is illustrated in Fig. 5 which shows a plot of τ_β vs φ_p for HEC with three different molecular weights. It can be clearly seen that φ_p^+ decreases with increase of M_w, as expected.

A summary of φ_p^+ values obtained with the two systems studied, as well as some of the physical parameters of the polymer is given in Table 2. φ_p^* was calculated using the following equation

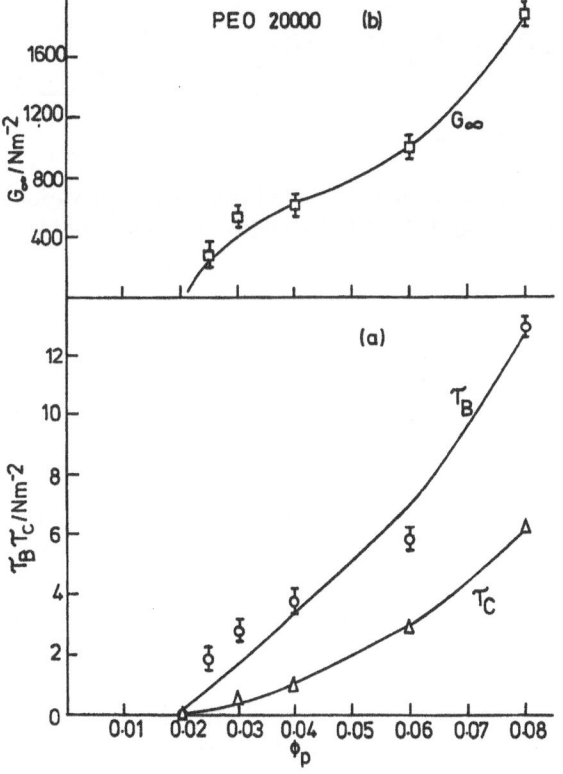

Fig. 4. G_∞, τ_β and τ_c vs free polymer volume fraction φ_p (PEO, M_w = 20000) for a polystyrene latex dispersion at φ = 0.3

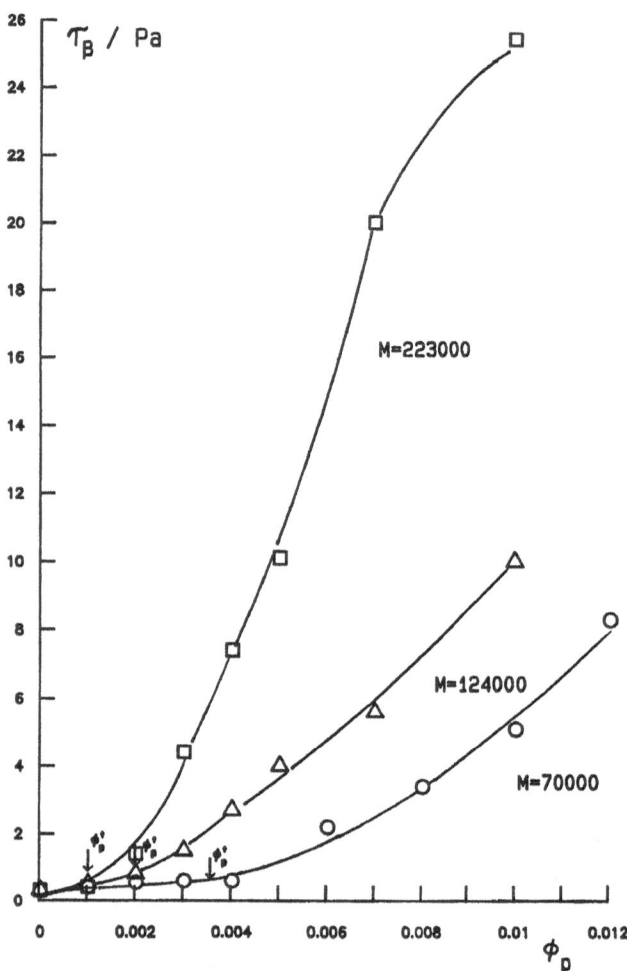

Fig. 5. τ_β vs φ_p of hydroxy ethyl cellulose with various molecular weights for a polystyrene latex dispersion $\varphi = 0.3$

Table 2. Summary of φ_p^+ values and physical characteristics of the polymers

(a) PEO			
M_w	φ_p^+	φ_p^*	Rg/nm
20000	0.02	0.035	5.52
35000	0.01	0.024	7.59
90000	0.005	0.012	12.9

(b) HEC			
M_w	φ_p^+	φ_p^*	Rg/nm
70000	0.0035	0.012	11.9
124000	0.002	0.007	17.1
223000	0.001	0.004	24.8

As discussed before [13, 25—27], it is possible to relate the Bingham yield stress, τ_β, to the interparticle interaction. The latter may be equated to the amount of energy required to separate the flocs into single units [26, 27], i.e.

$$\tau_\beta = N E_{Sep} , \qquad (12)$$

where N is the total number of contacts between particles in flocs and E_{Sep} is the energy required to break each contact. The total number of contacts N may be related to the volume fraction φ, and the average number of contacts per particle, n, by [27]

$$N = 1/2 \left(\frac{3\varphi n}{4\pi R^3} \right) . \qquad (13)$$

Combining Eqs. (12) and (13) gives

$$\tau_\beta = \left(\frac{3\varphi n E_{Sep}}{8\pi R^3} \right) . \qquad (14)$$

Thus, E_{Sep} can be calculated from τ_β, provided a value can be assigned for n. The maximum value for n is probaly 8, which is the average number of contacts in a floc for random close packing. However, recent work on the structure of aggregates indicates that quite open structures often arise, in which case n would be lower than 8. Thus, two values were assumed for n, namely 8 and 4, which is probably

$$\varphi_p^* = \frac{M_w}{b R_g^3 N_{av} \rho} , \qquad (11)$$

where R_g is the radius of gyration of the polymer that was calculated using the Stockmayer-Fixman relationship [24], b is a constant that is equal to 5.63 for hexagonal close-packing of the polymer coil, N_{av} is Avagadro's constant and ρ is the density of the polymer.

It can be seen from Table 2 that φ_p^* values are higher than φ_p^+. This is particularly the case with HEC. It should be mentioned, however, that calculation of φ_p^* using Eq. (11) is very approximate and based on a crude model of hexagonally close-packed polymer coils behaving as hardspheres.

the upper and lower limits of n. Moreover, one has to assume that above the yield point all contacts are broken. This assumption is justified if the flocculation is weak. Above the flocculation point the system probably consists of a network of weakly flocculated latex. These flocs can be reversibly broken under high shear, resulting in the formation of primary units. Evidence that this is the case is obtained from the relatively small dependence of plastic viscosity on free polymer concentration. Thus, at relatively high shear rates most of the flocculated structure is broken down.

The results of the calculation of E_{Sep} from φ_β on the basis of the above assumptions are given in Table 3 for the PEO system and in Table 4 for the HEC system. These values of E_{Sep} may be equated to the free energy of flocculation due to depletion. The latter is basically the free energy minimum in the free energy-distance curve in the presence of free polymer. Several models are available for calculation of G_{dep}, most of which are based on the concept of depletion of chain polymers near a non-adsorbing hard surface, leading to an osmotic attraction between two particles when two such depletion layers overlap. The earliest model is due to Asakura and Oosawa [5], who arrived at the following expression for the depletion free energy of attraction,

$$G_{dep}/kT = -(3/2)\varphi_2 X^2 \; ; \quad 0 < X < 1 \, , \qquad (15)$$

where φ_2 is the volume concentration of the polymer that is given by

$$\varphi_2 = \frac{4\pi\Delta^3 N_2}{3v} \, , \qquad (16)$$

where Δ is the depletion thickness that is equal to R_g, N_2 is the number of polymer molecules, and v is the total volume of the solution; $\beta = R/\Delta$ and X is given by the expression

$$X = \left(\Delta = \frac{h}{2}\right)\Big/\Delta \, , \qquad (17)$$

where h is the distance of separation between the outer surfaces. Clearly, when $h = 0$, i.e. at the

Table 3. Summary of the results of E_{Sep} calculated from the experimental τ_β values and G_{dep} calculated using Asakura and Oosawa's [5] and Fleer, Scheutjens and Vincent (FSV) [14] models

φ_p	τ_p/Nm^{-2}	E_{Sep} (kT) ($n = 8$)	E_{Sep} (kT) ($n = 4$)	G_{dep} (kT) (Asakura and Oosawa's model)	G_{dep} (kT) (FSV model)
(a) PEO, $M_w = 20000$					
0.025	2.0	9.1	18.2	25.3	81.9
0.03	2.8	12.7	25.4	30.3	102.9
0.04	3.8	17.3	34.6	40.5	149.5
0.06	5.8	26.4	52.8	60.7	271.1
0.08	13.1	59.6	111.2	80.9	397.4
(b) PEO, $M_w = 35000$					
0.015	2.3	10.5	21.0	16.4	54.1
0.02	4.4	20.0	40.0	21.8	78.0
0.03	7.0	31.9	63.8	32.7	135.0
0.04	11.7	53.2	106.0	43.6	203.0
(c) PEO, $M_w = 90000$					
0.01	1.2	5.5	11.0	12.3	48.4
0.015	2.8	12.7	24.5	18.4	85.6
0.02	4.4	20.0	40.0	24.5	131.3
0.025	5.9	26.9	53.8	30.6	185.7

Table 4. Summary of results of E_{Sep} calculated from the experimental τ_β values and G_{dep} calculated from theory

φ_p	τ_p/Nm^{-2}	E_{Sep} (kT) $(n = 8)$	E_{Sep} (kT) $(n = 4)$	G_{dep} (kT) (Asakura and Oosawa's model)	G_{dep} (kT) (FSV model)
(a) PEO, M = 70000					
0.002	0.5	2.3	4.6	2.7	28.6
0.003	0.6	2.8	5.6	4.0	43.4
0.004	0.7	3.2	6.4	5.4	58.5
0.006	2.2	10.0	20.0	8.0	89.7
0.008	3.4	15.5	30.9	10.7	122.2
0.010	5.1	23.2	46.4	13.4	156.0
0.012	8.3	37.7	75.4	16.0	191.1
(a) M = 124000					
0.001	0.5	2.3	4.6	1.6	17.0
0.002	0.9	4.1	8.2	3.1	34.6
0.003	1.6	7.3	14.5	4.7	53.0
0.004	2.7	12.3	24.5	6.2	72.0
0.005	4.0	18.2	36.4	7.8	91.8
0.007	5.7	25.9	51.8	10.9	133.3
0.010	10.0	45.5	90.9	15.6	200.7
(a) M = 223000					
0.001	0.4	1.8	3.6	1.8	20.7
0.002	1.9	8.7	17.3	3.7	42.9
0.003	4.9	22.3	44.5	5.5	66.7
0.004	7.9	35.9	71.9	7.3	91.8
0.005	10.7	48.7	97.3	9.2	118.5
0.007	20.0	90.9	181.8	12.8	176.3
0.10	25.4	115.5	231.0	18.3	274.2

point where the free polymer oils are "squeezed out" from the region between the particles, $x = 1$.

The thoery of Asakura and Oosawa [5] has been modified by several authors, for example by Vrij [28] and Sperry [6]. More recently, a general approach of the interaction of a hard sphere in the presence of non-adsorbing polymer has been introduced by Fleer et al. (FSV model) [4] who arrived at the following expression for G_{dep}, (when $h = 0$)

$$G_{dep} = \frac{2\pi R}{v_1}(\mu_1 - \mu_1^0)\Delta^2 \left(1 + \frac{2\Delta}{3a}\right), \quad (18)$$

where v_1 is the molecular volume of the solvent, μ_1 is the chemical potential of the solvent at a volume fraction of free polymer φ_p and μ_1^0 the corresponding value at $\varphi_p = 0$. $(\mu_1 - \mu_1^0)$ can be calculated from φ_p using the expression (19).

$$\frac{\mu_1 - \mu_1^0}{kT} = -\frac{\varphi_p}{n} - (1/2 - \chi)\varphi_p^2, \quad (19)$$

where n is the number of polymer segments and χ is the polymer — solvent interaction parameter, which for PEO is equal to 0.473, and for HEC it is 0.47.

Value of G_{dep} at various φ_p values are given in Table 3 for PEO and in Table 4 for HEC. For the PEO system it can be seen from the data in Table 3 that the calculated G_{dep} values deviate from those of E_{Sep} obtained from the experimental τ_β value. The closest agreement between the theoretical and experimental values of G_{dep} is obtained using Asakura and Oosawa's model.

With HEC (Table 4), deviation between E_{Sep} and G_{dep} is also obtained, although some agreement is obtained in certain cases, e.g. agreement between

E_{Sep} (assuming $n = 8$) with G_{dep} using Asakura and Oosawa's model for $M_w = 70000$. One should remember that all theories are based on a hard-sphere model for both particles and polymer coils, a clearly unrealistic assumption. Moreover, several assumptions have to be made for calculating E_{Sep} from τ_β. Thus, at best, agreement between E_{Sep} and G_{dep} is expected to be qualitative.

References

1. Napper DH (1985) "Polymeric stabilisation of colloidal dispersions, Academic Press, London, N.Y.
2. Prestige C, Tadros TF (1988) J Colloid Interface Sci 12:660
3. Ottewill RH (1982) In: Goodwin JW (ed) Concentrated Dispersions, Royal Society of Chemistry, London, Chapter 3, No 43
4. Fleer GJ, Scheutjens JHMH, Vincent B (1984) ACS Symp Ser 240:245
5. Asakura S, Oosawa F (1954) J Chem Phys 22:1255; (1958) J Polym Sci 37:183
6. Sperry PR (1982) J Colloid Interface Sci 87:375
7. Vincent B, Luckham PF, Waite F (1980) J Colloid Interface Sci 73:508
8. Vincent B, Edwards J, Emmett S, Jones A (1986) Colloids Surfaces 17:261
9. Krieger IM (1972) Adv Colloid Interface Sci 3:111
10. Krieger IM, Dougherty M (1959) Trans Soc Rheol 3:137
11. Costello BA de L, Luckham PF, Tadros TF (1988/1989) 34:301
12. Bromley C (1985) Colloids Surf 17:1
13. Prestidge C, Tadros TF (1988) Colloids and Surfaces 31:325
14. Casson N In: Mill CC (ed) (1959) Rheology of Disperse Systems, Pergamon Press, Oxford, pp 84—104
15. Van Olphen H (1956) Clay Miner 4:68; (1958) 6:106
16. Goodwin JW, Smith RW (1974) Faraday Discuss Chem Soc 57:126; Goodwin JW, Khider AH (1976) In: Kerker M (ed) Colloid and Interface Sci, Academic Press, New York, Vol IV, p 529
17. Cosgrove T, Crowley TL, Vincent B, Barnett KG, Tadros TF (1981) Faraday Symp Chem Soc 16:101
18. Ottewill RH, private communication
19. Costello BA de L, Ph. D. Thesis in preparation
20. Costello BA de L, Luckham PF, Tadro TF, to be published
21. Hopkinson A, Tadros TF, to be published
22. Van Den Boomgaard T, Change HC, King TA, Tadros TF, Vincent B (1978) J Colloid & Interface Sci 66:68
23. Gover M, Tadros TF, to be published
24. Stockmayer WH, Fixman M (1963) J Polym Sci Part C 1:137
25. Heath D, Tadros TF 76 (1983) Faraday Disc Chem Soc 76:203
26. Luckham PF, Vincent B, Tadros TF (1983) Colloids Surfaces 6:101
27. Gillespie (1960) J Colloid Sci 5:219
28. Vrij A (1976) Pure Appl Chem 48:471
29. Flory PJ (1953) "Principle of Polymer Chemistry", Cornell University Press, p 511

Author's address:

Dr. Th. F. Tadros
ICI Agrochemicals
Jealotts Hill Research Station
Bracknell
Berks, RG12 6EY, United Kingdom

Interfacial electrochemistry of oxides: Recognition of common principles

J. Lyklema, L. G. J. Fokkink and A. de Keizer

Department of Physical and Colloid Chemistry, Wageningen Agricultural University, Wageningen, The Netherlands

Abstract: Experimental data on the surface charge and enthalpy of double-layer formation are compared between a number of oxides and under a variety of conditions. It is expedient to introduce the concepts "specific" and "generic", in order to distinguish between properties referring to a specific oxide, electrolyte, or experimental condition and features of general applicability. In this way it becomes possible to recognize common backgrounds which, in turn, contribute to the understanding and can be advantageously used in a variety of applications, such as colloid stability and double-layer modelling.

Key words: Oxides, electrical double layer; haematite, electrical double layer; rutile, electrical double layer; oxides, interfacial double layer; electrical double-layers, common principles

Statement of the problem

The formation of electrical double-layers around dispersed particles involves electrical and chemical contributions and the accompanying change in Gibbs energy therefore has two terms. Generally, this can be written as

$$\Delta G = \Delta G(\text{el}) + \Delta G (\text{chem}) . \tag{1}$$

By a number of techniques, ΔG can be measured. For instance, ΔG can be obtained by analyzing the adsorption equilibrium of electrolytes, containing charge-determining (also called potential-determining) ions (c.d. ions). However, there is no operational way to subdivide ΔG into its constituents. In other words, model assumptions must be invoked to realize this. This issue is reminescent of the inoperationality of splitting electrochemical potentials ($\tilde{\mu}_i$) into a chemical (μ_i) and an electrical part ($z_i F \psi$, where ψ represents the electrical potential at the position where $\tilde{\mu}_i$ is measured and where F is the Faraday constant).

Similar problems are encountered with the enthalpy of double-layer formation which can be measured microcalorimetrically:

$$\Delta H = \Delta H (\text{el}) + \Delta H (\text{chem}) . \tag{2}$$

Accepting the thermodynamic impossibility of breaking down ΔG and ΔH into their constituent parts, the desirability of having insight into the electrical and chemical contributions is clear. Such insight is, for instance, very relevant in interpreting colloid stability: in the DLVO picture interaction at constant potential involves chemical and electrical contributions, whereas that at constant charge is of a purely electrical nature. Automatically, the problem arises as to what extent the required division can be realized with a minimum of non-thermodynamical assumptions.

Below, we propose an approach to achieve this. It is based on a comparison of double-layer properties obtained with different systems under differing conditions, whereby a distinction is made between *specific* properties (i.e., properties applying only to a specific electrolyte or to a specific colloid) and properties that the systems have in common, which we shall call *generic*.

Apart from the general consideration that the intended distinction is helpful in classifying systems (it has, for instance, been used to distinguish between ideal and non-ideal gases), it is likely to contribute to the separations envisaged in (1) and (2), because electrical contributions are generic, whereas chemical ones are specific. More precisely, $z_i F \psi$ is identical for different ions of the same valency, whereas chemical contributions are typically size-dependent and, hence, specific. It is appreciated that as soon as $z_i Fc$ is taken as the definition of "electrical" energy, automatically all other contributions are *by definition* "chemical", even if on closer inspection they have electrical origins, as is the case with chemical bonds or hydrogen bridges. Any difference between the mean potential and the potential of mean force is also counted as "chemical".

Comparison between double layers on different materials

Over the past decades the technique of potentiometric colloid titrations, together with the development of better controlled systems, has made enough progress to allow a significant comparison between double-layer properties on different substrates. Figure 1 gives a first example, dealing with mercury, silver iodide, and ruthenium dioxide.

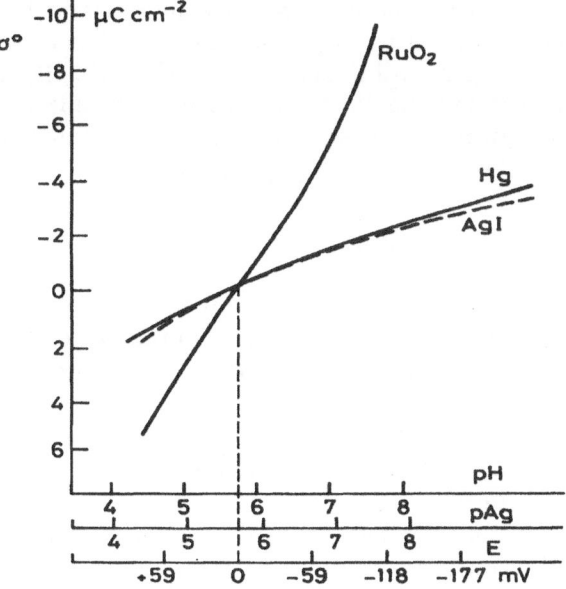

Fig. 1. Comparison between the electrical double layer on mercury [1], silver iodide [2], and ruthenium oxide [3]. Electrolyte, KCl or KNO$_3$, 0.1 M. Temperature 25°C. The pH, pAg, and E axes apply to RuO$_2$, AgI and Hg, respectively

The surface charge σ^0 for mercury was found by integration of Grahame et al.'s directly determined differential capacitances [1]: those on AgI and RuO$_2$ by potentiometric colloid titration with AgNO$_3$ and KI or HCl and KOH, respectively [2, 3]. For technical details, see the original papers.

In order to allow comparison, the abscissae axes have been normalized with respect to their points of zero charge (p.z.c.) or, in the case of mercury, its electrocapillary maximum (e.c.m.). Moreover, for AgI and RuO$_2$ one unit of pAg or pH, respectively, was set equal to 59 mV. That is, for these surfaces Nernstian behavior was presumed. This identity is today well-documented, mainly by comparison with the behavior of the corresponding electrode material. For the AgI-system this correspondence has been discussed before [4]; for oxides Nernstian behavior has been found to apply over several decades for, e.g. hematite [5], zirconium oxide [6], and titanium dioxide [7]. In fact, our analysis expounded below, confirms this indirectly.

As Fig. 1 demonstrates, mercury and silver iodide behave similarly, at least as to their interfacial electrochemistry. This correspondence recurs in such features as the adsorption of organic molecules (on both systems, butanol adsorbs with its hydrophobic moiety towards the surface, and in both systems the surface concentration Γ_{BuOH} passes through a maximum as a function of σ^0, which is situated to the right of the point of zero charge [8]) and lyotropic sequences (on both surfaces σ^0 increases from Li$^+$ to Rb$^+$ as the counterion, where the ion specificity is more pronounced on AgI [9, 10]). The differential capacitances (slopes of the curves in Fig. 1) are also similar; at the p.z.c. they are even identical. On the other hand, RuO$_2$ behaves very differently. The most striking feature is that its capacitance is higher than that on the other two surfaces by about a factor of ten. As several other oxides do also have high capacitance double layers the question automatically arises whether or not σ^0 (pH) curves for various oxides are as similar as those between Hg and AgI.

In order to allow for such a comparison, normalization of the pH axes with respect to the p.z.c.'s is required. Zero points are oxide-*specific* properties, reflecting the chemical affinity of protons for the surface: the higher this affinity, the higher the p.z.c. However, if the required normalization is carried out it is concluded that for at least three different oxides, notwithstanding substantial differences in p.z.c., an almost perfect congruence is obtained [11, 12]. In Fig. 2 the band width is indicated between

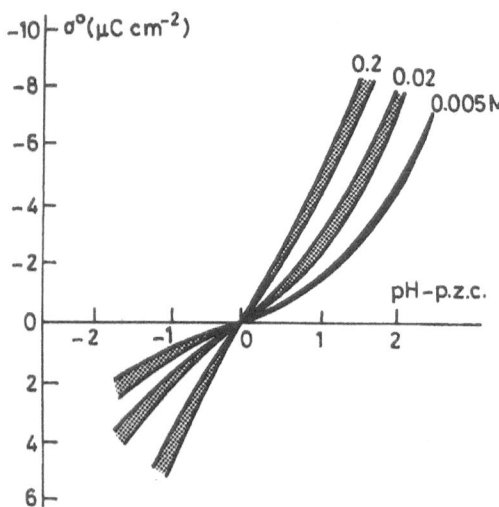

Fig. 2. Congruence of σ^0 (pH-p.z.c.) curves. For RuO$_2$, α-Fe$_2$O$_3$ and TiO$_2$ (rutile) at 25°C all data fall within the band width of the plotted master curves

which the curves for RuO$_2$, TiO$_2$ (rutile) and α-Fe$_2$O$_3$ (hematite) are found. It is concluded that the charge formation outside the p.z.c. is an oxide-independent, i.e., a *generic* phenomenon.

The observed genericity is a pabulum for further thought. Fistly, there is the question of representativity: do all σ^0 (pH-p.z.c.) curves on all oxides coincide? Before answering this question it is noted that a number of practical problems are inherent in the measurement of these curves. Measurements done by different investigators in different laboratories and on diffently prepared oxides are likely to show some differences. One of them is the specific surface area of the sample, of which the choice is proportionally reflected in the curves. The data on the bases of which the curves of Fig. 2 are based have been obtained by different authors in the same department, which may be a guarantee for uniformity in sample preparation and quality of the precautions taken. Against this background, it is striking that silica is an exception: its σ^0 (pH-p.z.c.) curve has a very different shape [13]. The colloidal stability of silica sols is also unusual: the sols tend to be stable around the p.z.c. (about 2), becoming less stable if the particles are (negatively) charged [14, 15]. Alumina appears to agree, at least in part [18], and there are more examples of systems obeying our congruence [17].

Disregarding the idiosyncratic behavior of silica, and allowing for more exceptions, it is concluded that, as rule, the σ^0 (pH) curves rest on two foundations:

i) an oxide-*specific*, chemically determined p.z.c.;
ii) a *generic* charge formation mechanism.

It is likely that this latter process is essentially electrostatical and determined by the solution side of the double layer. In fact, the master curves of Fig. 2 can be very well reproduced by a simple Gouy-Stern picture [18], assuming a) Nernst behavior of the surface potential, and b) a constant, oxide-independent Stern layer capacitance $C^i = \varepsilon^i \varepsilon_0 / d$ of about 450 µF cm^{-2}, where ε^i is the relative dielectric permittivity of the Stern layer, d is its thickness, and ε_0 the permittivity of vacuum. The precise value of C^i is not critical, provided it is high. (By comparison, on AgI and Hg, C^i is around 30 µF cm^{-2} at the p.z.c.) From our observation that C^i is a generic parameter, it is concluded that the water structure adjacent to the surfaces must be very similar between the various oxides, and from the fact that C^i is high and constant within experimental error, it is inferred that ε^i is high and insensitive to dielectric saturation, and that d is low. In turn, the high invariant value of ε^i suggests that the bulk water structure persists down to the very surface. All of this is in contrast to AgI and Hg, which behave as hydrophobic surfaces; ε^i is about 15 at the p.z.c. and dependent on σ^0. In this light, the deviating behavior of silica may be attributed to a very strong water structure formation which is broken down upon charging the surface.

Effect of temperature

Normalized surface charge curves are not only generic with respect to the nature of the oxide, they also exhibit similar behavior as a function of temperature. Figure 3 shows results for hematite and rutile; all observations fall within the hatched band width [18]. We call this phenomenon *temperature congruence*.

Mathematically, temperature congruence implies that the integral

$$\int_0^{\sigma^0} (\text{pH}' - \text{p.z.c.}) d\sigma^{0'} = B \qquad (3)$$

is independent of temperature. The electrical part of the Gibbs energy of a double layer is given by

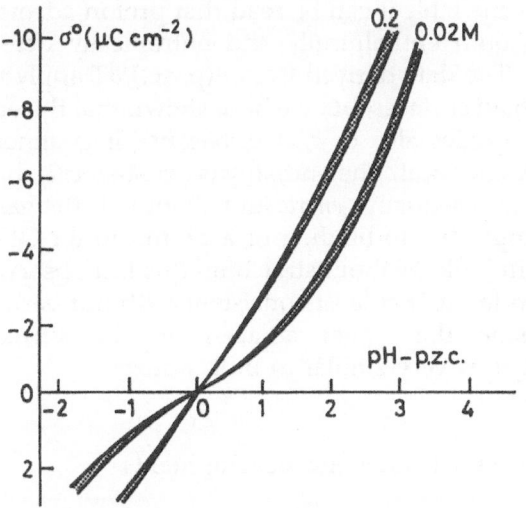

Fig. 3. Temperature congruence. For α-Fe$_2$O$_3$ and TiO$_2$ (rutile) all data between 5°C and 50°C coincide within the indicated band width

$$G^\sigma(\text{el}) = \int_0^{\sigma^0} \psi^{0'} d\sigma^{0'}, \qquad (4)$$

which, because of the validity of Nernst's law, may be combined with (3) to give

$$G^\sigma(\text{el}) = -\frac{2.3\,RT}{F} \int_0^{\sigma^0} (\text{pH}' - \text{p.z.c.}) d\sigma^{0'}$$

$$= \frac{2.3\,RT}{F} B, \qquad (5)$$

which is proportional to T. From this, using the Gibbs-Helmholtz relation

$$H^\sigma(\text{el}) = RT^2 \frac{\partial[G^\sigma(\text{el})/T]}{\partial T} \approx 0. \qquad (6)$$

In other words: the electrical contribution to the enthalpy of double layer formation is very small and perhaps negligible. For two reasons $H^\sigma(\text{el})$ need not be *exactly* zero. First, the curves of Fig. 3 have a certain band width, and secondly, there is some difficulty about the proper way to execute the differentiation in (3). We return to this conclusion and its inherent uncertainty below.

It would be very interesing to investigate the temperature effect more systematically on more oxides, silica among them.

Double-layer microcalorimetry

Enthalpies of double-layer formation can also be directly obtained in a sufficiently accurate microcalorimeter [19]. Figure 4 gives the acid branches of microcalorimetric titration experiments in which the cumulative enthalpy is measured as a function of σ^0. These titrations are reversible in that back-and-forth titration to the same σ^0 only produces additional water (apart from an insignificant dilution) of which the corresponding enthalpy agreed with the standard value. The data are given per unit of mass of the oxides; however, the interesting parameter is the slope of the curve, i.e., the enthalpy per unit of charge, H_c (in J μC^{-1}). As the curves are linear, there is no need to distinguish between integral and differential enthalpies per unit of charge.

The following observations can be made:

i) The lines are linear, there is no break at the p.z.c.;
ii) The lines are independent of the concentration of electrolyte (not shown in the figure);
iii) H_c is higher for hematite.

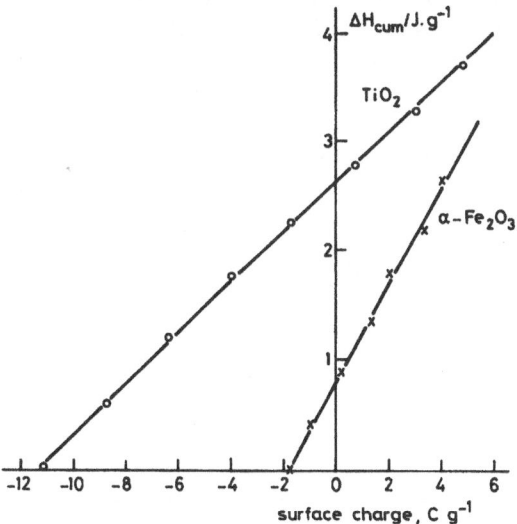

Fig. 4. Cumulative adsorption enthalpy for acid titration of rutile and hematite. Electrolyte: 0.02 M KNO$_3$

In our terminology, H_c is oxide-*specific*, but *generic* with respect to the sign of σ^0 and c_{salt}. All these observations are consistent with (6): H_c is essentially identical to $H_c(chem)$ because $H_c(el) \approx 0$, so that also $H^\sigma(el) \approx 0$. The absence of a break (or change of slope) around $\sigma^0 = 0$ indicates that the charge forming process is identical on (+) and (−) oxides; it is likely that this process is just adsorption and desorption of protons. In turn, this implies that only one pK_a suffices to define completely the acidity/basicity of an oxide surface. The same has been suggested in a different context by Bolt et al. [18].

That $H_c \approx H_c(chem)$ is reflected in iii): protons have a higher affinity for hematite, and this is the reason for its higher p.z.c. (≈ 9, as compared with 5.4 for rutile).

Thermodynamic data for double layer formation

In Table 1 we summarize thermodynamic functions for the charge formation, expressed in kJ per mole of protons adsorbed. Besides the microcalorimetry there is another source: from the absolute value of the p.z.c. and its variation with T it is possible to derive $\Delta_{ads}G^0_{H^+}$, $\Delta_{ads}H^0_{H^+}$ and $\Delta_{ads}S_{H^+}$ [18]. As the thermodynamic analysis is phenomenological, it gives $\Delta_{ads}G^0$, etc., only per unit of charge; but we interpreted this as due to protons, for reasons given above. Further, $\Delta_{ads}H^0_{H^+}$ equals H_c except for a factor of F. There are some uncertainties in the precise establishment of $\Delta_{ads}H^0_{H^+}$ from $\partial(p.z.c.)/\partial T$, to be discussed elsewhere (see [19]), but the agreement between the enthalpies obtained in this way and those from microcalorimetry is sufficiently close to accept the data as a good first approximation.

From the table it can be read that proton adsorption is both enthalpically and entropically determined. The data derived from $\partial(p.z.c.)/\partial T$ apply to uncharged surfaces, but we have shown that the enthalpy applies also to $\sigma^0 \neq 0$, because it is almost entirely chemical. The enthalpy is oxide-*specific*, but $T\Delta_{ads}S$ is essentially *generic* and about half the gain in entropy due to binding of a proton to an OH^- group in bulk. Without stretching this last observation too far, it is at least consistent with our earlier conclusion that water adjacent to the surface hydroxyls is very similar to bulk water.

Further conclusions and developments

In the present study we have shown that by comparing data on different systems and under different conditions considerable insight is gained on the formation and structure of electrical double layers on oxides, allowing only few model assumptions. The key to the analysis is the distinction between *specific* and *generic* features, allowing us to recognize common principles. Having gone so far, a number of elaborations and extensions present themselves. Among them, we mention the following.

i) Having established that $\Delta_{ads}H^\sigma(el) \approx 0$, the question must be asked why this is so. Answering this problem requires a thorough theoretical analysis of the various contributions to the formation of double layers. In a forthcoming publication [19] we shall demonstrate that, indeed, $\Delta_{ads}H^\sigma(el)$ is very low, but non-zero for high-capacitance double layers. Basically, this can be understood, thus, in such double layers a large fraction of the counterions adsorbs very close to the surface ions, and hence, at almost the same potential. In this way the electrical contributions of surface ions and counterions compensate each other to a large extent.

ii) Considering the previous remark, it follows that a different behavior must be expected for low-capacitance double layers, such as AgI and Hg (see Fig. 1).

iii) We mentioned already that a systematic temperature study of silica (the "exceptional" oxide) might well yield interesting results.

iv) Our approach may be extended to more difficult systems. Anticipating a forthcoming study on this matter [20] we note that also for the binding of Cd^{2+} ions to surface hydroxyls root principles may

Table 1. Thermodynamic functions for the binding of protons to oxides. Data in kJ mole^{-1}. Temperature 20 °C

	$\Delta_{ads}G^0_{H^+}$	$\Delta_{ads}H^0_{H^+}$		$T\Delta_{ads}S$
		Micro-cal.	$\dfrac{\partial(p.z.c.)}{\partial T}$	
Rutile	−31.0	−21	−17.6	+13.4
Hematite	−48.8	−36	−36.3	+12.5
(bulk OH⁻)	(−79.5)	(−57.5)		(+22.5)

be recognized. One of them is that this binding is essentially entropical, just like the formation of $CdOH^+$ in solution from Cd^{2+} and OH^-.

The overall conclusion may be that the potentialities of this thermodynamic approach are by no means exhausted.

References

1. Integration of differential capacitance curves by Grahame DC, Poth MA, Cummings JI (1952) J Am Chem Soc 74:4422
2. Lyklema J, Overbeek JTG (1961) J Colloid Sci 16:595
3. Interpolated data from Kleijn JM, Lyklema J (1987) J Colloid Interface Sci 120:511
4. Bijsterbosch BH, Lyklema J (1978) Adv Colloid Interface Sci 9:147
5. Penners NHG, Koopal LK, Lyklema J (1986) Colloids Surf 21:457
6. Ardizzone S, Radaelli M (1989) J Electroanal Chem 369:461
7. Hepel T, Hepel M, Osteryoung RA (1982) J Electrochem Soc 129:2132
8. Bijsterbosch BH, Lyklema J (1965) J Colloid Sci 20:665
9. Lijklema J (1961) Kolloid-Z 175:129
10. Grahame DC (1951) J Electrochem Soc 98:343
11. Fokkink LGJ, de Keizer A, Kleijn JM, Lyklema J (1986) J Electroanal Chem 208:401
12. Lyklema J (1987) Chem & Ind 21:741
13. Tadros TF, Lyklema J (1968) J Electroanal Chem 17:267
14. Depasse J, Watillon A (1970) J Colloid Interface Sci 33:430
15. Allen LH, Matijevic E (1969) J Colloid Interface Sci 31:287 and (1970) ibid. 33:420
16. Fokkink LGJ, de Keizer A, Lyklema J (1989) J Colloid Interface Sci 127:116
17. Stumm W (1989) private communication
18. Bolt GH, van Riemsdijk WH (1982) In: Bolt GH (ed) Soil Chemistry, B Physicochemical Models, 2nd ed. Elsevier, Amsterdam, pp 457; van Riemsdijk WH, de Wit JCM, Koopal LK, Bolt GH (1987) J Colloid Interface Sci 116:511
19. Fokkink LGJ, de Keizer A, Lyklema J (1990) Colloids Surfaces, in press
20. Fokkink LGJ, de Keizer A, Lyklema J, in course of preparation

Authors' address:

Prof. Dr. J. Lyklema
Department of Physical and Colloid Chemistry
Wageningen Agricultural University
P.O. Box 8038
6700 EK Wageningen, The Netherlands

Progress in Colloid & Polymer Science   Progr Colloid Polym Sci 83:52—55 (1990)

Molecular wetting films

A. M. Cazabat, F. Heslot and N. Fraysse

Collège de France, Physique de la matière condensée, Paris, France

Abstract: The growth of precursor films of non-volatile liquids results from the balance between the disjoining pressure gradient and the dissipative processes. — From the profile and the extension of films developing on solid surfaces, information can be obtained on the disjoining pressure and on the friction terms. For relatively thin films, the viscous friction in the film is the dissipative process. For ultrathin, possibly monomolecular films, it is the friction of the molecules on the solid which plays the role.

Key words: Wetting; thin films; film pressure; monomolecular films; disjoining pressure

Introduction

For non-volatile liquids, i.e., when the transport of molecules through the gas phase is negligible, precursor films may grow at the macroscopic liquid edges and spread over the bare solid surface.

The driving parameter for spreading [1] is the initial spreading coefficient S_0 [2]:

$$S_0 = \gamma_{SO} - \gamma - \gamma_{SL} \, ,$$

where γ_{SO} is the solid-vacuum interfacial tension (bare surface); γ is the liquid-vacuum (or liquid-gas), and γ_{SL} is the solid-liquid interfacial tension.

S_0 is positive for complete wetting. In this case, a precursor film actually spreads over the solid. On a horizontal surface, a thick film of bulk liquid will ultimately be formed. In the capillary rise geometry, the equilibrium film thickness is limited, because of the gravity [3]. However, this final state is approached for geological times and plays no role in the spreading dynamics within the time of the experiments.

S_0 is negative for partial wetting. An (usually molecular) film may spread on the surface, but the rest of the edge is static. Here, the final state is a liquid wedge with a finite contact angle in equilibrium with the ultrathin film.

The dynamics of growth of precursor films depends on the motion of the macroscopic edge. In order to be free of macroscopic dynamic parameters, we shall only consider here the capillary rise geometry, where the liquid meniscus is static at long times. Gravity effects are always negligible.

The experimental studies are performed by ellipsometry. For thicker films, the ellipsometric profiles are compared to the predictions of the hydrodynamic theory of de Gennes [4]. For molecular films, we propose a simple theoretical analysis which follows roughly the same lines [5].

Theoretical models

In the capillary rise geometry, the film grows at the edge of a static meniscus. Let z be the local film thickness, x the longitudinal coordinate, t the time and U the average velocity. The continuity equation is:

$$\frac{\partial z}{\partial t} + \frac{\partial}{\partial x} [zU] = 0 \, . \tag{1}$$

The velocity U results from the balance between the driving force, which is proportional to the disjoining pressure gradient, and the friction.

In the hydrodynamic case, the velocity field in the film is a Poiseuille flow. If $\pi(z)$ is the disjoining pressure, the velocity U is given by [4]

$$U = \frac{z^2}{3\eta} \frac{\partial \pi}{\partial x} .$$ (2)

This allows to rewrite Eq. (1) as a diffusion-like equation:

$$\frac{\partial z}{\partial t} = \frac{\partial}{\partial x}\left[D(z) \frac{\partial z}{\partial x} \right] ,$$ (3)

with

$$D_h(z) = -\frac{z^3}{3\eta} \frac{d\pi}{dz} .$$ (4)

Here, η is the bulk viscosity of the liquid.

In the monomolecular case, the velocity U is just the velocity of the molecules.

If a is the friction coefficient between molecules and solid surface, U can be written as [5]

$$U = +\frac{v_0}{a} \frac{\partial \pi}{\partial x}$$ (5)

$$D_m(z) = -\frac{v_0 z}{a} \frac{d\pi}{dz} .$$ (6)

Here, v_0 is the molecular volume.

The local "thickness" z is an average over a large number of molecules and becomes less than the molecule size for non-compact monolayers. In this case, the number of molecules per unit surface, c, and the film presure P are commonly used. The corresponding formulae are

$$z = cv_0$$

$$P(z) = -z\pi(z) + \int_0^z \pi(h)dh$$ (7)

$$D_m(c) = \frac{1}{a} \frac{dP}{dc} .$$

The spreading condition is $P(z) > 0$. For thick films (complete wetting), $z\pi(z)$ vanishes and the second term is just S_0:

$$S_0 = \int_0^\infty \pi(h)dh .$$

But thin films can spread ($P(z) > 0$), even for negative S_0. As a matter of fact, extremely thin, submolecular films always spread as $P(z)$ is just the perfect gas law at vanishing thickness:

$$P(z) \to \frac{kT}{v_0} z \qquad P(z) > 0$$

$$z \to 0$$

$$D_m(z) \to \frac{kT}{a} .$$

Thin films, a few molecular layers thick, form an intermediate case. Only qualitative predictions can be done, but not without importance.

What is still valid is the proportionality between $D(z)$ and $d\pi/dz$. This means that any transition betwen films of different thicknesses ($d\pi/dz = 0$) will produce a vertical part in the film profile.

For volatile liquids such transitions are known to produce step-like adsorption isotherms [6, 7]. Here, they will lead to step-like thickness profiles for non-volatile liquids.

It is worth noting that the corresponding range of thicknesses is not easily studied by equilibrium methods: non-volatile liquids are not convenient for adsorption isotherms and direct disjoining pressure measurements cannot reach these molecular thicknesses [8].

Detailed information can be obtained by systematically studying the dynamics of the last (n^{th}) layer when the spreading of the subjacent layers has been stopped [9]. The last layer spreads approximately as a monolayer. Thus,

$$D_{n,n-1} \cong \frac{1}{K_{n,n-1}} \frac{dP_{n,n-1}}{dc} ,$$

$K_{n,n-1}$ being a friction coefficient between layers n and $n - 1$. $P_{n,n-1}$ is the film pressure of the nth layer over the n-1th one. Spreading is achieved for $P_{n,n-1} > 0$.

Comparison with experiments

We shall not describe our ellipsometric set-up, or discuss the relevance of the measurements at submolecular thicknesses and on surfaces that are not atomically flat; this can be found in the literature [10—12].

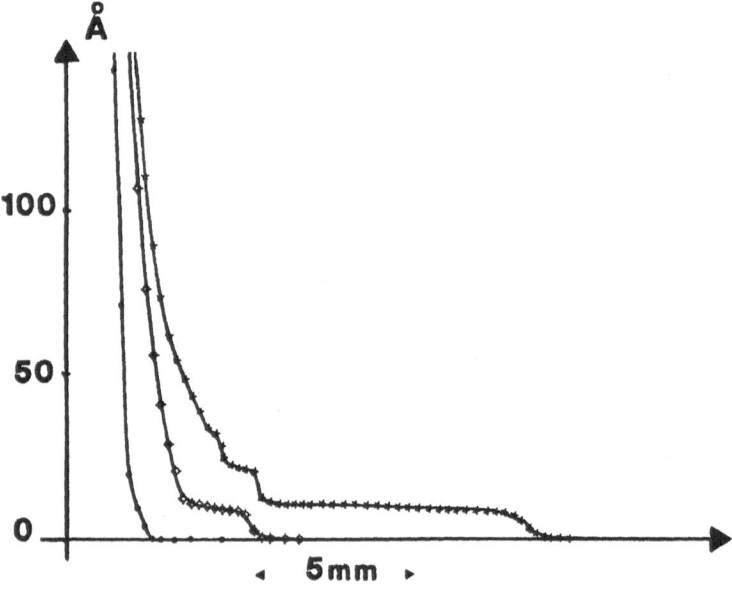

Fig. 1. Ellipsometric profiles of TK in the capillary rise geometry: after 3 h, after 24 h, after 215 h

Ellipsometric profiles recorded at increasing times in the capillary rise geometry for tetrakis (2-ethyl-hexoxy)silane (TK) are shown in Fig. 1. TK is a star-like molecule with a central silicon atom and four aliphatic arms. The molecule diameter h_0 is about 10 Å.

Successive molecular layers can be distinguished on the profile. The first layer forms a tongue of practically constant thicknes ($z \approx h_0$) terminated by a steeper edge. This means that the film pressure has a sigmoidal shape (Fig. 2) with possibly a two-dimensional (2D) phase transition between a 2D gas ($z \ll h_0$) and a 2D liquid ($z \lesssim h_0$). The diffusion coefficient for the steep part of $P(z)$ (i.e., $z \approx h_0$) can be calculated from the length of the tongue, which grows approximatively as

$$L(t) \approx \sqrt{D(h_0)t} \ .$$

Then, one obtains

$$D(h_0) \approx 1.6 \ 10^{-10} \ \mathrm{m^2 \ s^{-1}} \ .$$

At smaller thicknesses the diffusion coefficient is smaller. An average value over the range $z < h_0$ is of the order of $2 \ 10^{-11} \ \mathrm{m^2 \ s^{-1}}$. However, the relevance of this value is highly questionable:

— first, if there is a phase transition in the film the profile is composed of a (quasi) vertical part followed by a smoother tail;

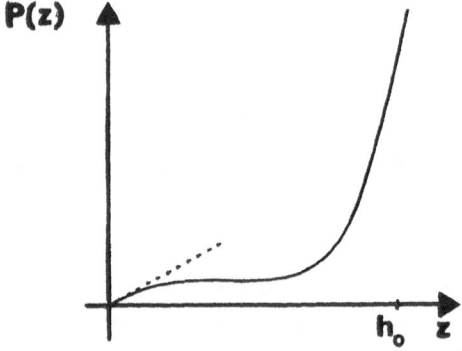

Fig. 2. Schematic behavior of the film pressure $P(z)$ for TK and PDMS

— second, if the surface is not perfectly homogeneous chemical heterogeneities or geometrical steps produce a scattering in the values of the disjoining pressure. This is well known for adsorption isotherms. A scattering in the value of $D(h_0)$ will result, which in turn contributes to the apparent width of the tip of the tongue. Very roughly, this contribution can be written as

$$\Delta \left[\sqrt{D(z < h_0}\right] \approx \sqrt{D_{\max}(h_0)} - \sqrt{D_{\min}(h_0)} \ .$$

In the present case, where $D(h_0) \ggg D(z < h_0)$, this term might well be the dominant one. Thus, the only relevant information on D at very low thickness ($z < h_0$) must be obtained by studying the spreading of submolecular droplets [10].

By comparison of the values of $D(z < h_0)$ for films or for droplets, some information on the surface heterogeneity can even be deduced.

Profiles similar to that of Fig. 1 have been obtained for light silicone oils (polydimethylsiloxane, PDMS) [11—12]. Here again, the presence of a monomolecular tongue terminated by a steep edge suggests that attractive interactions are present in the monolayer, with possibly a 2D-phase transition.

On the contrary [10], ellipsometric profiles are rather smooth in squalane, where the 2D perfect gas behavior seems to be observed as soon as $z < h_0$. The corresponding D value is found to be

$$D \approx (kT)/a \approx 7 \ 10^{-11} \ \text{m}^2 \ \text{s}^{-1} \ .$$

This is in perfect agreement with the value calculated by Teletzke [13] from a hopping model of molecular diffusion.

Conclusion

From the theoretical analyses and the first experimental investigations, it appears that the dynamic profiles of growing precursor films contain a lot of information. For non-volatile liquids, the study of these profiles is the only direct way to investigate the properties of the disjoining pressure at the molecular thicknesses. It avoids the difficulties and ambiguities of the force machine measurements by really looking at the solid/film/gas system.

Acknowledgments

Fruitful discussions with P. G. de Gennes, G. Findenegg, and J. Scheutjens are gratefully acknowledged.

References

1. de Gennes PG (1985) Rev Mod Phys 57:827—863
2. Cooper WA, Nuttall WA (1915) J Agr Sci 7:219
3. Padday JF (1971) In: "Thin liquid films and boundary layers". Special discussions of the Faraday Society, Academic Press, London and New York 64
4. Joanny JF, de Gennes PG (1986) J Phys, Paris 47:121
5. Cazabat AM, submitted
6. Bassignana IC, Larher Y (1985) Surface science 147:48
7. Ball PC, Evans R (1988) J Chem Phys 89:4412
8. Deryaguin BV, Churaev NV, Muller VM (1987) Surfaces forces — Consultant Bureau — New York and London, and references
9. Heslot F et al.: unpublished results
10. Heslot F, Cazabat AM, Levinson P (1989) Phys Rev Lett 62:1286
11. Heslot F, Fraysse N, Cazabat AM (1989) Nature 338:640
12. Heslot F, Cazabat AM, Fraysse N (1989) to appear in "Liquids — J Phys D"
13. Teletzke GF (1983) Thesis — University of Minnesota

Authors' address:

A. M. Cazabat
Collège de France
Physique de la matière condensée
11, place Marcellin-Berthelot
75231 Paris Cedex 05, France

Progress in Colloid & Polymer Science   Progr Colloid Polym Sci 83:56—58 (1990)

Construction of an atomic force microscope and application of atomic force microscopy and scanning tunneling microscopy in surface chemistry

M. Pitsch, O. Metz, J. Strnad and H.-H. Kohler

Institut für Physikalische und Makromolekulare Chemie, Universität, Regensburg, Regensburg, FRG

Abstract: A new force detector for atomic force microscope measurements is presented. A tip carrying an organic monolayer is covered by a thin metal layer that forms a tunneling gap which can be modulated by the surface forces. This arrangement eliminates the adjustment problems of the usual atomic force microscopy devices. Several experimental methods of tip preparation are discussed. — Scanning tunneling microscope images of adsorbed monolayers of *n*-octadecyl — trichlorosilane, tetraethyleneglycol laurylether, and *n*-hexadecane are presented.

Key words: Atomic force microscopy; scanning tunneling microscopy, atomic force tip; octadecyltrichlorsilane; MIM tunneling junction

Introduction

In 1986, an instrument known as the atomic force microscope (AFM) was proposed by Binnig et al. [1], which images surface topography on the atomic scale by detecting the interatomic force between a tip and the sample surface. The AFM consists of a tip, a flexible cantilever, and some device to measure the deflection of the spring. The spring deflection sensor can be based on electron tunneling between the back of the spring and a rigid reference electrode [1—6]. Images can be obtained by measuring either in the attractive or the repulsive range of the surface forces. We propose here a new type of force detector which can be used in a conventional scanning tunneling microscope set-up. AFM measurements are simplified, and a combination of STM and AFM is possible.

Construction of an atomic force microscope

A critical point of AFM devices with scanning tunneling microscope (STM) detection is the adjustment between lever and STM tip, which has not been sufficient up to now. A simple force detection

arrangement that allows use of a conventional STM as force detector has been developed by Bryant and coworkers [6]. Their combined system may be described as an AFM which can also serve as an STM.

In order to simplify and miniaturize this system, we developed a force detector which measures the deflection of a thin aluminum membrane covering an insulated STM tip. The base of an electropolished tungsten tip is insulated with a thick layer of varnish while the apex is insulated with a thin Langmuir-Blodgett film of cadmium arachidate. Thus, a well defined, stable metal-insulator-metal tunneling gap between the aluminium membrane and the tungsten tip is produced. The tunneling distance is modulated by surface forces acting between the membrane and the surface of the sample [7].

There are considerable difficulties in insulating a tip with varnish and then with a Langmuir-Boldgett film. Therefore, an electrochemically modified aluminium tip was used to avoid this insulation difficulties. The shaft of the electrolytically etched tip [8] is insulated by a layer of aluminium oxide produced by anodic oxidation [9]. During oxidation the tip is pressed against a silicon rubber, so that in this stage, no oxide forms around the apex. The tip is dipped in a 2.0 mM solution of *n*-octadecyl-trichlorosilane (OTS, $C_{18}H_{37}SiCl_3$) in *n*-hexadecane. The free aluminum of the apex region thus is coated with an adsorbed monolayer of OTS [10, 11]. It is known that OTS forms an ideal monolayer of hydrocarbon chains on aluminum. Actually, the adsorption takes place on the thin aluminum oxide layer spontaneously formed in ambient air. An evaporated aluminum layer with a thickness of about 50—100 nm forms the second electrode for the tunneling junction. These tips show insufficient resolution; probably, aluminum is too soft to form good tips.

Therefore, we adapted this method to a tungsten tip (Fig. 1). For coarse insulation polyvinylalcohol is used; then the tip is coated with OTS and aluminum in the same manner as with the aluminum tip. Again, OTS forms an ideal closed monolayer on the thin tungsten oxide layer, as can be seen from the STM image. (Fig. 2) Rows of hydrocarbon chainends are seen; within the rows single hydrocarbon chainends are recognizable. The distance between these chains is approximately 0.5 nm, so that the area required by a molecule is 0.25 nm².

With these new tips atomic resolution could not be achieved as of yet. Another disadvantage is that the tips are very delicate — at direct contact with the surface of the sample they are usually irreparably damaged, contrary to STM tips, which after contact with the surface sometimes work better than before.

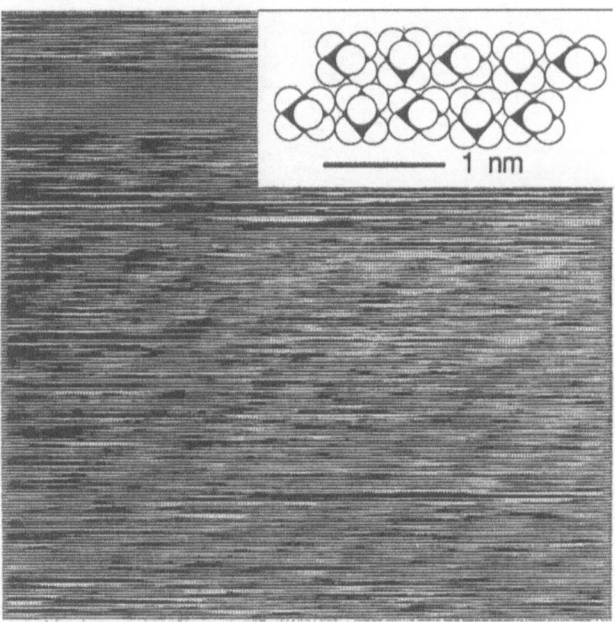

Fig. 2. STM image of *n*-octadecyl-trichlorosilane on tungsten. Scan area: 5 nm × 5 nm, x-scan frequency: 50 Hz, sample bias: —90 mV, tunneling current 10 pA. A model of hydrocarbon chainends in the same scale is inserted in the upper right corner

AFM images of adsorbed molecules with a conventional AFM were already made by some groups [12—14], but there is a problem with reproducibility. Once this problem is solved, AFM tips will be an indispensable tool for the investigation of adsorption on nonconductors as well as conductors.

STM images

STM can only be used to study the arrangement of single organic molecules on conducting surfaces. We will show here a few STM images of organic adsorbates deposited on a graphite surface.

An STM image of a tetraethyleneglycol lauryl ether ($C_{12}h_{25}(OCH_2CH_2)_4OH\%$ [15], a nonionic surfactant that is adsorbed at highly oriented pyrolytic graphite (HOPG) from a 0.1 mM aqueous solution is shown in Fig. 3. Measurements were made in the presence of the aqueous solution. Near a step in the graphite surface individual molecules of the polyglycol ether can be seen that are closely packed. Both the hydrophobic and hydrophilic parts of the molecule lie flat on the surface. This behavior was

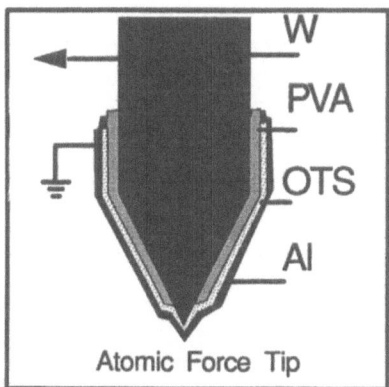

Fig. 1. Schematic view of an AFM tip. W: tungsten tip, PVA: polyvinylalcohol, OTS: *n*-octadecyl-trichlorosilane, Al: evaporated aluminum layer

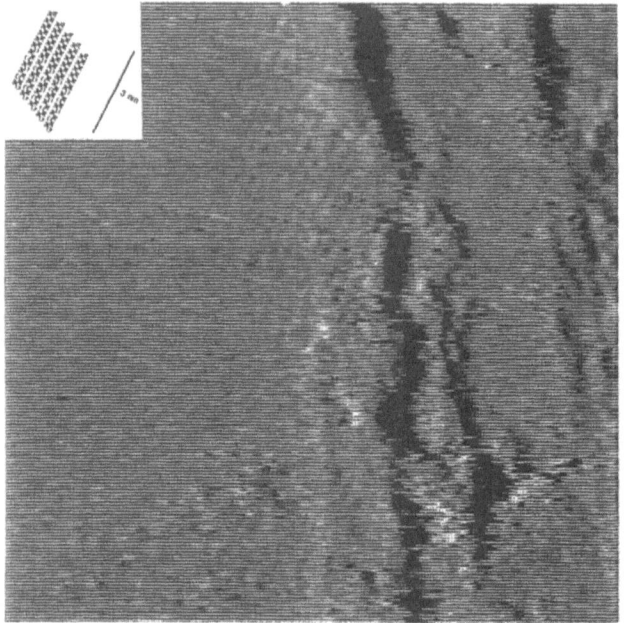

Fig. 3. STM image of tetraethyleneglycol lauryl ether molecules adsorbed from a $1 \cdot 10^{-4}$ M aqueous solution on HOPG. Scan area: 20 nm × 20 nm, x-scan frequency: 20 Hz, sample bias: —20 mV, tunneling current 200 pA. A model of tetraethyleneglycol lauryl ether in the same scale is inserted in the upper left corner

predicted by Findenegg [16] from caloric measurements. It can be seen that the rows fit perfectly with a structural model of the polyglykol ether.

Figure 4 shows *n*-hexadecane, recorded in a drop of *n*-hexadecane on HOPG. Only the first layer of *n*-hexadecane is adsorbed strongly enough to be seen by STM. Again, highly ordered rows are seen; this pattern, too, fits well with the structural model of *n*-hexadecane.

References

1. Binnig G, Quate CF, Gerber C (1986) Phys Rev Lett 56:930
2. Binnig G, Gerber C, Stoll E, Albrecht TR, Quate CF (1987) Europhys Lett 3:1281
3. Marti O, Drake B, Hansma PK (1987) Appl Phys Lett 51:484
4. Heinzelmann H, Grütter P, Meyer E, Hidber H, Rosenthaler L, Ringger M, Güntherodt H-J (1987) Surface Science 189:29
5. Hansma PK, Elings V, Marti O, Bracker CE (1988) Science 242:209
6. Bryant PJ, Miller RG, Yang R (1988) Appl Phys Lett 52:2233
7. Pitsch M, Metz O, Kohler H-H, Heckmann K, Strnad J (1989) Thin Solid Films 175:81
8. Jaquet PA (1956) Metal Rev 1:157
9. Peters DJ, Blackford BL (1989) Rev Sci Instrum 60:138
10. Polymeropoulos EE, Sagiv J (1978) J Chem Phys 69:1839
11. Polymeropoulos EE (1977) J Appl Phys 48:2404
12. Marti O, Ribi HO, Drake E, Albrecht TR, Quate CF, Hansma PK (1988) Science 239:50
13. Dovek MM, Albrecht TR, Kuan SWJ, Lang CA, Emch R, Grütter P, Frank CW, Pease RFW, Quate CF (1988) J Microsc 152:229
14. Worcester DL, Miller RG, Bryant PJ (1988) J Microsc 152:817
15. Brij 30, Aldrich-Chemie, D-7924 Steinheim
16. Findenegg GH, private communication

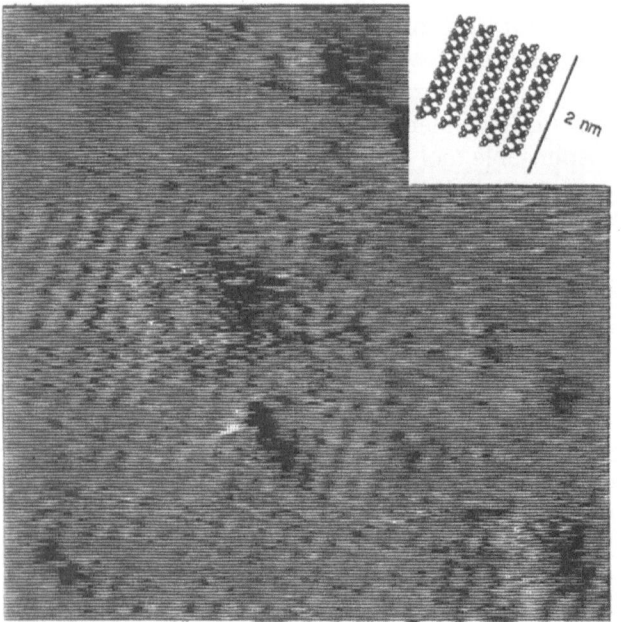

Fig. 4. STM image of *n*-hexadecane on HOPG. Scan area: 10 nm × 10 nm x-scan frequency: 20 Hz, sample bias: —20 mV, tunneling current 400 pA. A model of *n*-hexadecane in the same scale is inserted in the upper right corner

Authors' address:

Manfred Pitsch,
Diplom Chemiker
Institut für Physikalische und Markomolekulare Chemie
Universität Regensburg
Universitätsstraße 31
8400 Regensburg, FRG

Adsorption of short-chain polyoxyethylene alkylethers from aqueous and non-aqueous solutions onto graphite. A thermodynamic study using liquid-solid chromatography

Th. Rheinländer, H. Dropsch and G. H. Findenegg

Physikalische Chemie II, Ruhr-Universität Bochum, Bochum, FRG

Abstract: The adsorption of several oligo-oxyethylene alkylethers (C_nE_m, with $n = 4, 6, 8$ and $m = 2, 3, 4$) from acetonitrile (A) and methanol (M), and from binary mixtures of these solvents with water (W) onto graphite has been studied in the Henry's law region, using liquid-solid elution chromatography. The standard Gibbs energy of adsorption from solution ($\Delta_d G^0$) derived from the surface-specific retention volume of the amphiphilic solutes is analyzed in terms of group contributions for methylene groups of the alkyl chain (Δg_C^0) and oxyethylene groups (Δg_E^0). For all situations studied (binary eluents with up to 60 volume % W), Δg_C^0 and Δg_E^0 are negative, indicating that the two types of groups are both preferentially adsorbed at the graphite surface. The enthalpic and entropic group contributions Δh_C^0 and Δs_C^0 are negative, but Δh_E^0 and Δs_E^0 for the adsorption from M are positive. The composition dependence of Δg_C^0 and Δg_E^0 in the solvent system $A + W$ is analyzed on the basis of the parallel-layer model of adsorption from multicomponent systems. It is shown that in the present systems the nonideality of the bulk solution causes an enhancement of the adsorption of the methylene groups, but for the oxyethylene groups it has the opposite effect (i.e., adsorption becoming weaker as the concentration of water in the eluent mixture increases).

Key words: Adsorption; oxyethylene alkylethers; C_nE_m oligomers; liquid-solid chromatography; graphite; nonionic surfactants

Introduction

The adsorption of non-ionic surfactants of the n-alkylpoly(oxyethylene glykol) type, $CH_3(CH_2)_{n-1}(OCH_2CH_2)_mOH$ (abbreviated as C_nE_m), from aqueous solutions onto hydrophobic surfaces like graphite is dominated by the tendency to remove the hydrocarbon chains from the aqueous surroundings. The role played by the polar oxyethylene groups in the adsorption process is, however, not well understood. In a recent paper [1], we reported measurements of adsorption isotherms and enthalpies of displacement of C_8E_4 and several shorter C_nE_m oligomers onto graphitized carbon black. Strong adsorption up to a surface concentration corresponding to a complete monolayer of C_nE_m molecules oriented parallel to the graphite surface was observed. This strong adsorption is partly due to the attractive lateral interactions in a close-packed monolayer of higher n-alkanes and related molecules on graphite [2, 3]. In the present work, we studied the strength of adsorption of oligomeric C_nE_m amphiphiles in the region of high dilution (Henry's law region), where lateral interactions between adsorbed solute molecules are absent. The results have been analyzed in terms of group contributions for the n methylene groups of the alkyl chain and the m oxyethylene groups of the C_nE_m molecules.

Adsorption from solution represents a displacement reaction of adsorbed solvent molecules (S) by the solute (C_nE_m)

$$C_n E_m(sol) + rS(ads) = C_n E_m(ads) + rS(sol) \, , \quad (1)$$

where (sol) and (ads) refer to the dissolved and adsorbed state, respectively, and r is a stoichiometric number. In the case of adsorption from binary mixtures there is also competitive adsorption of the two solvents S_1 and S_2. The standard Gibbs energy of the displacement reaction (1) at infinite dilution of the solute is denoted as $\Delta_d G^0$ [4]; it can be expressed formally by group contributions from the n methylene (methyl) groups from the hydrocarbon chain (Δg_C^0) and from the m oxyethylene groups (OCH$_2$CH$_2$) of the hydrophilic chain (Δg_E^0):

$$\Delta_d G^0 = n \Delta g_C^0 + m \Delta g_E^0 + \Delta g_0 \, , \quad (2)$$

where Δg_0 is an end-group contribution. Quantitative estimates of the group increments Δg_C^0 and Δg_E^0 can be obtained on the basis of Eq. (2) from $\Delta_d G^0$ values for a series of $C_n E_m$ oligomers.

Standard Gibbs energies $\Delta_d G^0$ for the displacement reaction (1) can be determined by liquid-solid elution chromatography. The experimentally accessible surface-specific retention volume of the solute is $V_s = (V_R^0 - V_\mu)/A_s$, where V_R^0 is the retention volume of the solute extrapolated to infinite dilution in the solvent, V_μ is the hold-up volume of the column, and A_s is the surface area of the adsorbent in the column [5]; V_s is related to $\Delta_d G^0$ by

$$\Delta_d G^0 = -RT \ln(V_s/V^0) \, , \quad (3)$$

where V^0 is a standard volume per unit area, and T is the experimental temperature of the column.

In the present paper, measurements of the surface-specific retention volume V_s are reported for several oligomeric $C_n E_m$ amphiphiles in a column packed with mesoporous graphitic carbon, using mixtures of acetonitrile and water as the eluent. A few complementary measurements with mixtures of methanol and water are also presented for two different temperatures. The results are analyzed in terms of the group contributions Δg_C^0 and Δg_E^0. This analysis shows that the adsorption of the two types of groups from solution is dominated by different interactions which lead to a different dependence of these free energy increments on the polarity of the binary eluent.

Materials and methods

The alkyl(oxyethylene) oligomers were either gifts from Henkel (C_4E_2, C_6E_2, C_6E_3) or purchased from Bachem (C_8E_2, C_8E_3, C_8E_4) and had a specified purity of 99%. These materials were stored under nitrogen gas at $-10\,^\circ$C and used without further purification. Acetonitrile and methanol of p.a. grade (Riedel-de Haën, >99.5% purity by GLC) where dried with 3 Å molecular sieve and filtered through 0.45 μm PTFE membranes before use. Deionized water was passed through a Milli-Q reagent water system (Millipore), which leads to a total organic carbon (TOC) content of <20 ppb.

A HPLC column packed with porous graphitic carbon (PGC) was supplied by Chromatographite (Edinburgh, UK). This material (designation PGC 93) consists of spherical sponge-like particles (7 μm mean particle diameter) made up from graphite crystallites of around 10 nm in size, and its exposed surface consists of patches of graphite basal plane [6]. PGC 93 has a specific surface area of 105 m^2 g^{-1}, as determined by the standard BET method with nitrogen ($a_m = 0.162$ nm^2) at 77 K in our laboratory. The column contained 0.916 g PGC, hence the surface area was $A_s = 96$ m^2.

The chromatograph had a similar configuration as described elsewhere [7]. A Model 6000 A constant flow rate metering pump (Waters) was connected to a Model 7125 sampling valve (Rheodyne). The sample loop had a volume of 20 μl. The column had an inner diameter of 4.5 mm and a length of 100 mm; it was immersed in a liquid thermostat (Haake R 20). The column outlet was connected to a Model R401 differential refractometer (Waters) and a microburette flowmeter, which were both kept in an air thermostat. A flow rate of 24 ml/h was used. The column hold-up volume V_μ was calculated from the retention time of an unsorbed tracer and the flow rate. Measurements with deuterated acetonitrile d3 (Aldrich) in acetonitrile gave $V_\mu = 1.4277$ ml, in good agreement with results for D$_2$O in water (1.4282 ml). The extracolumn hold-up volume amounted to less than 15% of the total hold-up volume.

Results

The surface-specific retention volume of the $C_n E_m$ solutes at infinite dilution in the eluent was obtained by an extrapolation method based on the Langmuir equation which relates the surface concentration q of the solute to its equilibrium concentration in solution c by

$$q = \frac{ac}{1 + bc} \, ,$$

where a and b are constants. The net retention volume $V_N = V_R - V_\mu$ of the solute at a concentration c is proportional to the derivative $(dq/dc)_c$; hence

$$V_N = \frac{V_N^0}{(1 + bc)^2} \, . \quad (4)$$

For given compositions of the eluent mixture, V_N was measured for several solute concentrations in the range 0.2—10 mmol^{-1}, and the parameters V_N^0 and b were determined by fitting Eq. (4) to these data by a least-squares procedure. Figure 1 shows such a fit for C_8E_4 in pure acetonitrile. The surface-specific retention volume at infinite dilution, $V_s = V_N^0/A_s$, of C_nE_m oligomers eluted by acetonitrile + water mixtures at 25 °C is given in Table 1. Up to a mole fraction of water $x_1 \simeq 0.5$ (volume fraction $\Phi_1 \simeq 0.25$), V_s is a relatively weak function of the eluent composition. At higher concentrations of water, however, V_s rises markedly with x_1 and becomes excessively large for higher oligomers in the water-rich domain.

Table 1. Chromatographic retention of C_nE_m oligomers in acetonitrile + water mixed eluents on a column packed with porous graphitic carbon (PGC 93) at 25 °C: Surface-specific retention volume V_s as a function of the mole fraction of water (x_1) in the eluent

x_1	$V_s/(\text{mm}^3 \text{ m}^{-2})$					
	C_4E_2	C_6E_2	C_6E_3	C_8E_2	C_8E_3	C_8E_4
0	2.26	5.93	9.44	15.73	24.75	38.3
0.1	2.12	5.81	9.05	16.08	24.66	36.4
0.2	2.05	5.90	9.06	17.71	26.77	38.5
0.3	2.10	6.33	9.56	20.17	30.1	42.8
0.4	2.04	6.93	10.33	23.33	34.4	48.1
0.5	2.19	8.47	12.23	30.3	43.5	60.3
0.6	2.68	11.07	15.68	43.2	60.2	82.9
0.7	3.70	17.21	24.05	77.5	106.0	143.9
0.8	6.87	39.53	57.2	227.0		
0.9	30.9					

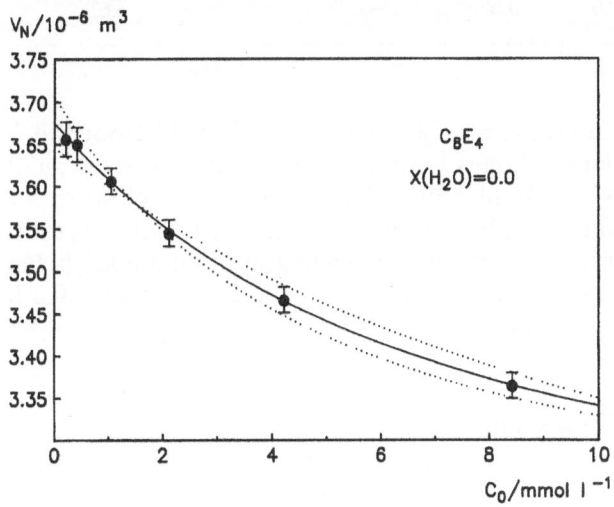

$V_N/10^{-6}\ m^3$

C_8E_4

$X(H_2O)=0.0$

$c_0/\text{mmol l}^{-1}$

Fig. 1. Net retention volume V_N as a function of the concentration of the solute in the sample loop c_0 for C_8E_4 eluted by pure acetonitrile. The full curve represents the fit of the experimental data by Eq. (4)

Table 2. Retention of C_nE_m oligomers eluted by methanol and methanol + water mixed eluents on PGC 93 at 25° and 45 °C: Surface-specific retention volume V_s for two different temperatures and solvent compositions

T	$V_s/(\text{mm}^3 \text{ m}^{-2})$				
	C_4E_2	C_6E_2	C_8E_2	C_8E_3	C_8E_4
Pure methanol ($x_1 = 0$):					
25 °C	3.72	11.05	33.31	53.21	87.47
45 °C	3.18	8.34	22.63	36.98	60.18
Methanol + water ($x_1 = 0.601$):					
25 °C	28.02	179.4			
45 °C	20.49	107.6			

Table 2 presents surface-specific retention volumes for the elution of C_nE_m oligomers by methanol at two temperatures (25° and 45 °C) and a few results for methanol + water mixtures. Here again, V_s becomes excessively large for the higher oligomers at higher water contents of the eluent. The thermodynamic standard quantities $\Delta_d G^0$, $\Delta_d H^0$ and $T\Delta_d S^0$ derived from the retention data in Table 2 are summarized in Table 3. As $\Delta_d G^0$ exhibits a weak temperature dependence and was determined only at two temperatures, the un-

Table 3. Standard thermodynamic quantities (based on $V^0 = 1\ \text{mm}^3 \text{ m}^{-2}$) for the adsorption of C_nE_m oligomers from solution onto the graphite surface at 25 °C (values in kJ mol^{-1})

Substance	$\Delta_d G^0$	$\Delta_d H^0$	$T\Delta_d S^0$
Solvent: Methanol			
C_4E_2	−3.26	−6.2	−2.9
C_6E_2	−5.95	−11.1	−5.1
C_8E_2	−8.69	−15.2	−6.5
C_8E_3	−9.85	−14.3	−4.4
C_8E_4	−11.08	−14.7	−3.6
Solvent: Methanol + water ($x_1 = 0.601$)			
C_4E_2	−8.26	−12.3	−4.0
C_6E_2	−12.86	−20.1	−7.2

certainty in $\Delta_d H^0$ and $\Delta_d S^0$ is rather large (10—20%).

The compounds studied in this work represent members of two homologous series, i.e., $C_n E_2$ ($n = 4, 6, 8$) and $C_8 E_m$ ($m = 2, 3, 4$). On the basis of Eqs. (2) and (3) one expects a linear dependence of $\ln(V_s/V^0)$ on n and m within the two series. Figure 2 shows examples of such logarithmic plots. Table 4 summarizes the group increments Δg_C^0, Δg_E^0, and the end-group contribution Δg_0 derived from the experimental data. The increment rule expressed by Eq. (2) with the parameters listed in Table 4 fits the

Table 4. Group contributions to the standard Gibbs energy $\Delta_d G^0$ for the displacement of solvent by $C_n E_m$ compounds based on Eq. (2) for the adsorption from mixtures of acetonitrile + water at 25°C. The increments Δg_C^0 and Δg_E^0 and the end-group term Δg_0 (based on $V^0 = 1$ mm^3 m^{-2}) are given in kJ mol^{-1}. The solvent composition is expressed as mole fraction x_1 and volume fraction Φ_1 of water

x_1	Φ_1	Δg_C^0	Δg_E^0	Δg_0
0	0	−1.20	−1.10	4.99
0.1	0.037	−1.26	−1.01	5.19
0.2	0.079	−1.34	−0.96	5.49
0.3	0.128	−1.40	−0.93	5.63
0.4	0.186	−1.51	−0.90	6.06
0.5	0.255	−1.63	−0.85	6.26
0.6	0.339	−1.73	−0.81	6.06
0.7	0.444	−1.89	−0.77	5.83
0.8	0.578	−2.17		5.61

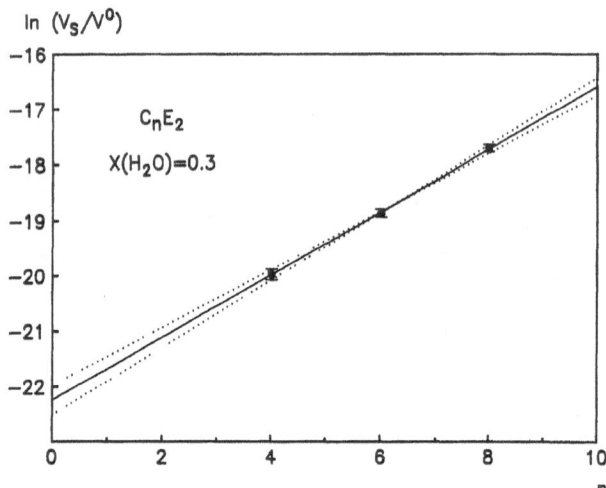

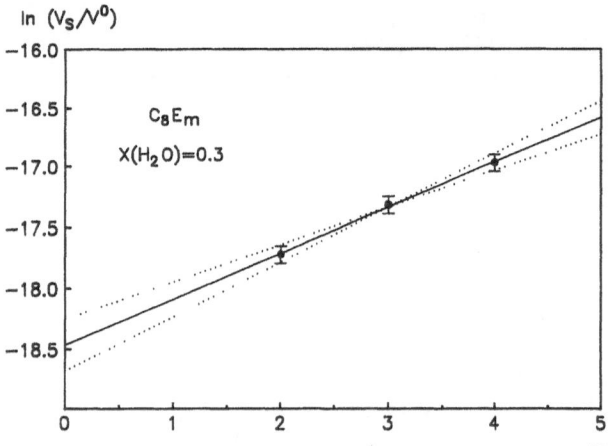

Fig. 2. Logarithm of the surface-specific retention volume V_s for the two homologous series of amphiphiles eluted by a binary eluent acetonitrile + water ($x_1 = 0.3$): $C_n E_2$ with $n = 4, 6, 8$ (upper graph); $C_8 E_m$ with $m = 2, 3, 4$ (lower graph). The standard volume V^0 is expressed in m^3 m^{-2} in these graphs

experimental $\Delta_d G^0$ values (as derived from the V_s data of Tables 1 and 2) within a maximum deviation of ± 0.05 kJ mol^{-1}. The estimated absolute uncertainty of the group increments (including systematic errors of the chromatographic method) is less than ± 0.1 kJ mol^{-1} for Δg_C^0 and less than ± 0.2 kJ mol^{-1} for Δg_E^0.

Discussion

Group contribution concept

The present work shows that for oligomeric $C_n E_m$ amphiphiles the standard Gibbs energy of adsorption from solution $\Delta_d G^0$ can be represented by group contributions, as defined in Eq. (2). The magnitude of $\Delta_d G^0$ and of the end-group contribution Δg_0 depends on the chosen standard state (in particular, on the standard volume per unit surface area V^0), but the group contributions Δg_C^0 and Δg_E^0 are independent of the chosen units. For all situations studied, Δg_C^0 and Δg_E^0 are negative, which implies that the nonpolar methylene groups and the polar oxyethylene groups are both preferentially adsorbed from solution onto the graphite surface.

For the adsorption from the pure organic solvents, the group increments Δg_C^0 and Δg_E^0 are of similar magnitude, but the constituent enthalpic and entropic parts of Δg^0 are quite different, at least in the case of methanol (see Table 5). The

Table 5. Group contributions to $\Delta_d G^0$, $\Delta_d H^0$, $\Delta_d S^0$ for adsorption of $C_n E_m$ oligomers from methanol (and methanol + water mixtures) on graphite (25 °C). Group contributions Δg_i^0 and Δh_i^0 in kJ mol^{-1}, Δs_i^0 in J K^{-1} mol^{-1}

	Δg_i^0	Δh_i^0	Δs_i^0
Methanol:			
Methylene (i = C):	−1.35	−2.25	−3.0
Oxyethylene (i = E):	−1.20	+0.2	+4.5
Methanol + water (x_1 = 0.601):			
Methylene (i = C):	−2.3	−3.9	−5.5

adsorption of nonpolar methylene segments is exothermic and is connected with a decrease of entropy ($\Delta h_C^0 < 0$, $\Delta s_C^0 < 0$), whereas the adsorption of the polar oxyethylene groups from methanol is weakly endothermic, but is connected with a significant entropy increase $\Delta h_E^0 > 0$, $\Delta s_E^0 > 0$). These positive values of Δh_E^0 and Δs_E^0 can be rationalized by asserting that in solution the oxyethylene groups are solvated by methanol molecules via hydrogen bonds, and that adsorption at the graphite surface involves a desolvation, i.e., a breaking of hydrogen bonds. Unfortunately, Δh^0 and Δs^0 is not available for the adsorption from acetonitrile, where no solvation via hydrogen bonds is possible.

In water-containing solvent mixtures the group contributions Δg_C^0 and Δg_E^0 exhibit a pronounced but qualitatively different dependence on the composition. As illustrated in Fig. 3 for the solvent system acetonitrile + water, an increase of the water contents of the solvent causes an enhancement of the adsorption of the non-polar CH$_2$ groups (Δg_C^0 becoming more negative), but a weakening of the adsorption strength of the polar oxyethylene groups (Δg_E^0 becoming less negative). To understand these opposite trends it is useful to separately consider the "ideal" displacement process and the nonideality effects in the bulk solution.

For the adsorption of solutes from a binary solvent mixture, in the case of ideal mixing behavior of the bulk phase, one expects that the standard Gibbs energy of displacement becomes more negative as the concentration of the less strongly adsorbed component of the solvent mixture increases. In the binary solvent systems of the present study, the organic component (acetonitrile and methanol, respectively) is preferentially adsorbed at the

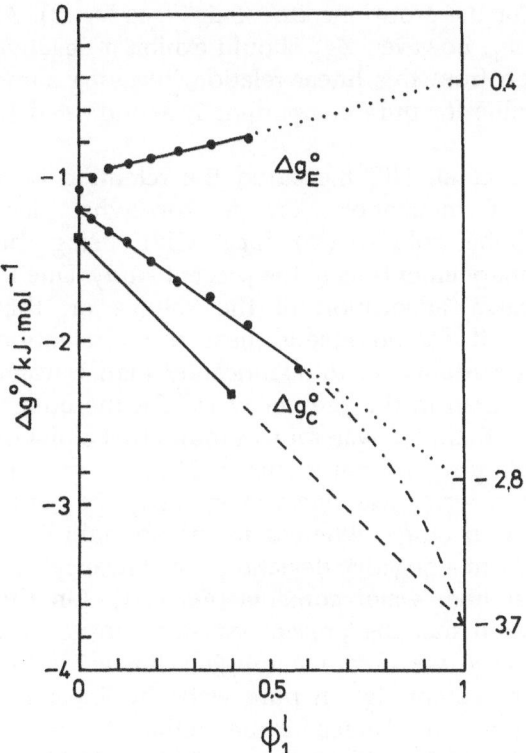

Fig. 3. Group increments Δg_C^0 and Δg_E^0 as a function of the volume fraction of water Φ_1^l in the binary eluents acetonitrile + water (●), and Δg_C^0 for methanol + water (■) at 25 °C. Linear extrapolations up to pure water are indicated for the two eluents: ..., acetonitrile + water, – – –, methanol + water. A more realistic extrapolation for Δg_C^0 data in acetonitrile + water is marked by dashed-dotted symbols (see text)

graphite surface [8, 9]. Moreover, an analysis of the surface excess isotherms of the (acetonitrile + water)/graphite system shows [8] that in the composition range of the present study the adsorbed layer next to the graphite surface consists of nearly pure acetonitrile. In this limiting case of strong preferential adsorption of one of the solvents, and in absence of significant nonideality effects in the bulk solution, one expects that for a given solute the standard Gibbs energy of displacement will vary proportional to $r \ln \Phi_2$, where Φ_2 is the volume fraction of the preferentially adsorbed solvent in the binary eluent and r is the number of adsorbed solvent molecules displaced by one solute molecule (see next section). In the composition range near pure solvent 2, when $\ln \Phi_2 \simeq 1 - \Phi_2 = \Phi_1$, this ideal displacement process leads to a linear decrease of $\Delta_d G^0$ with increasing Φ_1, as is indeed found for the group increment Δg_C^0 (see Fig. 3). At higher Φ_1, however, Δg_C^0 should exhibit a negative deviation from this linear relation, reaching some finite value for pure component 1, as indicated in Fig. 3.

Karger et al. [10] measured the retention of a series of n-alkanols on a somewhat less hydrophobic column (Bondapak-C18), using the same binary eluents as in the present study. Due to the weaker adsorption of the solutes on that substrate, it was possible to measure the retention in those systems up to significantly higher water contents than in the present work. For the binary eluent methanol + water it was found that a plot of a variable proportional to our $\Delta_d G^0$ vs the volume fraction of water was linear over nearly the entire composition range, whereas in the acetonitrile + water eluent a negative deviation from linearity was found at high water contents ($\Phi_1 \rightarrow 1$). On the assumption that the present systems behave in a similar way, we obtain a rough estimate of the group increment Δg_C^0 in pure water by linear extrapolation of the values for the methanol + water mixtures, which yields $\Delta g_C^0 = -3.7$ kJ (mol CH$_2$)$^{-1}$. Accodingly, in the plot of Δg_C^0 vs Φ_1 for the solvent system acetonitrile + water, a pronounced negative deviation from linearity must occur at high volume fractions of water, in order to meet the same Δg_C^0 value in pure water (see Fig. 3). The resulting estimate for the standard Gibbs energy of adsorption of methylene segments from water onto graphite, $\Delta g_C^0 = -3.7$ kJ mol^{-1}, agrees in magnitude with the corresponding group increment for the transfer of alkanes from aqueous solutions into liquid hydrocarbons (-3.7 kJ mol^{-1} [11]); it is somewhat greater than the corresponding value for the adsorption from aqueous solutions at the oil/water interface (-3.2 to -3.4 kJ mol^{-1} [12]), and significantly greater than the corresponding value for the adsorption from aqueous solutions at the air/water interface (-2.5 to -2.9 kJ mol^{-1} [12, 13]).

Whereas the concentration dependence of Δg_C^0 in binary solvent mixtures conforms, at least in a qualitative way, with the predictions for an ideal displacement reaction, the opposite concentration dependence found for Δg_E^0 can be rationalized only in terms of nonideal mixing behavior in the bulk phase. One expects that on adding water to an aprotic eluent like acetonitrile, the oxyethylene groups of C_nE_m molecules will be hydrated, and thus the tendency of these groups to be removed from the solvent will decrease. This effect may ex-plain the rather pronounced variation of Δg_E^0 in the acetonitrile + water system in the region of small water concentrations (see Fig. 3). Note that a solvation of the oxyethylene groups via hydrogen bonds was already implied by the positive values of Δh_E^0 and Δs_E^0 in the adsorption from pure methanol (Table 5). Unfortunately, the corresponding enthalpy and entropy data for the adsorption from acetonitrile + water mixtures are not available.

Application of a theoretical model

We will now analyze the observed dependence of the adsorption parameters of solutes on the composition of the solvent mixture in terms of a simple theoretical model, based on the parallel-layer model of adsorption from solution [14]. In this theory the adsorbed layer is taken as an autonomous phase, and the activity coefficients of the individual components in both the adsorbed phase and the bulk phase are defined on the basis of a Flory-Huggins type lattice model. If the solute is present at infinite dilution in a binary solvent mixture (components 1 and 2) from which component 2 is preferentially adsorbed at the solid/liquid interface, the following relation for the retention of the solute (either the capacity ratio k' or the surface-specific retention volume V_s) applies [15]:

$$\frac{V_s}{V_{s(2)}} = \left(\frac{\Phi_2^a}{\Phi_2^l}\right)^r \exp\left[-\frac{1}{4} r S_{21}(1 + 2\Phi_2^a - 3\Phi_2^l)\right],$$
(5)

where $V_{s(2)}$ is the retention volume in pure solvent 2, Φ_i^a and Φ_i^l represent the volume fractions of i in the adsorbed phase and bulk phase ($\Phi_1^l + \Phi_2^l = 1$), and r is the molecular size ratio of solute and component 2. S_{21} represents a linear combination of Flory-type interaction energy parameters for molecules of solvent i with each of the r chain segments of a solute molecule (χ_{is}), and between unlike solvent molecules (χ_{12}), viz. $S_{21} = \chi_{2s} - \chi_{1s} + \chi_{12}$. For an ideal mixture of the three components, $S_{21} = 0$.

In the case of strong preferential adsorption of component 2 from the solvent mixture, $\Phi_2^a \simeq 1$ for a wider composition range $\Phi_2^l \leqslant 1$, and for these compositions of the eluent, Eq. (5) reduces to

$$\frac{V_s}{V_{s(2)}} = \left[\frac{1}{\Phi_2^l} \exp\left\{-\frac{3}{4} S_{21}(1 - \Phi_2^l)\right\}\right]^r.$$
(6)

Equation (6) was applied to the present systems by choosing realistic values for the size ratio r and using a non-linear least-square fitting procedure to obtain S_{21} from the experimental retention data $V_s(\Phi_2^l)$. The cross-sectional area of the C_nE_m molecules in the adsorbed state on graphite was estimated by assuming a similar configuration as for long-chain n-alkanes, for which there is evidence from scanning tunneling microscopy [16] that each CH_2 group occupies the area of one carbon hexagon of the graphite basal plane ($a_h = 0.0524$ nm^2); assuming for C_nE_m molecules that the ether oxygens are geometrically equivalent to CH_2 and the end-groups (CH_3 and OH) equivalent to two CH_2, we obtain $a_s(C_nE_m) = (n + 3m + 3)a_h$ [1]. The cross-sectional area of solvent 2 was taken as $a_2 = 0.33$ nm^2 (acetonitrile) and $a_2 = 0.21$ nm^2 (methanol). In Table 6, the resulting values of $r = a_s/a_2$ are summarized, together with the best-fit values for S_{21} and $V_{s(2)}$. Generally, a good fit of the experimental data by Eq. (6) was found in the whole (or nearly the whole) experimental composition range, as shown, for example, in Fig. 4. Furthermore, the fitted values of $V_{s(2)}$ agree with the experimental retention volumes V_s in pure solvent 2 within the limits of experimental error.

The interaction parameters S_{21} given in Table 6 reflect the different balance of hydrophilic and hydrophobic interactions of the individual C_nE_m oligomers with the solvent mixture. This is shown in Fig. 5, where S_{21} is plotted against the HLB value (hydrophile-lipophile-balance) of the amphiphiles,

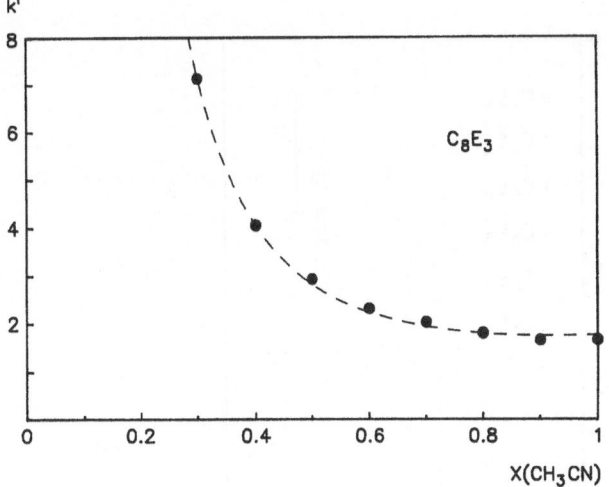

Fig. 4. Capacity ratio k' of C_8E_3 vs mole fraction of acetonitrile in the binary eluent acetonitrile + water at 25°C: ●, experimental data; —— fit by Eq. (6). The capacity ratio k' is linearly related to the surface-specific retention volume by $k' = A_s V_s / V_\mu$

defined as HLB $= 20\, M_E/M$, (where M and M_E represents the molar mass of the entire molecule and its hydrophilic oligooxyethylene chain, respectively) [17]. Within the given limits of error, these results can be represented by the linear relation

$$S_{21} = 0.16\ \text{HLB} - 0.37 \ .$$

Accordingly, S_{21} can be expressed formally in terms of the corresponding interaction parameters S_C and S_E for the v_C segments occupied by the n methylene groups, and the v_E segments occupied by the m oxyethylene groups, respectively, viz.

$$S_{21} = \zeta_C S_C + \zeta_E S_E = S_C + \zeta_E (S_E - S_C) \ , \tag{7}$$

where $\zeta_C = v_C/r$, $\zeta_E = v_C/r$, and $\zeta_C + \zeta_E = 1$. By comparing the two expressions for S_{21}, we find $S_C = -0.37$ and $S_E = 2.83$.

The following relation for the group contributions Δg_i^0 is obtained from Eqs. (2), (3), (5), and (7):

$$\Delta g_i^0 = \Delta g_{i(2)}^0 - RT \ln \left(\frac{\Phi_2^a}{\Phi_2^l} \right)$$

$$+ \frac{1}{4} RT\, S_i (1 + 2\Phi_2^a - 3\Phi_2^l) \ , \tag{8}$$

Table 6. Fit of the surface-specific retention volume of C_nE_m oligomers by Eq. (6) for binary eluents acetonitrile + water (A + W) and methanol + water (M + W) at 25°C: Estimated values of the size ratio r and best-fit values of S_{21} and $V_{s(2)}$ (mm^3 m^{-2})

Oligomer	r	S_{21}	$V_{s(2)}$
A + W:			
C_4E_2	2.09	1.80	2.23
C_6E_2	2.41	1.31	5.35
C_6E_3	2.89	1.66	9.67
C_8E_2	2.73	1.17	15.3
C_8E_3	3.22	1.46	26.2
C_8E_4	3.70	1.65	42
M + W:			
C_4E_2	3.25	0.65	
C_6E_2	3.75	0.38	

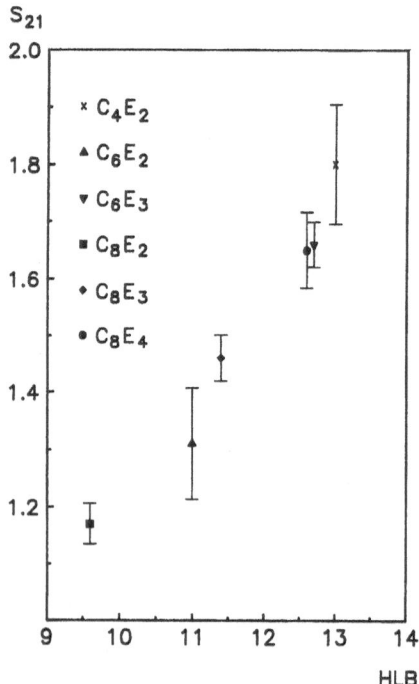

Fig. 5. Interaction parameter S_{21} for C_nE_m solutes in acetonitrile + water mixed eluents (25 °C) as a function of the HLB value of the amphiphiles. The error bars of the best-fit S_{21} values are indicated

where index i stands for the methylene groups (C) or oxyethylene groups (E), and $\Delta g^0_{i(2)}$ is the group contribution in pure solvent 2. The logarithmic term is positive ($\Phi^a_2 > \Phi^l_2$) and accounts for the lowering of Δg^0_i due to a weaker binding of the solvent to the surface, as the concentration of the preferentially adsorbed component decreases. The third term on the r.h.s. of Eq. (8) is due to the nonideality of the intermolecular interactions and may cause an increase or a decrease of Δg^0_i for decreasing Φ^l_2, depending on the sign of the parameter S_i. According to Eq. (8), two regimes are of special interest:

i) *Low concentrations of water:*

When component 2 (organic solvent) is strongly preferentially adsorbed from the binary solvent mixture, we have $\ln\left(\dfrac{\Phi^a_2}{\Phi^l_2}\right) \simeq -\ln(\Phi^l_2) \simeq -\Phi^l_1$ for small Φ^l_1, and thus, Eq. (8) yields

$$\Delta g^0_i = \Delta g^0_{i(2)} - RT\left(1 - \frac{3}{4}S_i\right)\Phi^l_1 , \qquad (9)$$

which is consistent with the observed linear dependence of the group contributions on Φ^l_1 in acetonitrile-rich eluent mixtures (Fig. 3). Moreover, as the parameter S_C is negative (see above), it follows for the methylene groups that the effect of the nonideal intermolecular interactions acts in the same direction as the weakening of the solvent binding to the surface, both causing a lowering of Δg^0_C (i.e., an increasing strength of adsorption of methylene groups) with increasing water concentration in the solvent mixture.

Conversely, in the case of the oxyethylene groups, since S_E is positive, the intermolecular interactions and the surface interactions act in opposite directions, and because S_E is large ($\frac{3}{4}S_E > 1$), the concentration dependence of Δg^0_E is dominated by the favorable interactions with the solvents so that Δg^0_E increases with increasing water concentration in the solvent mixture (corresponding to a weaker adsorption of oxyethylene groups), as indeed observed (Fig. 3).

ii) *High concentration of water:*

In the limit $\Phi^l_1 \rightarrow 1$, we have $\Phi^l_2 \rightarrow 0$ and $\Phi^a_2 \rightarrow 0$, but $\lim\left(\dfrac{\Phi^a_2}{\Phi^l_2}\right) = K > 1$. For pure solvent 1, we obtain

$$\Delta g^0_{i(1)} = \Delta g^0_{i(2)} - RT\left(\ln K - \frac{1}{4}S_i\right) . \qquad (10)$$

Accordingly, $\Delta g^0_{i(1)}$ will be more negative or less negative than $\Delta g^0_{i(2)}$, depending on whether or not the effect of the surface interactions ($\ln K$) is overcompensated by favorable interactions of the segments i with solvent 1 (large positive S_i). Moreover, the value of $\Delta g^0_{i(1)}$ will be more negative than the linear extrapolation of the expression in Eq. (9) to $\Phi^l_2 = 1$ if $\ln K - S_i/4$ is greater than $1 - (\frac{3}{4})S_i$. This condition is met when K is large (strong preferential adsorption of solvent 2) and S_i is either positive or not too negative; the latter situation appears to apply for methylene segments in acetonitrile + water mixtures (Δg^0_C in Fig. 3).

Conclusions

The factors affecting the adsorption of hydrocarbon chains and oxyethylene groups of nonionic surfactants from polar solutions onto an energetically homogeneous hydrophobic surface (graphite) have been analyzed in terms of a group contribution concept. For the adsorption of hydrocarbon (methylene and methyl) groups from solvents like methanol, the enthalpy and entropy of adsorption (displacement) is negative, and becomes more negative as water is added to the solvent. No evidence of a hydrophobic effect is seen in the experimental composition range (up to 40 volume-% water in methanol). However, the principal factor in the adsorption of oxyethylene groups from methanol is a large positive Δs_E^0, while Δh_E^0 is nearly zero. It would be useful to determine these enthalpic and entropic group contributions to the adsorption of $C_n E_m$ surfactants for several protic and aprotic solvents and their mixtures with water.

The composition dependence of the group contributions Δg_i^0 in binary mixtures of acetonitrile and water can be understood in terms of the Flory-Huggins-type lattice model for adsorption from multicomponent mixtures. Although the highly directional interactions involved in these aqueous systems are not accounted for by this model, it is, nevertheless, capable of describing the differences in the adsorption of the two types of groups resulting from nonideality effects of the bulk phase in a qualitative manner. In particular, it follows that the different composition dependence of the group contributions Δg_C^0 and Δg_E^0 is a consequence of the different sign of the interaction parameters S_i of Eq. (8) for the hydrophobic and hydrophilic groups of $C_n E_m$ amphiphiles.

Acknowledgement

Financial support from the Fonds der Chemischen Industrie is gratefully acknowledged.

References

1. Findenegg GH, Pasucha B, Strunk H (1989) Colloids and Surfaces 37:223
2. Findenegg GH, Liphard M (1987) Carbon 25:119
3. Bien-Vogelsang U, Findenegg GH (1986) Colloids and Surfaces 21:469
4. Davies J, Everett DH (1983) In: Everett DH (ed) Colloid Science, Vol 4, Royal Society of Chemistry, London, p 101
5. Riedo F, Kováts E (1982) J Chromatogr 239:1
6. Knox JH, Kaur B, Millward GR (1986) J Chromatogr 353:3
7. Köster F, Findenegg GH (1982) Chromatographia 15:743
8. Pesch W, Findenegg GH, to be published
9. Everett DH, Fletcher AJP (1986) J Chem Soc Faraday Trans 1, 82:2605
10. Karger BL, Grant JR, Hartkopf A, Weiner PH (1976) J Chromatogr 128:65
11. McAuliffe C (1966) J Phys Chem 70:1267
12. Tanford C (1980) The Hydrophobic Effect, 2nd ed., Wiley, New York
13. Clint JH, Corkill JM, Goodman JF, Tate JR (1968) J Colloid Interface Sci 28:522
14. Ash SG, Everett DH, Findenegg GH (1968) Trans Faraday Soc 64:2639
15. Findenegg GH, Köster F (1986) J Chem Soc Faraday Trans 1, 82:2691
16. McGonigal GC, Bernhardt RH, Thomson DJ (1990) Appl Phys Lett, in press; Pitsch M, Metz O, Strnad J, Kohler HH (1990) Progr Colloid Polymer Sci 83 (this volume)
17. Shinoda K, Friberg S (1986) Emulsions and Solubilization, Wiley, New York, p 66

Authors' address:

Prof. Dr. G. H. Findenegg
Physikalische Chemie II
Ruhr-Universität Bochum
Postfach 102148
4630 Bochum 1, FRG

Progress in Colloid & Polymer Science

Progr Colloid Polym Sci 83:68—74 (1990)

Non-equilibrium effects in liquid-solid chromatography and flow sorption microcalorimetry

Z. Király and I. Dékány

Department of Colloid Chemistry, Attila József University, Szeged, Hungary

Abstract: Several flow techniques were studied to obtain adsorption and enthalpy data at liquid/solid interfaces. n-Butanol (solute), water (solvent) and graphitized Printex-300 carbon black (adsorbent) were selected for use. Adsorption and desorption paths in the one-step and step-by-step (cumulative) methods were investigated. Although the adsorption-displacement process was found to be reversible in nature, the adsorption and enthalpy curves of the one-step method eventually lie well above or below the corresponding cumulative isotherms due to non-equilibrium effects occurring in the one-step experiments.

Key words: Adsorption from solution; liquid chromatography; enthalpy of displacement; flow calorimetry

Introduction

The flow technique proved to be a powerful method for the investigation of adsorption phenomena at solid/solution interfaces. During the flow replacement experiment, a stream of pure liquid or of a solution is percolated through a column loaded with the adsorbent. In the one-step method the column inlet is switched to solutions of increasing concentration, starting from pure solvent in each case (adsorption path), which is followed by elution with pure solvent (desorption path). The step-by-step (cumulative) method is based on a finite sequence of closely spaced concentrations, starting from pure solvent initially (adsorption path); measurements are then made in the reverse direction (desorption path). If the solid sample is held in a chromatographic column and the outlet concentrations are continuously monitored and recorded by a detector, the adsorption excess isotherm (mass exchange) can be deduced from the breakthrough curves (flow frontal analysis liquid-solid chromatography, LSC). If the sorption vessel of a flow microcalorimeter is used as the adsorption column, the enthalpy isotherm of displacement (heat exchange) can be obtained (flow sorption microcalorimetry, FSM). In a more sophisticated procedure (which involves some practical dif-

ficulties), the detector is connected to the exit port of the calorimeter (on-line fit); hence, LSC and FSM measurements are made simultaneously on the same solid sample. In LSC, the one-step method has been applied by, among others, Sharma and Fort [1], while the cumulative method was preferred by Wang et al. [2] and Koch et al. [3]. In FSM, both methods were used by Findenegg et al. [4] and Woodbury and Noll [5]. Simultaneously LSC and FSM experiments were performed by Jednacak-Biscan and Pravdic [6, 7], Noll and Burchfield [8] and Denoyel et al. [9]. Questions arise of wehther the one-step and step-by-step methods lead to the same adsorption and enthalpy isotherms and how they correlate with the corresponding immersion experiments. Different calorimetric protocols led to different enthalpy isotherms [4, 5, 10], and efforts have been made to convert one of them to another [5]. Mixing between surface and bulk phases, mixing at the interface between replacing and replaced solutions, and non-equilibrium (kinetic) effects were considered to be the most likely reasons for such differences [5]. To obtain adsorption excess isotherms, it has been concluded that flow frontal analysis LSC is not restricted to instantaneous equilibrium chromatography [2]. Glückauf's method is based on the special evaluation of a single, high-step elution chromatogram. When it was applied, however, non-equilibrium effects were found to play an important role and a correction procedure was presented to eliminate them [11]. It may be inferred from the above-mentioned dynamic studies that a careful analysis of the methodology and experimental conditions should precede any theoretical conclusions drawn from the experimental results. The aim of this paper is to compare the adsorption and thermal data obtained from various flow methods and to interpret them solely from an experimental point of view.

Theoretical background

A schematic diagram of the experimental design, applicable to simultaneous FSM and LSC measurements, is shown in Fig. 1. The apparatus consists of a set of liquid sample holders, two pumps, and a six-way valve to select the liquid sample of interest, the sorption vessel of the microcalorimeter, and a connected analytical device (e.g., a differential refractometer) supplied with a recorder. Theoretical (ideal) and experimental chromatograms are presented in the schematic Fig. 2. The destribution function of the concentration, $F(t)$, is recorded con-

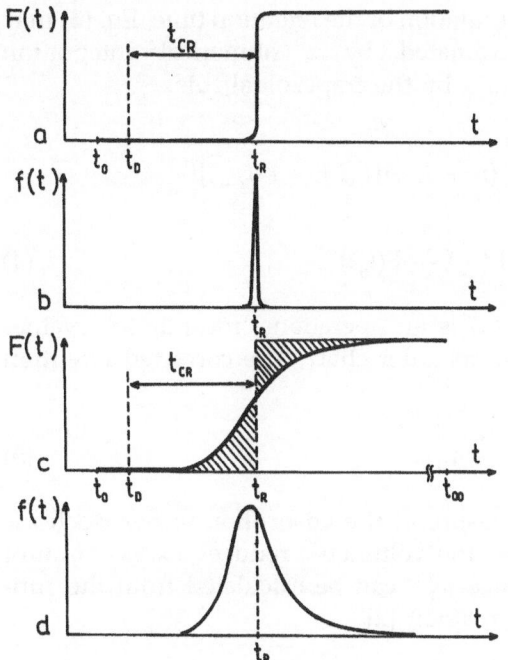

Fig. 2. Concentration distributions and the corresponding (probability) density functions for ideal (Figs. 2a and 2b) and experimental (Figs. 2c and 2d) chromatography

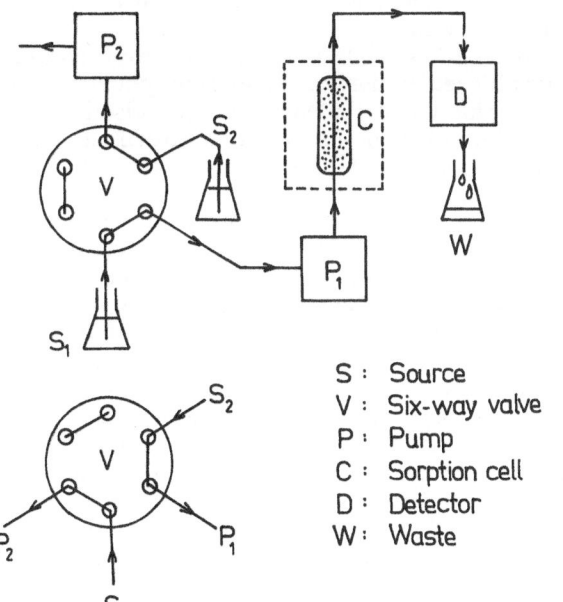

S : Source
V : Six-way valve
P : Pump
C : Sorption cell
D : Detector
W : Waste

Fig. 1. Experimental design for a simultaneous calorimetric-chromatographic assay of adsorption in solid/ binary liquid systems

tinuously as a function of time t. The corresponding density function is

$$f(t) = dF(t)/dt .$$ (1)

The extent of the system is defined as starting at the selection valve and ending at the detector. The liquid flow begins at t_0 and terminates at t_∞, after which a solution of new composition is started. The dead volume of the system V_D is the sum of the volumes of the tubing and fittings and the void volume of the column. If the volumetric flow rate Q is maintained constant, the dead time can be calculated as

$$t_D = V_D/Q .$$ (2)

The experimental quantity t_R is referred to as the retention time. It is located on the time scale at the point where the two shaded areas in Fig. 2 are equal. It can be expressed as a mean value:

$$t_R = \int_{t_0}^{t_\infty} tf(t)dt \bigg/ \int_{t_0}^{t_\infty} f(t)dt .$$ (3)

For computation of the retention time, Eq. (3) may be approximated by a numerical integration routine, e.g., by the trapezoidal rule:

$$t_R = \sum_{i=1}^{\infty} (t_i + t_{i-1})[F(t_i) - F(t_{i-1})]/$$

$$2[F(t_\infty) - F(t_0)] . \tag{4}$$

Equation (4) is an operational formula for evaluation of the recorder chart. The corrected retention time is

$$t_{CR} = t_R - t_D . \tag{5}$$

As a measure of the adsorption under isochoric conditions, the volumetric reduced excess amount of substance $n_1^{\sigma(v)}$ can be calculated from the fundamental relation [2]

$$dn_1^{\sigma(v)}/dc_1 = Qt_{CR}/m , \tag{6}$$

where c_1 is the concentration of component 1, and m is the mass of the adsorbent bed. In the step-by-step method, the adsorption excess isotherm $n_1^{\sigma(v)}$ vs c_1 can be constructed as

$$[n_1^{\sigma(v)}]_K = \sum_{k=1}^{K} [\Delta n_1^{\sigma(v)}]_k = \frac{Q}{m} \sum_{k=1}^{K} [t_{CR}\Delta c_1]_k . \tag{7}$$

Equation (7) involves the evaluation of a number K of S-shaped chromatograms. In the one-step method, the following relationship is applied:

$$\left[n_1^{\sigma(v)} = \frac{Q}{m} t_{CR}c_1 \right]_k , \tag{8}$$

where $k = 1, 2, ..., K$. Similarly, in the cumulative and one-step FSM experiments:

$$[\Delta_{21}H]_K = \sum_{k=1}^{K} \Delta H_k \tag{9}$$

$$[\Delta_{21}H]_k = [H(c_1) - H(c_1 = 0)]_k , \tag{10}$$

where $\Delta_{21}H$ is the integral enthalpy of displacement, which in part involves the integral enthalpy of adsorption [12], and ΔH_k is the enthalpy of displacement during a concentration step $[\Delta c_1]_k = [c_1]_k - [c_1]_{k-1}$. The corresponding quantities in the

immersion experiment, if a volume V^0 of solution of composition c_1^0 is equilibrated with a mass m of solid, are given by

$$[n_1^{\sigma(v)}]_k = \left[\frac{V_0}{m} (c_1^0 - c_1) \right]_k \tag{11}$$

$$[\Delta_{21}H]_k = [\Delta_w H]_k - [\Delta_w H]_{k=0} , \tag{12}$$

where c_1 is the equilibrium concentration in the bulk solution and $[\Delta_w H]_{k=0} = \Delta_w H(c_1 = 0)$ is the enthalpy of wetting in pure component 2. If component 1 has the greater affinity for the solid surface, the adsorption path is defined by proceeding from pure component 2 towards component 1, and the desorption path is related to the reverse direction. If a particular initial state and a particular final state of the system are considered, Eqs. (7), (8), and (11) are expected (at least in principle) to be equivlaent to each other, as are Eqs. (9), (10), and (12).

Experimental

Liquids: Double distilled water was saturated with n-butanol (Reanal) in a specially built separatory funnel thermostated at $25 + 0.1°C$. The saturation concentration of n-butanol in water was determined with an LCD 201 differential refractometer to be 0.968 mol dm^{-3}. Solutions of the desired concentrations were obtained by diluting with water.

Adsorbent: Printex-300, a carbon black from Degussa, was graphitized under an argon atmosphere by Sigri Elektrographit GmbH, Meitingen. The adsorbent was treated with n-heptane for 50 h in a Soxhlet apparatus. The B.E.T. specific surface area was found to be 56 m^2 g^{-1}.

Methods: In the LSC experiments, a chromatographic precolumn was used, loaded with 0.3965 g of adsorbent and held in a water-bath at $25 + 0.1°C$. The flow rate was 0.429 cm^3 min^{-1}. The concentration profiles were monitored with an LCD 201 differential refractometer. The dead volume V_D was determined by noting the time required for the solvent to run through the dry column at a fixed flow rate. Via Eq. (4), V_D was also calculated from the breakthrough curve of a 0.04% NaCl solution. For this electrolyte solution neither adsorption in a separate immersion experiment nor any heat effects in flow calorimetry could be detected.The average value of V_D was 0.756 cm^3. For comparison, the adsorption excess isotherm was determined by the conventional batch method, too. FSM experiments were performed with an LKB 2107 isothermal microcalorimeter at $25 + 0.02°C$. The sorption vessel contained 0.2020 g of adsorbent and a liquid flow rate of 0.361 cm^3 min^{-1} was applied. Blank runs were also made to estimate heats of mixing at the boundary between replacing and replaced solutions. The column-packing of carbon beads was modelled by a column-packing of teflon beads as inert material [13]. No heats of mixing could be detected in the cumulative experiments, but exothermal heats appeared in the one-step replacements (in both the adsorption and desorption directions), the magnitude of which increased with increasing concentration steps.

Results

The volumetric-reduced excess isotherms of the adsorption of n-butanol from the aqueous phase onto graphitized Printex-300 are shown in Fig. 3 on a relative concentration scale, $c_r = c_1/c_{sat}$. In the step-by-step method, the adsorption path was found to be reproducible by the desorption path (reversible displacement process) and only the average values of the corresponding experimental points are indicated in the figure. The static (immersion) isotherm coincides with the cumulative curve within experimental error. The agreement of the one-step adsorption and desorption data with the cumulative and static isotherms is also satisfactory up to a concentration of about $c_r = 0.20$. Beyond this, however, the adsorption and desorption curves incline more and more down and up, respectively, relative to the cumulative curve. The integral

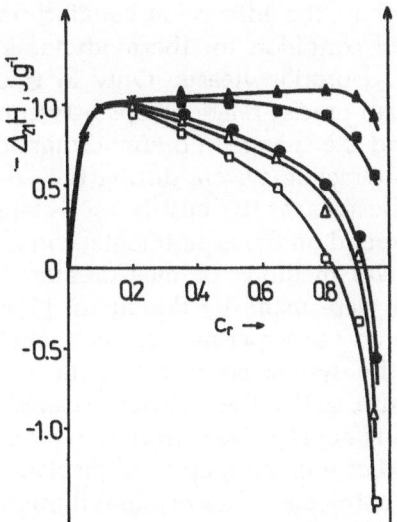

Fig. 4. Integral enthalpy isotherms of the displacement of water (2) by n-butanol (1) on graphitized Printex-300 at 25°C. (●) reversible cumulative; (▲) one-step adsorption; (△) one-step desorption; (■) one-step adsorption and (□) one-step desorption corrected by heats of mixing between solutions on teflon; (∗) coinciding points

enthalpies of the displacement of water on the graphitized carbon by n-butanol are shown in Fig. 4. Again, the cumulative isotherm proved to be reversible and, relative to this curve, considerable deviations of the one-step thermal data arose, even if heat of mixing corrections (between solutions) were applied. Heat of immersion measurements were also performed by using the batch unit of the calorimeter, but the results are not reported here because of the relatively high scattering of the data. In fact, immersion calorimetry is less sensitive than flow calorimetry and is, therefore, less applicable for the determination of small enthalpy differences (as in the present case) with high precision. However, a thermodynamic study on the system under discussion showed that the agreement between the cumulative and immersion enthalpies becomes satisfactory if they are corrected by heats of mixing between the surface and bulk phases [12].

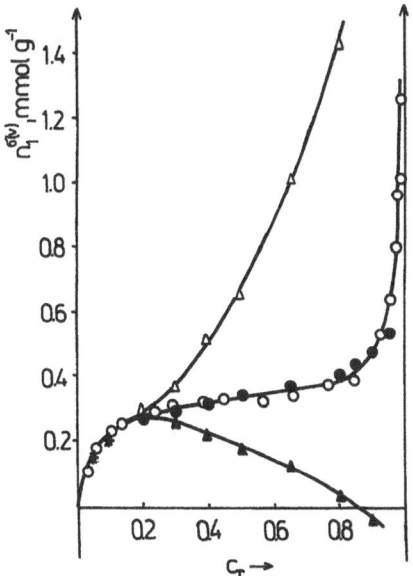

Fig. 3. Volume-reduced excess isotherms of the adsorption of n-butanol (1) from water (2) onto graphitized Printex-300 at 25°C. (○) static; (●) reversible cumulative; (▲) one-step adsorption; (△) one-step desorption; (∗) coinciding points

Discussion

In making LSC and FSM measurements it is important to check that the experiment is performed under well-defined conditions. Otherwise, caution is required when the results are interpreted. Della

Gatta argued [14] that "the adsorption equilibrium is the fundamental condition for thermodynamic definition of the adsorption heasts. Only in this case can we establish precise relationships between measured heats and the change in thermodynamic functions of the interacting system during the adsorption process. Besides, as the heat is not a state function, it will depend on the experimental conditions as well as the methods of measurement". Similar comments were made by Rouquerol [15]: "The heats measured are 'experimental heats' that depend on both the system studied and the experimental procedure, so that they cannot be called, in the absence of any appropriate correction (which must be clearly indicated), 'enthalpies of displacement', or still less, 'enthalpies of adsorption', though this is sometimes dones".

A more detailed analysis of the problem was given by Woodbury and Noll [5]. They concluded that "since the time scale for the cumulative method is several times longer than it is for the one-step method, one would expect any kinetic (non-equilibrium) effect present to be larger for the latter: such an effect would lead to a difference between the enthalpies of the two methods".

The above remarks are confirmed by the calorimetric data (Fig. 4), but they can be extended to the chromatographic measurements, too (Fig. 3). We find that if the displacement process takes place via non-equilibrium steps, the measured heats and the recorded breakthrough curves become ill-defined measures of the adsorption and, under such circumstances, it is illusory to regard the measured heats as enthalpies of displacement and the quantities deduced from Eq. (3) as excess amounts of substance. The recorder charts of the refractometer for two different concentration steps, $\Delta c_r = 0.05$ (adsorption: $0 \rightarrow 0.05$; desorption: $0.05 \rightarrow 0$) and $\Delta c_r = 0.80$ (adsorption: $0 \rightarrow 0.80$; desorption: $0.80 \rightarrow 0$), are shown in Figs. 5a and 5b, respectively, and the corresponding detector signals of the calorimeter in Figs. 6a and 6b. It may be concluded from the figures that the displacement process becomes reversible if, at a given column-packing and a given liquid-flow rate, the concentration increments are chosen to be small enough for the liquid front to pass through and the displacement process to take place via quasi-equilibrium steps throughout the column. A large concentration difference between the replacing and replaced solutions may conduce to the appearance of some non-equilibrium effects (e.g., back-mixing, mass transfer resistance, channelling and other complexities) which is clearly indicated by the irregular peaks in Figs. 5b and 6b. Although small concentration steps are preferred, the detectability decreases with decreasing concentration difference and, therefore, only instruments of high senstivity are suitable for the measure-

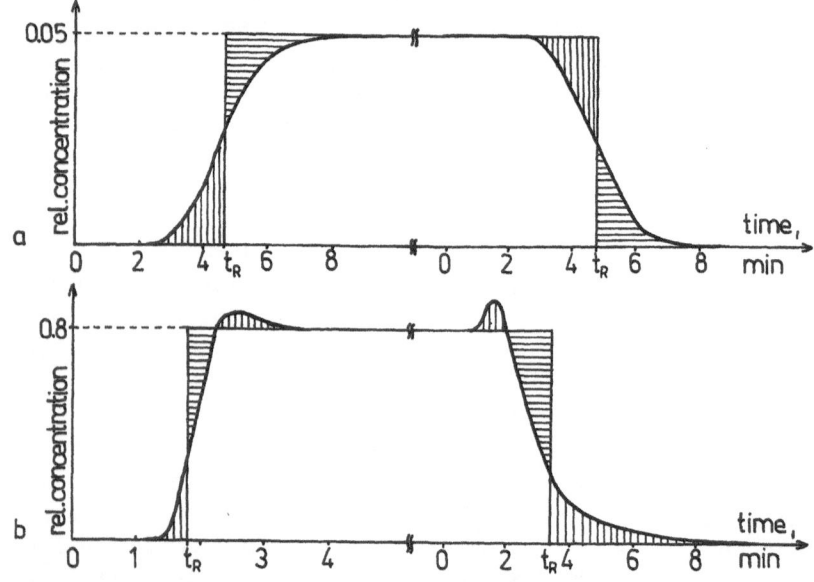

Fig. 5. Adsorption-elution breakthrough curves for relative concentration steps $0 \rightarrow 0.05 \rightarrow 0$ (Fig. 5a) and $0 \rightarrow 0.80 \rightarrow 0$ (Fig. 5b). The oppositely shaded areas are equal for each curve

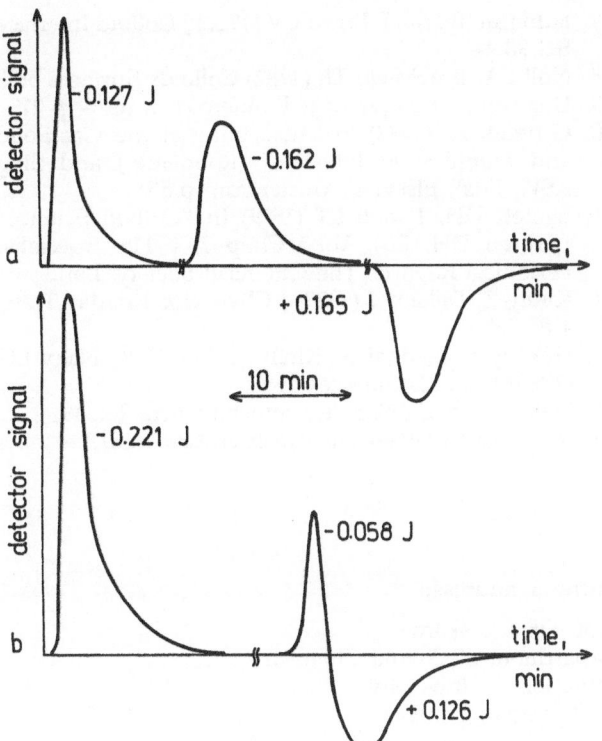

Fig. 6. Detector signals of the calorimeter for the same concentration steps as in Fig. 5. The first peak in Fig. 6a is a reference (calibration) peak

ments. A tight column-packing and a low liquid-flow rate are also expected to improve the experimental conditions, but they require an excessive pumping pressure and a very stable, drift-free detector signal. It can be seen in Fig. 4 that the one-step data still lie far from the cumulative enthalpy isotherm, even if they are corrected by heats of mixing on teflon, as described in the experimental section. What is more, the agreement may become even poorer. Consequently, a heat correction of this kind is unsuitable for this system, at least under the present experimental conditions (liquid flow rate, column parameters). It is interesting to note that the one-step "enthalpies" of desorption, without corrections, run quite close and parallel to the cumulative enthalpy curve. The reason for this is that, to reach the equilibrium state, a longer time is required for the elution process than for the reverse adsorption path (see the time scale in Fig. 6b). To a certain extent, this kinetic effect can be related to the thermodynamic driving force of the displacement: the desorption is less favored energetically than the adsorption. The one-step excess "iso-

therms" (Fig. 3) also indicate that the one-step desorption takes a longer time to equilibrate than the one-step adsorption: the desorption curve lies well above the adsorption curve, as a consequence of the longer retention times (see the time scale in Fig. 5b). Although the step-by-step method provides more accurate data, due precaution is required when the concentration steps are selected. If they are not small enough, the procedure may become a series of single steps, as in the one-step method. To test and eliminate non-equilibrium effects, it is advisable:

i) to perform multiple runs (adsorption-desorption cycles);

ii) to reduce the flow rate or to increase (say, to double) the number of steps by reducing (halving) the concentration intervals. For a reversible displacement process (physisorption), both versions are applicable. If the isotherms are reproducible, the results are not "damaged" by the experimental procedure. For an irreversible process (e.g., chemisorption), variant i) can be applied only if chemical reactions are completed. Therefore, variant ii) is preferred, with the restriction that fresh solid sample should be used in each adsorption-desorption run. Finally, it is to be noted that a heat of mixing correction can be made in a very elegant manner, as proposed by Woodbury and Noll [5]. This correction is based on a combination of the broadening of the breakthrough curves with the excess enthalpy function of the bulk liquid. Unfortunately, this procedure was not applicable in the present work, because it requires simultaneous FSM and LSC measurements.

Conclusions

The appearance of non-equilibrium effects in frontal analysis liquid/solid chromatography and flow sorption microcalorimetry makes it difficult to deduce adsorption excess isotherms and enthalpy isotherms of displacement from the experimental results. The reproducibility of the measurements under the same experimental conditions does not necessarily mean that the experimental results can be associated with the desired adsorption equantities (excess amounts of substance and enthalpies of displacement). To eliminate non-equilibrium effects, quasi-equilibrium experimental conditions are required. Besides comparisons with static (immersion) measurements, dynamic data can be tested with regard to non-equilibrium effects:

i) via adsorption-desorption cycles, and

ii) by reducing the liquid-flow rate or the concentration increments to increase the time for the displacement process.

If chemisorption occurs, variant ii) is preferred, a fresh solid sample being used in the test experiment.

References

1. Sharma SC, Fort T Jr (1973) J Colloid Interface Sci 43:36
2. Wang HL, Duda JL, Radke CJ (1978) J Colloid Interface Sci 66:153
3. Koch CS, Köster F, Findenegg GH (1978) J Chromatogr 406:257
4. Findenegg GH, Koch C, Liphard M (1983) In "Adsorption from Solution" (Ottewill RH, Rochester CH and Smith AL, Eds), Academic Press, London, p 87
5. Woodbury GW Jr, Noll LA (1987) Colloids Surfaces 28:233
6. Jednacak-Biscan J, Pravdic V (1980) J Colloid Interface Sci 75:322
7. Jednacak-Biscan J, Pravdic V (1982) J Colloid Interface Sci 90:44
8. Noll LA, Burchfield TE (1982) Colloids Surfaces 5:33
9. Denoyel R, Rouquerol F, Rouqerol J in ref. 4 p 225
10. Groszek AJ (1982) In "Adsorption at the Gas-Solid and Liquid-Solid Interface" (Rouqerol J and Sing KSW, Eds), Elsevier, Amsterdam, p 55
11. Everett DH, Podoll RT (1979) In "Colloid Science" "Everett DH, Ed), Vol 3 Chap 2, p 115. Specialist Periodical Reports, The Chemical Society, London
12. Király Z, Dékány I (1989) J Chem Soc Faraday Trans 1 85:3373
13. Dékány I, Zsednai Á, Király Z, László K, Nagy LG (1986) Colloids Surfaces 19:47
14. Della Gatta G (1985) Thermochim Acta 96:349
15. Rouquerol J (1985) Thermochim Acta 96:377

Authors' address:

Prof. Dr. I. Dékány
Department of Colloid Chemistry
Attila József University
6720 Szeged, Hungary

Progress in Colloid & Polymer Science   Progr Colloid Polym Sci 83:75—83 (1990)

On the nature and role of water structuring in the adsorption of nonionic surfactants on hydrophobic surfaces

B. Kronberg and R. Silveston

Institute for Surface Chemistry, Stockholm, Sweden

Abstract: High pressure liquid chromatography (HPLC) results were used to calculate the temperature dependence of the hydrocarbon-water interaction parameter. The temperature dependence has been interpreted in terms of water structuring around the hydrocarbon. The role of water structuring is to enhance the solubility of hydrocarbons in water, which is equivalent to a decrease of the hydrocarbon-water interaction parameter. The poor solubility of hydrocarbons in water is not due to the water structuring, but to other factors such as the large energy required to form a cavity in the water for the hydrocarbon. This is a reflection of the very large cohesive forces in liquid water. These results, together with the temperature dependence of the polyethylene oxide-water interaction parameter, have been used to calculate the temperature dependence of the adsorption of a nonionic surfactant, viz. an ethoxylated nonyl phenol with 20 ethylene oxide units (NP-EO$_{20}$). It is shown, that water structuring causes, an increased adsorption with temperature. Without water structuring a decreased adsorption is predicted.

Key words: Adsorption; nonionic surfactants; water structuring; interaction parameters

Introduction

Adsorption isotherms of surfactants on hydrophobic surfaces are known to correlate with the solution properties of the surfactants, i.e., plateau adsorption is obtained at the critical micelle concentration (CMC) of the surfactant, regardless of i) the surface polarity (provided the surface is not so polar that specific interactions occur with the head group of the surfactant), ii) surfactant structure, ionic or nonionic, and iii) solvent composition, such as additions of salt or water soluble organic additives. This fact indicates that the driving force of surfactant adsorption on hydrophobic surfaces is to be found in the surfactant-solvent interaction.

The purpose of this paper is to discuss the origin of the main driving force of the adsorption of nonionic surfactants on hydrophobic surfaces. By nonionic surfactants, we mean chain molecules with two blocs: one which is a hydrocarbon moiety, and another which is a polyethylene oxide chain.

Adsorption model

In order to understand the adsorption isotherms of surfactants, we need a model that somewhat realistically describes the adsorbed state of the surfactants on a hydrophobic surface. Most commonly, surfactant adsorption isotherms are analyzed in terms of the Langmuir equation. However, such an analysis does not give any relevant molecular information since two very important assumptions are made using the equation. These assumptions are i) that the solute and solvent molecules should be of the same size, and ii) that there should not be any molecular interaction between the solute and solvent. Thus, the use of the Langmuir equation implies a model where noninteracting molecules of the same size compete for the surface, and where the adsorption is governed by their adsorption free energy.

It is, therefore, of interest to investigate other, perhaps more realistic models of the adsorption of nonionic surfactant molecules. Since these surfactants are chain molecules the Flory-Huggins description of the system is applicable. Thus, the chemical potential of the solution phase is easily described, where the surfactant-water interaction is determined by both the hydrocarbon-water and ethylene oxide-water interactions. The difficulty lies in the description of the chemical potential of the surface phase. Adsorption of chain molecules on solid surfaces has previously been described by, for example, Prigogine and Marechal [1], Silberberg [2], and Roe [3]. None of these descriptions fit the adsorption of nonionic surfactants where the hydrocarbon moiety is assumed to adsorb, at least partly, in contact with the surface, and the ethylene oxide chain is assumed to protrude into the solution. Great improvements have been achieved through the Scheutjens-Fleer theory on the adsorption of polymers [4] and lately, by Koopal et al. [5] on the adsorption of surfactant molecules. These theories suffer from the disadvantage, however, of being quite complex, disallowing the fitting of experimental results. We, therefore, simply used the Flory-Huggins description of the chemical potential for the surfactant in the surface phase [6]. It is assumed that the surfactant molecules adsorb with their hydrocarbon moiety in contact with surface. This fact will, of course, affect the combinatorial entropy of the surfactant. Instead of calculating this effect, we put all such effects into the interaction parameter χ^s in the surface phase. Thus, the interaction parameter in the surface phase should be considered as a residual of the chemical potential once a rough calculation of the combinatorial entropy has been performed using the Flory-Huggins expression.

Thus, the chemical potential μ of the surfactant (component 2) in the solution and surface phase is written as

$$\frac{\mu_2 - \mu_2^0}{RT} = \ln(\varphi_2) + (1 - r)\varphi_1 + r\chi(\varphi_1)^2 \quad \text{(1a)}$$

$$\frac{\mu_2^s - \mu_2^{0,s}}{RT} = \ln(\varphi_2^s) + (1 - r_1^s)$$

$$+ r\chi^s(\varphi_1^s)^2 + \frac{\gamma a_2^0}{kT}, \quad \text{(1b)}$$

where superscript s indicates the surface phase. Here φ is the volume fraction, r is the size ratio of

the solute to the solvent, χ is the interaction parameter, a_2^0 is the surface area occupied by a surfactant molecule, and γ is the interfacial tension between the solution and the solid. Similar equations can be written for the chemical potential of the water in the two phases [6]. Equating the chemical potential of the two components in the two phases gives the following expression at low solution concentrations of the surfactant,

$$\ln(\varphi_2) = \ln\left(\frac{\varphi_2^s}{(1 - \varphi_2^s)^r}\right) + r\chi^s(1 - 2\varphi_2^s)$$

$$- r\chi - \frac{qa_1}{RT}(\gamma_1 - \gamma_2), \quad \text{(2)}$$

where q is the number of segments of the surfactant that are in contact with the surface and the γ's are the interfacial tensions between the respective pure component and the surface.

Equation (2) is an explicit expression of the solution composition as a function of the surface composition. From the equation, we see that plotting $\ln\left(\frac{\varphi_2^s}{(1 - \varphi_2^s)^r}\right) - \ln(\varphi_2)$ vs $(1 - 2\varphi_2^s)$ should give a straight line with a slope of $-r\chi^s$ and an intercept of $r\chi + \frac{qa_1}{RT}(\gamma_1 - \gamma_2)$. Analysis of experimental isotherms [6, 7] shows that there are two contributions to the adsorption free energy. The first contribution, $\frac{qa_1}{RT}(\gamma_1 - \gamma_2)$, is due to the exchange of surface-water molecular contacts for contacts between the surface and the hydrocarbon moiety of the surfactant. The second driving force is $r(\chi - \chi^s)$, i.e., it originates from a difference in the surfactant-water interaction parameter in the surface phase compared with that of the solution phase. When $\chi > \chi^s$ there will be an increased adsorption.

Table 1 shows the two driving forces for the adsorption of ethoxylated nonyl phenols, NP-EO$_n$, where n is equal to 10, 20 or 50, on a polystyrene latex [6]. The table clearly shows that the dominating driving force is the difference between the χ and χ^s parameters. This difference originates in the fact that the whole of the adsorbed surfactant molecule is not "seen" by the water molecules. The surfactant molecules adsorb with their hydrocarbon part in contact with the surface and the ethylene

Table 1. The two driving forces for the adsorption of NP-EO$_n$ on polystyrene latex (calculated for $q = 4$). (From [6])

Surfactant	$r(\chi - \chi^s)$	$\dfrac{qa_1}{RT}(\gamma_1 - \gamma_2)$
NP-EO$_{10}$	5	1.6
NP-EO$_{20}$	7	1.6
NP-EO$_{50}$	6	1.6

oxide chains in contact with the aqueous solution. This surfactant orientation decreases the probability of finding hydrocarbon-water molecular contacts. Thus, many of the unfavorable hydrocarbon-water contacts, that are found in the aqueous phase are lost in the surface phase. Thus, Table 1 reveals that the adsorption is dominated by the hydrocarbon-water interaction. Hence, the major driving force for the adsorption of nonionic surfactants on hydrophobic surfaces is akin to that of micellization.

HPLC measurements

Recently, we utilized the high pressure liquid chromatography (HPLC) technique in order to obtain a deeper understanding of the hydrocarbon-water interaction [8]. The HPLC system studied consisted of a stainless steel column packed with glass beads (20—40 μm), which were coated with a polymer, viz. polydimethyl siloxane (PDMS). The mobile phase was pure water. The difference in retention volume between nonpolar molecules, such as toluene, and a nonretarded salt, viz. NaNO$_3$, was measured as a function of temperature. This set-up gives the partitioning of the probe between the aqueous phase and the nonpolar PDMS phase at infinite dilution of the probe. Thus, from the partition coefficient K, which is the ratio of probe concentration in the aqueous phase to that in the PDMS phase, one obtains the free energy of transfer of the probe from the aqueous phase to the nonpolar polymer phase, $G^{tr}_{(w \to polymer)}$. Thus the chromatographic process mimics that of the adsorption of nonionic surfactants, viz. the removal of hydrocarbon-water contacts to form hydrocarbon-hydrocarbon contacts. We will, therefore, denote the free energy of transfer by $G^{tr}_{(w \to hc)}$.

The partition coefficient is directly related to the retention volume of the probe V^R and the marker salt V^{mk} through

$$\frac{G^{tr}_{(w \to hc)}}{RT} = \ln(K)$$

$$= \ln\left(\frac{V^R - V^{mk}}{V^p}\right) + \ln\left(\frac{V_1^p(\infty)}{V_1^w(\infty)}\right), \quad (3)$$

where V^p is the volume of the polymer phase, $V_1^p(\infty)$ and $V_1^w(\infty)$ are the partial molar volumes of the probe at infinite dilution in the polymer and aqueous phase, respectively. Recently, we studied the transfer of toluene and ethyl benzene [8] and benzene and propyl benzene [9].

Analysis of the temperature dependence of the free energy of transfer gives the enthalpy and entropy of transfer of the probe between the two phases. Figure 1 shows the quantities $G^{tr}_{(w \to hc)}$, $H^{tr}_{(w \to hc)}$ and $TS^{tr}_{(w \to hc)}$ for the transfer of toluene from water to the nonpolar PDMS phase. The figure reveals that the free energy of transfer is large and negative, which is analogous to the large $r(\chi - \chi^s)$ contribution to the adsorption (Table 1). The large and negative free energy of transfer $G^{tr}_{(w \to hc)}$ is normally attributed to the water structuring around the hydrophobic probe. The analysis of the temperature dependence in terms of the transfer of the probe from the PDMS phase to the aqueous phase has been dealt with in detail in [8]. We will only briefly repeat the main conclusions in terms of the transfer

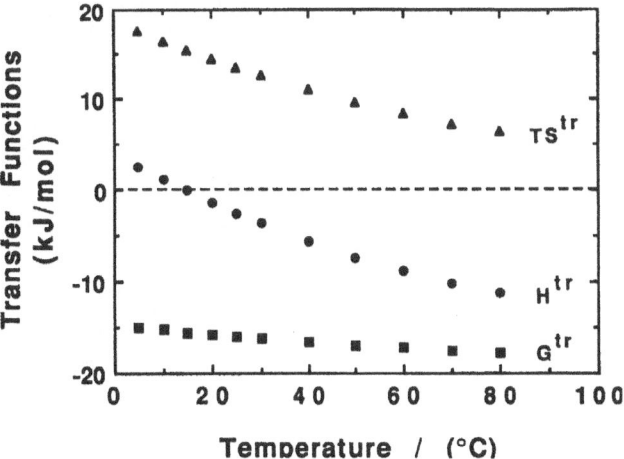

Fig. 1. The temperature dependence of the transfer functions $G^{tr}_{(w \to hc)}$, $H^{tr}_{(w \to hc)}$ and $TS^{tr}_{(w \to hc)}$ for the transfer of toluene from water to PDMS

of the probe from the aqueous solution to the PDMS phase, which is akin to the process of adsorption of nonionic surfactants from solution, i.e., a process where hydrocarbon-water interactions are exchanged for hydrocarbon-hydrocarbon interactions. Strictly speaking, the HPLC results should be corrected for the transfer of the probe from the PDMS phase to an aliphatic hydrocarbon phase. This correction is very small and does not have any appreciable temperature dependence. We will, therefore, restrict ourselves to discussing the primary results as they turn out from the chromatography experiments.

Water structuring

The entropy of transfer contains two contributions. The first contribution is due to the difference in combinatorial entropy of the probe in the two phases. This contribution is shown to be close to zero [8]. The second contribution is due to the structuring of water molecules in the proximity of the nonpolar probe. This contribution was found to have an exponential temperature dependence according to

$$S^{tr}_{(w \to hc)} = A e^{-T/\tau} , \qquad (4)$$

where A is a constant depending on the probe size (or more realistically the surface area of the probe), and τ is a universal constant, independent on probe, describing the rate of the build up of water structure with decreasing temperature. It was found that $\tau = 60 \pm 5$ K. From elementary thermodynamics it can be shown that the structuring contribution to the free energy $G^{tr}_{s(w \to hc)}$ and enthalpy $H^{tr}_{s(w \to hc)}$ is

$$G^{tr}_{s(w \to hc)} = \tau S^{tr}_{(w \to hc)} = \tau A e^{-T/\tau} \qquad (5)$$

$$H^{tr}_{s(w \to hc)} = (T + \tau) S^{tr}_{(w \to hc)} = (T + \tau) A e^{-T/\tau} . \qquad (6)$$

These quantities differ from those experimentally obtained by a constant $G^{tr}_{0(w \to hc)}$ and $H^{tr}_{0(w \to hc)}$, respectively. Since there is no other contribution to S^{tr} besides the structuring contribution we have

$$G^{tr}_{0(w \to hc)} = H^{tr}_{0(w \to hc)} . \qquad (7)$$

Thus, the experimentally obtained $G^{tr}_{(w \to hc)}$ and $H^{tr}_{(w \to hc)}$ are composed of two contributions,

$$G^{tr}_{(w \to hc)} = G^{tr}_{0(w \to hc)} + G^{tr}_{s(w \to hc)}$$

$$= G^{tr}_{0(w \to hc)} + \tau A e^{-T/\tau} \qquad (8)$$

$$H^{tr}_{(w \to hc)} = H^{tr}_{0(w \to hc)} + H^{tr}_{s(w \to hc)}$$

$$= H^{tr}_{0(w \to hc)} + (T + \tau) A e^{-T/\tau} . \qquad (9)$$

$G^{tr}_{0(w \to hc)}$ and $H^{tr}_{0(w \to hc)}$ are ascribed to the difference in chemical nature that the probe experiences during the transfer process. This term is due to differences in van der Waals forces and the large energy required to form a cavity in the water for the hydrocarbon. This latter contribution is a reflection of the large cohesive forces in liquid water. The temperature dependence of $H^{tr}_{0(w \to hc)}$ should thus be of the same order of magnitude as the temperature dependence of the enthalpy of vaporization, which, when compared to the magnitude of the temperature dependence of $H^{tr}_{s(w \to hc)}$, is negligible. Therefore, $H^{tr}_{0(w \to hc)}$ is considered to be independent on temperature.

Figure 2 displays the structuring contributions to the thermodynamic functions, showing that $G^{tr}_{s(w \to hc)}$ is positive, i.e., water structuring around the nonpolar probe increases the stability of the solution. Thus, water structuring also increases the solubility of nonpolar molecules. The reason for the very low water solubility of nonpolar molecules in water is thus found in the large values of $H^{tr}_{0(w \to hc)}$ or $G^{tr}_{0(w \to hc)}$. Figure 2 also shows that the enthalpy and entropy are almost compensating, resulting in a small magnitude of the free energy. The reason for

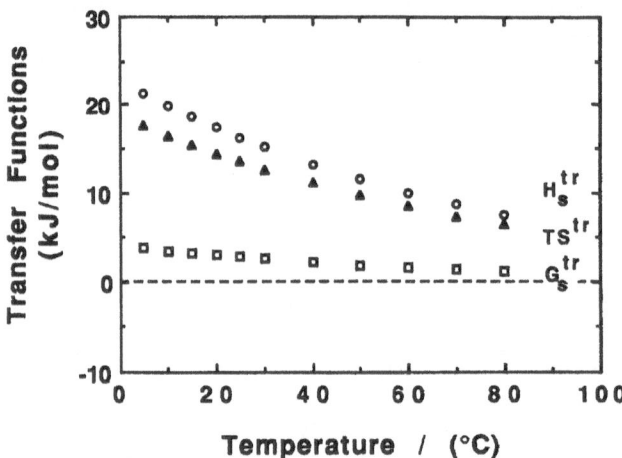

Fig. 2. The temperature dependence of the water structuring contribution to the transfer functions (i.e., $G^{tr}_{s(w \to hc)}$, $H^{tr}_{s(w \to hc)}$, $TS^{tr}_{(w \to hc)}$) for the transfer of toluene from water to PDMS

this compensation is found in the small value of τ compared to T, in the temperature range where water is a liquid, as can be seen from Eqs. (5) and (6). The interpretation of the water structuring adopted in this paper was previously discussed by Shinoda [10—13], Patterson [14, 15], and Hvidt [16—18].

Interaction parameters

Before applying these results to the adsorption of nonionic surfactants, we need to discuss the interaction parameters. In solution thermodynamics the most common way to describe the solute-solvent interaction is through the χ parameter. In the Flory-Huggins interpretation of the chemical potential, the χ parameter is considered to include all contributions other than the calculated combinatorial entropy of mixing. The χ parameter is thus considered to be an excess quantity containing both entropic and enthalpic contributions. Costas and Patterson [15] have shown that in aqueous solutions, another contribution to the excess entropy arises since all the water volume is not accesible to the solute, due to the hydrogen-bonded network existing in liquid water. Such contributions, not covered by the calculated combinatorial entropy of mixing, are included in the excess entropy term and are thereby included in the χ parameter. Thus, the description of the combinatorial entropy of the probe in aqueous and organic solutions should differ. We will not calculate any such differences, but rather will include it in the probe-polymer interaction parameter as an excess quantity. Thus, the relation between $G^{tr}_{(w \to hc)}$ and the χ parameters is [8]

$$G^{tr}_{(w \to hc)}/RT = (G^{tr}_{0(w \to hc)} + G^{tr}_{s(w \to hc)})/RT$$

$$= \chi_{hc,p} - \chi_{hc,w} , \qquad (10)$$

where $\chi_{hc,p}$ and $\chi_{hc,w}$ are the χ parameters for the probe (hydrocarbon)-polymer (PDMS) and probe-water interaction, respectively. From Eqs. (8) and (10), we conclude that the $\chi_{hc,w}$ interaction will have the following temperature dependence,

$$\chi_{hc,w} = (a_{hc} - b_{hc} e^{-T/\tau})/T , \qquad (11)$$

where both a_{hc} ($= T\chi_{hc,p} - G^{tr}_{0(w \to hc)}/R$) and b_{hc} ($= \tau A_c/R$) are constants. The constant A_c is equal to the A in Eq. (4), but is expressed per aliphatic

carbon atom (see below). We have here assumed that the $\chi_{hc,p}$ parameter has an inverse temperature dependence, since the major contribution to $\chi_{hc,p}$ comes from the van der Waals forces.

Both $G^{tr}_{(w \to hc)}/RT$ and solubility data of the probes in water show an extremum at around 15°C. It can be shown that such an extremum corresponds to the condition

$$\frac{d\chi_{hc,w}}{dT} = \frac{-1}{T_e^2} \left(a_{hc} - b_{hc} \left(1 + \frac{T_e}{\tau} \right) e^{-T_e/\tau} \right) = 0 .$$

$$(12)$$

Thus, given τ ($= 60$) and the temperature of the extremum, T_e ($= 288$ K), we obtain the ratio between a_{hc} and b_{hc}. An approximate value of b_{hc} is obtained from the hydrocarbon chain-length dependence of A in Eq. (4) (since $b_{hc} = \tau A_c/R$). In calculating the corresponding aliphatic carbon number of the alkyl benzenes, we assume that a benzene ring corresponds to 3.5 aliphatic carbons [19]. Thus, the slope of a plot of $\tau A/R$ as a function of the aliphatic carbon number of the probe gives a value of 14300 K for b_{hc}. Note that this value has been normalized to the size of a water molecule, since water is the solvent in our discussion. Thus, we find the following temperature dependence of a water-hydrocarbon interaction,

$$\chi_{hc,w} = (683 - 14300 \ e^{-T/60})/T . \qquad (13)$$

Figure 3 shows the temperature dependence of the $\chi_{hc,w}$ parameter. Included in the figure is also how the $\chi_{hc,w}$ parameter would have changed if there was no water structuring around the hydrocarbon, i.e., no exponential term in Eq. (13). We note that i) at high temperatures the two curves merge, i.e., when the water structuring has dissappeared, and ii) the water structuring decreases the $\chi_{hc,w}$ parameter. Thus, the water structuring acts favorably on the interaction between a hydrocarbon and water. This is contrary to the normally held opinion, viz., that water structuring is the cause of the poor hydrocarbon water interaction. We find, however, that the poor hydrocarbon-water interaction is due to the large and positive constant a_{hc} in Eqs. (11) and (13). The origin of this large constant lies in the difference in chemical nature of the two components and on the work of forming a cavity for the probe in the water, which is a reflection of the large cohesive forces in liquid water.

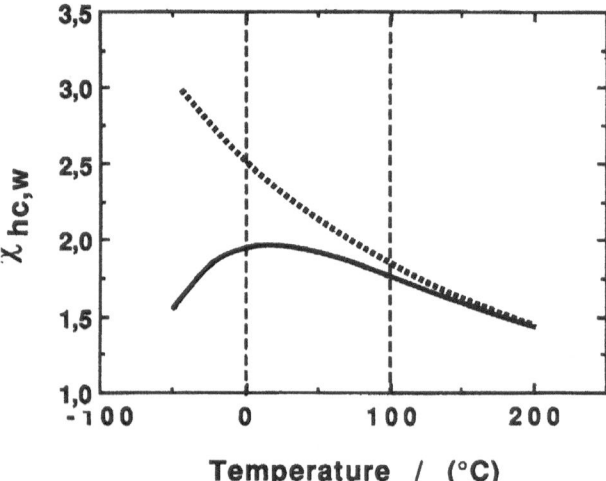

Fig. 3. The temperature dependence of the hydrocarbon-water interaction parameter, $\chi_{hc,w}$, (solid line). The dotted line represents the same interaction parameter, but calculated without the water structuring contribution

In order to predict the temperature dependence of the adsorption of nonionic surfactants, we also need the temperature dependence of the ethylene oxide-water interaction. This interaction is complicated by the fact that there is an ether group in the aliphatic chain, thus enabling hydrogen bonding with the water. We will, therefore, calculate Eq. (11) in terms of ethylene oxide-water interactions and add an extra term, $c_{eo}T$, where c_{eo} is a constant. Thus, we have for the partial molar excess free energy of mixing, G_E^M, of a polyethylene oxide chain with water,

$$\frac{G_E^M}{RT} = \chi_{eo,w} = \frac{1}{T}\left(a_{eo} - b_{eo}e^{-T/\tau} + c_{eo}T\right), \qquad (14)$$

where a_{eo} and b_{eo} are constants.

Phase diagrams show an upper critical solution temperature at roughly 373 K. However, increasing the temperature further reveals that there is a lower critical solution temperature. Thus, the temperature derivative of the $\chi_{eo,w}$ parameter must change sign somewhat in between these temperatures. Experimental phase diagrams (20) reveal that the middle of the two-phase region is found at approximately 470 K. Thus we have

$$\frac{d\chi_{eo,w}}{dT} = \frac{-1}{T_m^2}\left(a_{eo} - b_{eo}\left(1 + \frac{T_m}{\tau}\right)e^{-T_m/\tau}\right) = 0, \qquad (15)$$

where T_m is the temperature (470 K) at which the $\chi_{eo,w}$ is at a maximum. Keeping $\tau = 60$, Eq. (15) thus gives us the ratio between the a_{eo} and b_{eo}, viz., $a_{eo}/b_{eo} = 0.00350$. Substituting these results into Eq. (14) we have

$$\frac{G_E^M}{RT} = \chi_{eo,w} = \frac{b_{eo}}{T}\left(0.00350 - e^{-T/\tau}\right) + c_{eo}. \qquad (16)$$

Experimental $\chi_{eo,w}$ parameters have been obtained from osmotic pressure measurements at various temperatures (21) and from calculation of the critical $\chi_{eo,w}$ parameter from the phase diagrams (20) according to the scheme of Kjellander et al. (22). Thus, plotting $\chi_{eo,w}$ vs $(0.00350 - e^{-T/\tau})/T$ gives us $b_{eo} = 11800$ K and $c_{eo} = 0.502$ K. Thus, we have

$$\chi_{eo,w} = \frac{1}{T}\left(41.3 - 11800\,e^{-T/\tau} + 0.502\,T\right). \qquad (17)$$

The temperature dependence of the calculated $\chi_{eo,w}$ parameter is shown, together with the literature values in Fig. 4, displaying the maximum at 470 K. The $\chi_{eo,w}$ parameter, calculated without the water structuring contribution is also shown. As in Fig. 3, we see that the two curves merge at high temperatures, i.e., when the water structuring has disappeared.

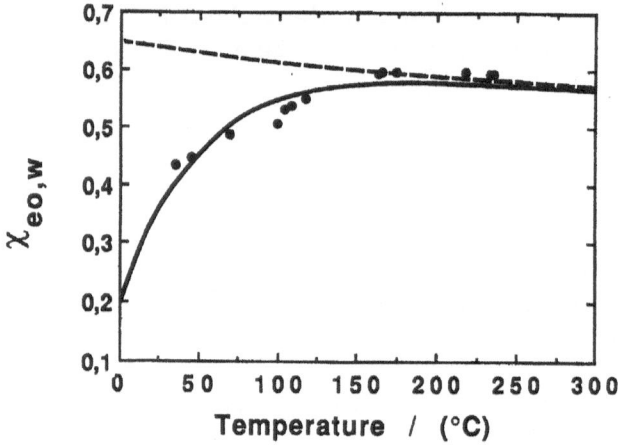

Fig. 4. The temperature dependence of the ethylene oxide-water interaction parameter, $\chi_{eo,w}$, (solid line). The dashed line represents the same interaction parameter, but calculated without the water structuring contribution. The points are literature values (see text)

We will use NP-EO$_{20}$, an ethoxylated nonyl phenol with 20 ethylene oxide units, to represent a model nonionic surfactant in our discussion. The surfactant-water interaction parameter χ(NP-EO$_{20}$, W) can be written in terms of the χ parameters of its individual constituents according to [6],

$$\chi(\text{NP-EO}_{20}, \text{W}) = (1 - \omega_{eo})\chi_{hc,w} + \omega_{eo}\chi_{eo,w}$$
$$- (1 - \omega_{eo})\omega_{eo}\chi_{hc,eo} , \qquad (18)$$

where ω_{eo} is the ethylene oxide weight fraction in the surfactant and $\chi_{hc,eo}$ is the hydrocarbon-ethylene oxide interaction parameter (normalized to the size of a water molecule). Since this latter interaction does not involve any interactions other than van der Waals forces, we will adopt an inverse temperature dependence of the $\chi_{hc,eo}$ parameter. We will use a value of 0.4 (at 25 °C) for this parameter [6].

Figure 5 shows the total surfactant-water interaction parameter χ(NP-EO$_{20}$, W) in the temperature range 0—200 °C. The figure shows that the χ(NP-EO$_{20}$, W) interaction parameter is monotonously increasing in the temperature range 0—100 °C and showing a maximum at temperatures just above 100 °C. The figure also shows how the χ(NP-EO$_{20}$, W) parameter would change if there were no water structuring. Thus, the water structuring lowers the surfactant-water interaction con-

siderably, especially at low temperatures. Again we conclude the water structuring favors the dissolution of hydrophobic substances.

Adsorption isotherms

Using these χ parameters, we are now able to calculate the temperature dependence of the adsorption isotherms using Eq. (2). In accordance with [6], we assume that the number of hydrocarbon-water contacts lost in the surface phase is a fraction λ of the number of contacts in the bulk liquid phase. Thus, we have [6],

$$\chi^s(\text{NP-EO}_{20}, \text{W}) = \lambda\chi(\text{NP-EO}_{20}, \text{W})$$
$$+ (1 - \lambda)\omega_{eo}\chi_{eo,w} . \qquad (19)$$

The parameters used in the following calculations are $\lambda = 0.5$, $r = 61.1$ and $(qa_1/RT)(\gamma_1 - \gamma_2) = 1.6$ at 25 °C. Adsorption isotherms have been calculated in the temperature range 0—100 °C. Figure 6a shows the predicted adsorption isotherms at selected temperatures. We see that the adsorption increases with temperature, reaching an almost constant value at higher temperatures and reflecting the change in the χ(NP-EO$_{20}$, W) parameter with temperature.

From the adsorption plateau values in Fig. 6 we calculated the area per surfactant molecule at close packing at the surface, A_2. Figure 7 shows that this quantity decreases with increasing temperature, reaching a constant value at high temperatures. Normally, such a behaviour would have been interpreted in terms of the extension of the ethylene oxide chain and its decrease in extension at higher temperatures. It is an interesting fact that the simple theory used here does not contain any assumptions as to the polyethylene oxide chain conformations. Figure 7 thus indicates that an interpretation in terms of polymer chain conformations is equivalent to an interpretation in terms if polymer-solvent interactions.

Figure 6b shows the calculated adsorption without the water structuring contribution. Here the adsorption decreases with temperature. Thus, the increasing adsorption with temperature in Fig. 6a is due to the water structuring in the system. Also, Fig. 7 shows how the area per molecule should change if there were no water structuring in the system. We see that we now obtain an increas-

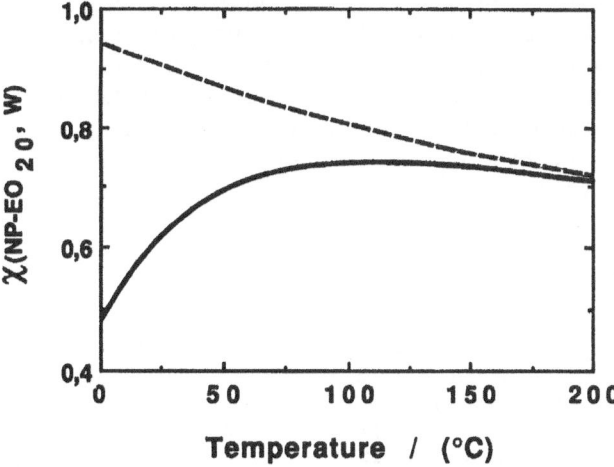

Fig. 5. The temperature dependence of the surfactant-water interaction parameter, χ(NP-EO$_{20}$, W), (solid line). The dashed line represents the same interaction parameter, but calculated without the water structuring contribution

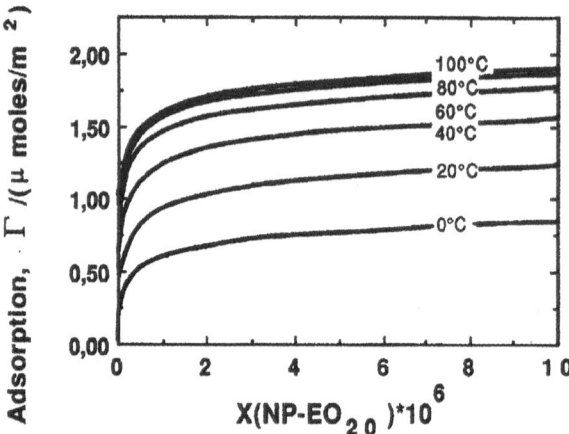

Fig. 6a. Calculated adsorption isotherms for the surfactant $NP-EO_{20}$, based on Eq. (2), at the indicated temperatures

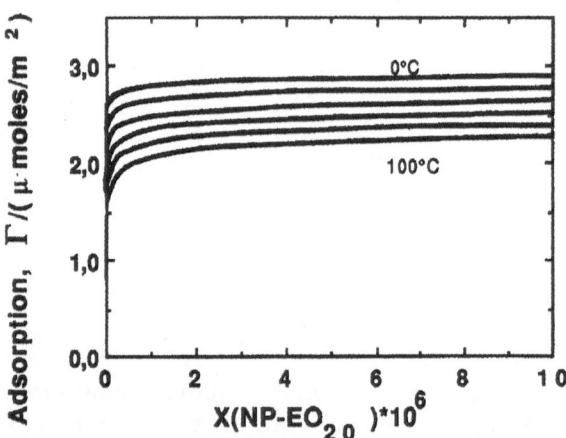

Fig. 6b. Calculated adsorption isotherms for the surfactant $NP-EO_{20}$, based on Eq. (2), at the indicated temperatures. The water structuring contribution has been excluded. The temperature difference between the curves is 20°C

ing temperature dependence, which is due to the inverse temperature dependence of the interaction parameters leading to a decreasing adsorption with increasing temperature.

Conclusions

We have shown in this work that water structuring is the reason for the abnormal temperature

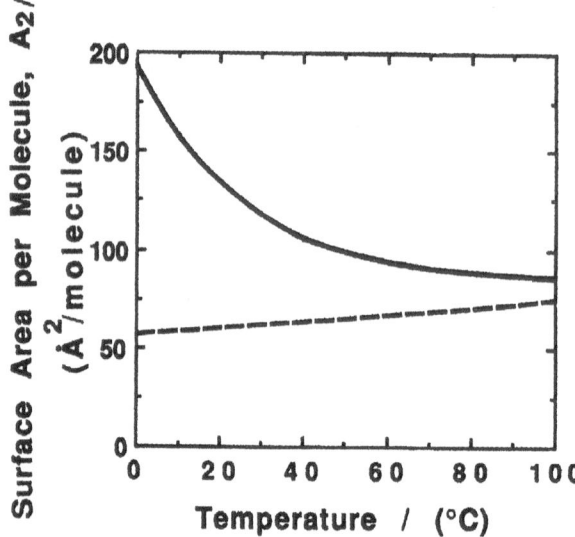

Fig. 7. The temperature dependence of the surface area per surfactant molecule calculated at close packing. The dashed line represents the same quantity, but calculated without the water structuring contribution

dependence of the interaction between a hydrocarbon, or an ethylene oxide chain, and water. The role of water structuring is to enhance the solubility of hydrocarbons in water, which is equivalent to a decrease of the hydrocarbon-water interaction parameter, $\chi_{hc,w}$. The poor solubility of hydrocarbons in water is not due to the water structuring, but due to other factors such as the large amount of energy required to form a cavity in the water for the hydrocarbon. This is a reflection of the very large cohesive forces in liquid water. The effect of water structuring plays a dominant role in the temperature dependence of the adsorption of nonionic surfactants, leading to an increase in adsorption with temperature — instead of a decrease, which is the normal behavior.

References

1. Defay R, Prigogine I, Bellemans A, Everett DH (1966) Surface Tension and Adsorption, chap 12, Longmans, London
2. Silberberg A (1962) J Phys Chem 66:1872, 1884; (1967) J Chem Phys 46:1105
3. Roe RJ (1965) J Chem Phys 43:1591; (1966) 44:4264
4. Scheutjens JMHM, Fleer GJ (1979) J Phys Chem 83:1619

5. Koopal LK, Wilkinson GT, Ralston J (1988) J Colloid Interface Sci 126:493
6. Kronberg B (1983) J Colloid Interface Sci 96:55
7. Kronberg B, Stenius P (1984) J Colloid Interface Sci 102:410; Kronberg B, Stenius P, Igeborn G (1984) J Colloid Interface Sci 102:418
8. Silveston R, Kronberg B (1989) J Phys Chem 93:6241
9. Silveston R, Kronberg B, in preparation
10. Shinoda K, Fujihira M (1968) Bull Chem Soc Jap 41:2162
11. Shinoda K (1977) J Phys Chem 81:1300
12. Shinoda K (1978) Principles of Solution and Solubility, Marcel Dekker: New York
13. Shinoda K, Kobayashi M, Yamaguchi N (1987) J Phys Chem 91:5292
14. Patterson D, Barbe M (1976) J Phys Chem 80:2345
15. Costas M, Patterson D (1985) J Chem Soc, Faraday Trans I, 81:2381
16. Hvidt B (1978) Biochem et Biophys Acta 537:374
17. Hvidt A (1983) Physiol Chem Phys Med NMR 15:501
18. Hvidt A (1983) Ann Rev Biophys Bioeng, 12:1
19. Rosen MJ (1978) Surfactants and Interfacial Phenomena, J Wiley & sons, New York
20. Saeki S, Nobuhiro K, Nakata M, Kaneko M (1976) Polymer 17:685
21. Rogers JA, Tam T (1977) Can J Pharm Sci 12:65
22. Kjellander R, Florin E (1981) J Chem Soc, Faraday Trans I, 77:2053

Authors' address:

Bengt Kronberg
Institute for Surface Chemistry
P.O. Box 5607
11486 Stockholm, Sweden

Progress in Colloid & Polymer Science   Progr Colloid Polym Sci 83:84—95 (1990)

Microcalorimetric adsorption enthalpies of ethylene oxide and propylene oxide copolymers from CCl_4 and H_2O on pyrogenic silica

E. Killmann and W. Melchior

Lehrstuhl für Makromolekulare Stoffe, Institut für Technische Chemie, Garching, FRG

Abstract: The adsorption enthalpies of block- and statistic copolymers of ethylene oxide (EO) and propylene oxide (PO), and of corresponding homopolymers on silica (Aerosil 200) from CCl_4 and H_2O are determined by microcalorimetry. With adsorption isotherms measured separately from the concentration difference in the solution fraction of adhered segments, p_{cal}, the binding enthalpies have been deduced. The p_{cal}-values from CCl_4 decrease with increasing amount adsorbed, molar mass, EO-fraction and -blocklength. The hydroxyl endgroups of the polymers strongly influence the adsorption enthalpies. With results on preferential adsorption of the EO-segments the dependences are interpreted by conformation models.

Key words: Polymer adsorption; blockcopolymers; silica; microcalorimetry

Introduction

Microcalorimetric measurements of adsorption enthalpies are a possible means of investigating the energetics of polymer adsorption [1—6]. Knowing the adsorption enthalpy per adsorbed segment and the adsorption isotherm, it is possible to determine the fraction of adhered segments, also in cases where no specific interactions of polymer- and surface-groups takes place and spectrometric methods (IR, NMR, ESR) fail. In addition, the segmental binding enthalpy can be evaluated under certain assumptions [6]. Beside these unique enthalpic determinations, a displacer method was recently developed in order to get the free enthalpy of adsorption [7—9].

Besides IR-spectrometric measurements made of the adsorption isotherms and of the fractions of adhered segments [10] in this study the integral adsorption enthalpies of homopolymers, PEO, PPO, block copolymers, EPE, PEP, and statistic copolymers, SEP, of ethylene oxide and propylene oxide on silica (Aerosil 200) in CCl_4- and H_2O-suspensions were determined calorimetrically. Two types of three-block copolymers with central propylene oxide block, EPE, and with central ethylene oxide block, PEP, were investigated. Previous investiga-

tions about the adsorption of EO-PO-block copolymers from the literature are discussed in [10].

This paper is mainly restricted to the measurements in apolar CCl_4. The influences of the chemical composition, block length, block position, molar mass and, especially, of the endgroups by chemical modification of the hydroxyl groups are reported. Measurements in the polar medium H_2O have demonstrated the decisive contributions of the solvent interactions and will be the subject of a future publication.

Experimental

The specification of the substances and their pretreatment have been reported [10]. In the microcalorimetric measurement the silica samples (Aerosil 200), evacuated, preheated and stored under dry nitrogen, are suspended in CCl_4 and H_2O, respectively, by shaking 20 min with 200 impulses/min in a three-dimensional shaking device and filled in the calorimetric vessel. All manipulations were accomplished under nitrogen atmosphere. The microcalorimetric measuring technique at 25 °C with the isoperibolic calorimeter LKB 8700 system, which introduced an improved titration procedure and a new evaluation method, is described in [6].

Results

Homopolymers

Integral adsorption enthalpy: The integral adsorption enthalpy ΔH is obtained with Eq. (1) from the measured heat ΔH_{meas} produced by the addition of polymer solution to the standard pretreated silica suspension in CCl_4 [4, 6]. The dilution heat ΔH_d in the corresponding experiment without silica has to be considered.

$$\Delta H = \Delta H_{meas} - \Delta H_d . \tag{1}$$

Measured values, related to 1 g adsorbent, are represented in Fig. 1 for polyethylene oxides PEO 400 and PEO 6000 in dependence of the polymer dosage concentration C_E.

A summarizing plot of the exothermic adsorption enthalpies of the homopolymers PEO and PPO is shown in Fig. 2. In CCl_4 the plateau values of the integral enthalpies ΔH decrease with increasing molar mass, whereas in H_2O the reverse dependence is found (Fig. 3). With the adsorption isotherms measured in the same systems [10] the enthalpies can be related to the adsorbed amounts A, given in segment mol per g silica. Especially the comparison of the adsorption enthalpies at the same molar number of adsorbed segments (ether units) is predicative. The plots ΔH vs A have a convex character, demonstrating a reduced increase of the adsorption enthalpy with increasing coverage (Fig. 4). The initial increase of the enthalpies of low molar mass PEO is in good accordance with that of PPO.

Influence of the endgroups: In order to evaluate the influence of the OH-endgroups on the adsorption enthalpy of PEO, samples with modified endgroups are measured. In Fig. 5 an evident variation in the plot of the adsorption enthalpies is shown. The adsorption enthalpy in the plateau decreases in CCl_4 from PEO 600 (with two hydroxyl endgroups over monomethylated PEO 550) to dimethylated PEO 500 by 2—2.5 J/g_{Ae} per methylation step at comparable molar masses. In contrast to this behavior, an increase of ΔH is observed in H_2O with the same substances. These dependencies are elucidated by the decrease of the plateau adsorption enthalpies with molar mass of PEO and PPO in CCl_4 (Fig. 6). The increasing values in H_2O are also plotted.

Fraction of adhered segments: Assuming the same adsorption enthalpy for an adsorbed polymer segment and for an adsorbed monomer molecule of similar chemical structure the fraction of adhered segments can be calculated [6] by the formula

$$p_{cal} = \frac{(\Delta H/A)_p}{(\Delta H/A)_M + \Delta H_{L,A,M} - \Delta H_{L,A,P}} , \tag{2}$$

with $\Delta H_{L,A,M} = \Delta H_{L,A,P}$ (Index: M = monomer, P = polymer, L = solvent $\Delta H_{L,A,M}$, $\Delta H_{L,A,P}$ = solvent-adsorbent interaction enthalpy per mol adsorption sites of the monomer and of the polymer segments respectively).

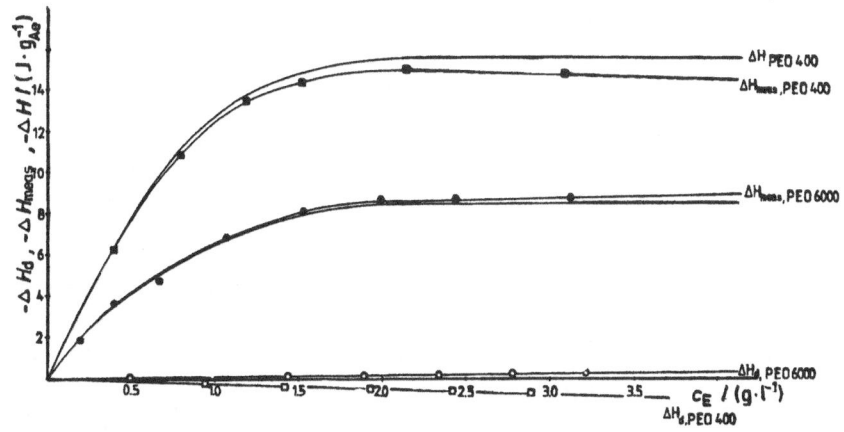

Fig. 1. Dependence of enthalpies measured ΔH_{meas} ●■, of dilution ΔH_d ○□ and of adsorption ΔH (without signs) on the dosage concentration c_E for PEO 400 and PEO 6000; CCl_4; 25 °C

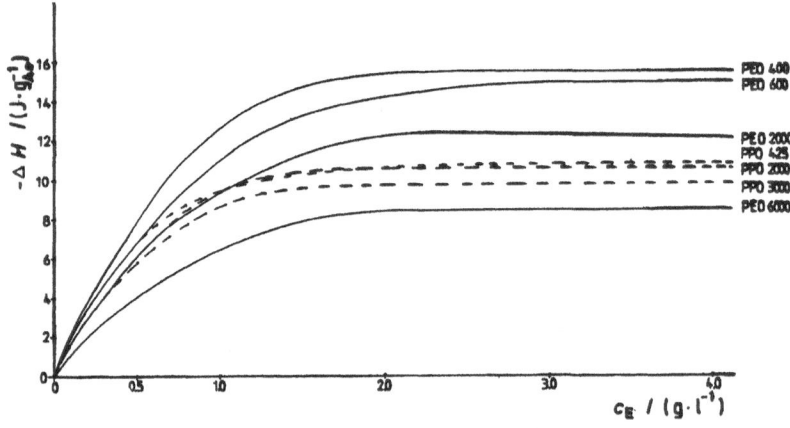

Fig. 2. Dependence of the adsorption enthalpy ΔH on the dosage concentration c_E for the homopolymers PEO and PPO; CCl_4; Aerosil 200; 25°C

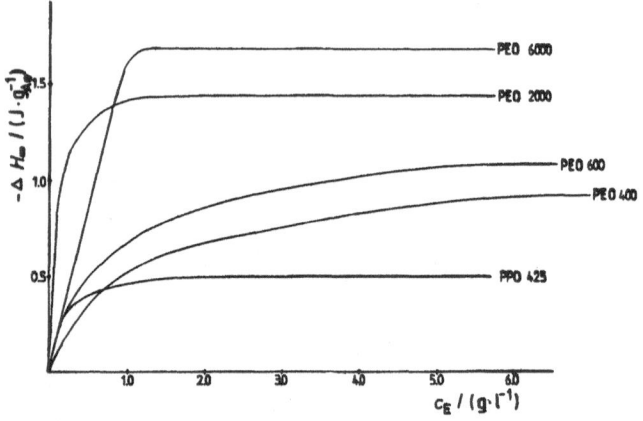

Fig. 3. Dependence of the adsorption enthalpy ΔH on the dosage concentration c_E for the homopolymers PEO and PPO; H_2O; Aerosil 200; 25°C

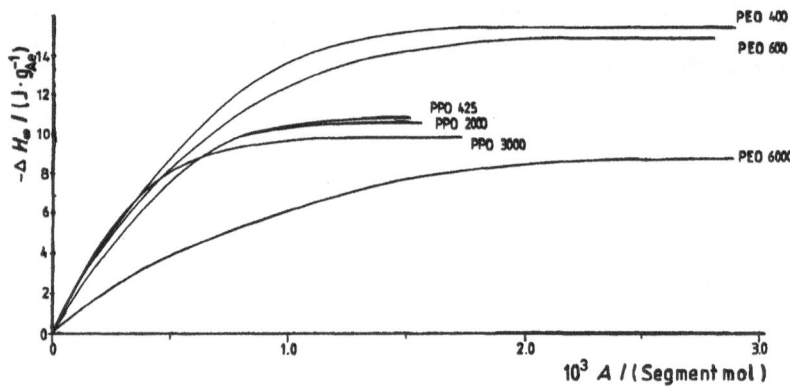

Fig. 4. Dependence of the adsorption enthalpy ΔH on the adsorbed amount A in segment mol for the homopolymers PEO and PPO; CCl_4; Aerosil 200; 25°C

With microcalorimetric measurements of mixing enthalpies it has been shown [12] that the enthalpic contributions of the polymer solvent and polymer polymer interactions, $|\Delta H_{P,L} + \Delta H_{P,P}| < 0.5$ kJ/mol, are negligible in comparison to those of the

other interactions. This is a necessary condition in the derivation of Eq. (2) [6].

The calculation has been accomplished with the plateau adsorption enthalpy $\Delta H_\infty = -7.94$ J/g_{Ae} and the adsorbed amount $A_\infty = 0.037$ g/g_{Ae} of the

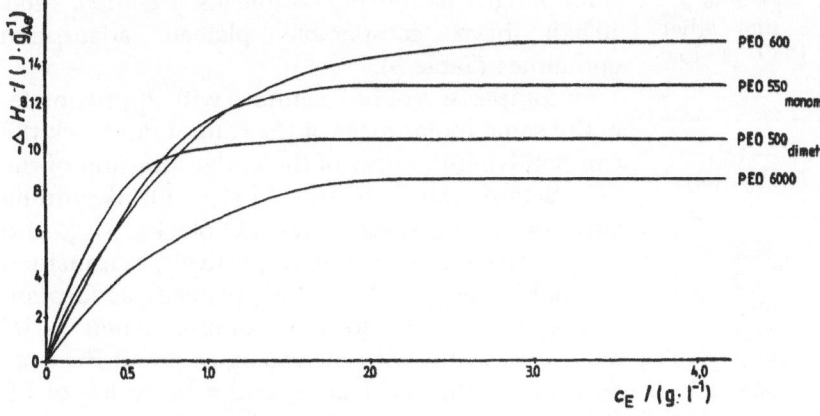

Fig. 5. Dependence of the adsorption enthalpy ΔH on dosage concentration c_E; comparison of dihydroxylated with mono- and dimethoxylated samples; CCl_4; Aerosil 200; 25°C

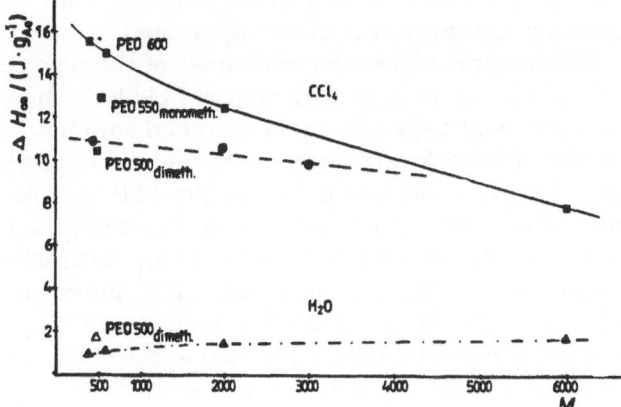

Fig. 6. Dependence of the plateau adsorption enthalpies ΔH_∞ on molar mass M of PEO ■ and PPO ● from CCl_4 and H_2O ▲, values for mono- and dimethoxylated PEO are signed ■▲; Aerosil 200; 25°C

monomer diethylether, as well as with the corresponding values $\Delta H_\infty = -20.93$ J/g_{Ae} and $A_\infty = 0.065$ g/g_{Ae} of ethanol [2]. The values calculated in both ways agree extremely well. In Table 1 these fractions p_{cal} are compared with p_{IR}-values, obtained by IR at full coverage [10].

At low coverage ($A = 0.02$ g/g_{Ae}) in the increasing region of the integral adsorption enthalpies the p_{cal} fractions of Table 2 have been evaluated with the same procedure using corresponding enthalpies for the monomers.

At full coverage the p_{cal}-values for PEO and PPO are smaller than at low coverage. The p_{cal}-values of PPO are essentially higher than those of PEO at full coverage and the fractions p_{cal} decrease with the molar mass of the polymers. The p_{cal} values are rather higher than the p_{IR}-values from IR-spectrometry.

Table 1. Adsorbed amount A_∞, enthalpies of adsorption ΔH_∞, fractions of adhered segments p_{cal} and p_{IR}, binding enthalpies $\Delta H_{P,A}$ per mol ether segments of homopolymers PEO and PPO at full coverage, $\Theta = 1$, from CCl_4 on Aerosil 200

Polymer	A_∞ (g/g_{Ae})	$-\Delta H_\infty$ (J/g_{Ae})	P_{cal} $\Theta = 1$	P_{IR}	$-\Delta H_{P,A}$ (KJ/mol)
PEO 400	0.167	15.55	0.28	0.14	52.6
PEO 600	0.162	15.0	0.27	0.15	51.5
PEO 1000	0.200	12.4*)	0.18*)	0.12	46.5
PPO 425	0.087	10.9	0.49	0.33	44.2
PPO 2000	0.09	10.65	0.47	0.32	44.0
PPO 3000	0.1	9.9	0.39	0.29	42.6

*) value of PEO 2000.

Table 2. Molar mass, fraction of adhered segments p_{cal} and p_{IR}, binding enthalpies $\Delta H_{P,A}$ per mol ether segments of homopolymers PEO and PPO at low coverage, $\Theta \to 0$, from CCl_4 on Aerosil 200

Polymer	M	P_{cal} $\Theta \to 0$	P_{IR}	$-\Delta H_{P,A}$ (KJ/mol)
PEO 400	400	0.7	0.66	43.3
PEO 600	600	0.66	0.88	51.2
PEO 1000	1000	0.85	0.53	49.3
PPO 425	425	0.65	0.82	44.3
PPO 2000	2000	0.65	0.61	51.5
PPO 3000	3000	0.74	0.58	56.8

Binding enthalpy: Under the same conditions as with Eq. (2) the binding enthalpy can be calculated from

$$\Delta H_{P,A} = \Delta H/Ap + \Delta H_{L,A} ; \qquad (3)$$

$\Delta H_{L,A}$ = molar wetting enthalpy of CCl_4 = -23.8 kJ/mol.

Formula (3) is derived in previous publications [2, 4, 5]. Assuming that one adsorbing polymer segment displaces one CCl_4-molecule from the surface with measured p_{IR}-values, the $\Delta H_{P,A}$ values are calculated at full coverage in Table 1 and at low coverage in Table 2. In comparison, at full coverage the binding enthalpies are for diethylether $\Delta H_{M,A}$ = -39.8 kJ/mol and for ethanol $\Delta H_{M,A}$ = -38.6 kJ/mol.

Block- (EPE, PEP) and statistic copolymers (SEP) of EO and PO

Integral adsorption enthalpy: The integral adsorption enthalpies ΔH of the block copolymers EPE and PEP show the same curvature in dependence of the dosage concentration c_E as do the homopolymers PEO and PPO (examples in Fig. 7). The plateau adsorption enthalpies are listed in Table 3. In Figs. 8 and 9 the integral adsorption enthalpies of EPE and PEP are plotted against the adsorbed amount A in segment mol/g_{Ae}, obtained from the adsorption isotherms [10]. EPE block copolymers of high molar mass (PE 10500, 6800) are characterized by a flat increase of ΔH with A, EPE of low molar mass by a steep increase, and EPE of middle molar mass (PE 10100, 9400) by an intermediate inclination. EPE-block copolymers with approximately the

same length of the EO-endblocks (PE 6100, 8100, 10100) have comparable plateau adsorption enthalpies (Table 3).

By comparison of EPE samples with approximately the same molar mass of the central PO-block the competitive influences of the endgroups and of the EO-fractions can be quantified (Fig. 10). Beginning with low molar mass of the EO-blocks, M_{EO}, and low EO-fraction (PE 6100), respectively, at transition to higher M_{EO} values the plateau adsorption enthalpy increases to a maximum. Then $-\Delta H$ decreases, again with increasing M_{EO} and EO-fraction (PE 6400), and falls to under the value of PE 6100 at the highest M_{EO} value (PE 6800). This decrease of the adsorption enthalpies corresponds to the reducing OH-endgroup fraction. By this influence the initial increase of the enthalpy with increasing EO-fraction is overcompensated.

The integral adsorption enthalpies of the reverse PEP block copolymers with central EO-block show an equivalent dependence on adsorbed amount, as do the EPE block copolymers with central PO-block (Fig. 9). With increasing M_{EO} of the PEP samples the plateau adsorption enthalpy increases (Fig. 11). But the values and the increase are distinctly smaller than with the comparable EPE polymers. The decrease of the enthalpy with molar mass observed with EPE (Fig. 10) related to the influence of the endgroups is not observed with PEP.

All statistic copolymers show an equivalent initial increase of the adsorption enthalpy with the adsorbed amount (in segment mol). The plateau adsorption enthalpies (Table 3) rise slightly with the EO-fraction (Table 4).

Fraction of adhered segments: In Table 3 the fraction of adhered segments P_{cal} at high coverage, $\Theta = 1$, calculated by Eq. (2) with adsorption enthalpies of the polymers and of the monomer diethylether, as well as the p_{IR}-values from IR experiments are listed. An average molar mass of the segments dependent on the EO-PO-composition is introduced to transfer the adsorbed amount in segment mol. At low molar mass the p_{cal}-values lie between those of the homopolymers PEO and PPO, and decrease with increasing molar mass under the values of the EO-homopolymers (Fig. 12). In Fig. 13 the p_{cal}-values are plotted against the EO-fraction of the EPE and PEP block copolymers and confronted with the corresponding p_{IR}-values already published in [10]. Although onyl values of constant molar mass should be compared, in Fig. 13 values of dif-

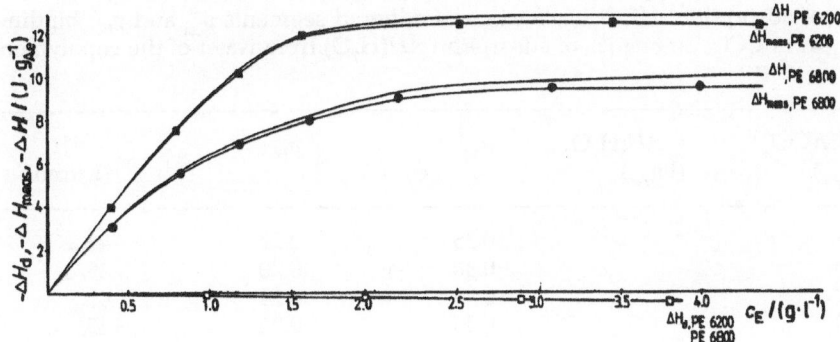

Fig. 7. Dependence of enthalpies measured ΔH_{meas} ●■, of dilution ΔH_{dil} □ and of adsorption ΔH (without signs) on the dosage concentration c_E for PE 6200 and PE 6800; CCl_4; Aerosil 200; 25 °C

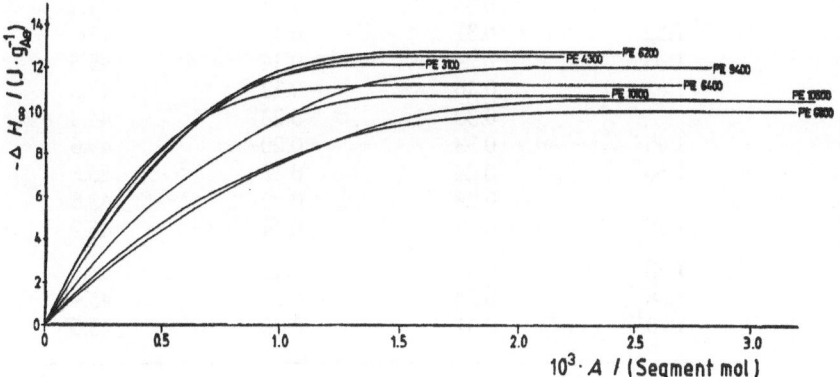

Fig. 8. Dependence of the adsorption enthalpy ΔH on the adsorbed amount A in segment mol for the block copolymers EPE; CCl_4; Aerosil 200; 25 °C

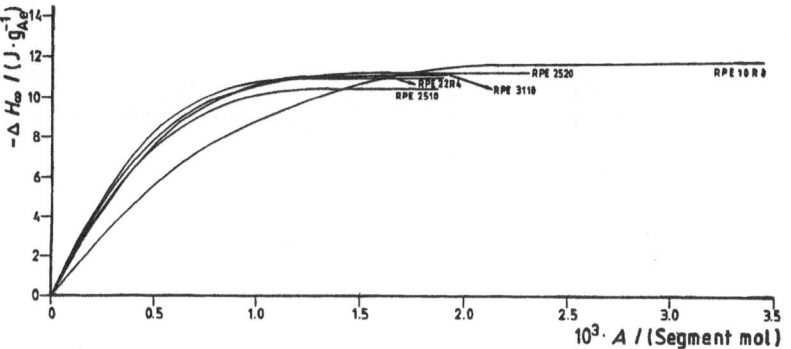

Fig. 9. Dependence of the adsorption enthalpy ΔH on the adsorbed amount A in segment mol for the block copolymers PEP; CCl_4; Aerosil 200; 25 °C

ferent molar masses ($1100 < M_{tot} < 8500$) are plotted; this is the reason for the strong scatter of the points. The p_{cal}-values are always higher than the p_{IR}-values — with increasing EO-fraction the p_{cal}-values and the p_{IR}-values decrease, and at 50% EO-fraction they tend to have constant values (Fig. 13).

As with the homopolymers the p_{cal}-values of the block copolymers at low coverage are essentially larger than at full coverage, and they are higher than the corresponding p_{IR}-values (Table 4).

Binding enthalpies: Using the fraction of adhered segments p_{IR} in Tables 3 and 4 the binding enthalpies have been calculated with Eq. (3) (Table 3). Because the values p_{IR} and ΔH are measured with different methods, a strong scatter of the $-\Delta H_{P,A}$

Table 3. Adsorbed amount A_∞, enthalpy of adsorption $\Delta H(CCl_4)$, fraction of adhered segments p_{cal} and p_{IR}, binding enthalpies per mol ether segments $\Delta H_{P,A}$ from CCl_4; enthalpies of adsorption $\Delta H(H_2O)$ from water of the copolymers at full coverage $\Theta = 1$ on Aerosil 200

Type	A_∞ (g/g$_{Ae}$)	$-\Delta H(CCl_4)$ (J/g$_{Ae}$)	$-\Delta H(H_2O)$ (J/g$_{Ae}$)	p_{cal} $\Theta = 1$	p_{IR}	$-\Delta H_{P,A}$ (KJ/mol)
RPE 2510	0.10	10.5		0.35	0.23	48.2
RPE 2520	0.12	11.3		0.30	0.20	48.2
RPE 3110	0.10	11.2		0.36	0.24	48.2
RPE 22R4	0.12	11.0		0.30	0.22	45.7
RPE 10R8	0.15	11.8		0.21	0.14	48.2
PE 3100	0.09	12.2		0.46	0.32	46.4
PE 4300	0.13	12.6	1.47	0.33	0.21	48.3
PE 6100	0.11	10.6		0.33	0.25	44.1
PE 6200	0.14	12.8	1.32	0.31	0.18	47.6
PE 6400	0.16	11.3	1.33	0.24	0.14	45.5
PE 6800	0.18	10.1	1.62	0.16	0.13	43.3
PE 8100	0.12	10.5		0.30	0.23	44.1
PE 9200	0.14	13.5	1.45	0.33	0.20	49.6
PE 9400	0.16	12.1	1.58	0.24	0.17	45.6
PE 10100	0.14	10.8		0.28	0.22	43.5
PE 10500	0.19	10.6	1.59	0.18	0.14	43.9
ZN 2364	0.10	11.7	1.68	0.33	0.23	45.8
ZN 1832	0.10	11.6	1.48	0.38	0.26	45.4
ZN 1833	0.10	10.4	1.70	0.31	0.23	43.5

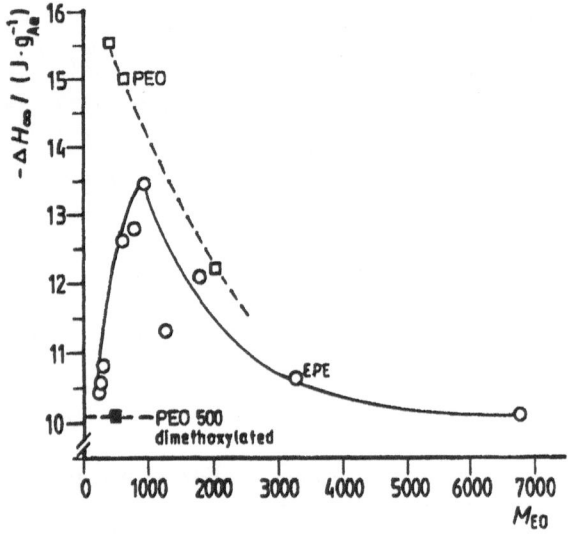

Fig. 10. Dependence of the plateau adsorption enthalpy ΔH_∞ on molar mass of the EO-fraction M_{EO}; PEO □, EPE ○, PEO 500 dimethoxylated ■; CCl_4; Aerosil 200; 25 °C

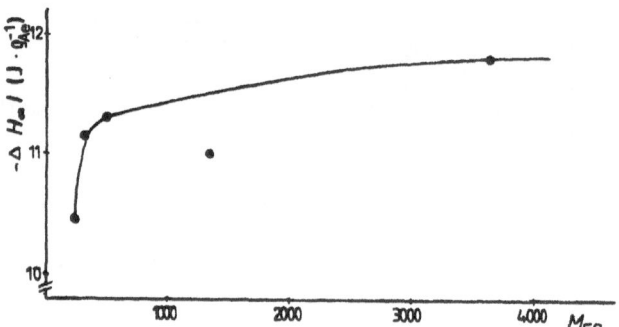

Fig. 11. Dependence of the plateau adsorption enthalpy ΔH_∞ on molar mass of the EO-fraction M_{EO} of the PEP block copolymers; CCl_4; Aerosil 200; 25 °C

values between 43 and 50 kJ/mol is observed. The $-\Delta H_{P,A}$ values agree with those of the homopolymers and can be related to the hydrogen bond formation. A systematic dependence on molar mass and EO-fraction of the blockcopolymers is not recognized.

Table 4. Total molar mass M_{tot}, molar mass of the EO-fraction M_{EO}, fraction of adhered segments p_{cal} and p_{IR} of the copolymers at low coverage, $\Theta \rightarrow 0$, from CCl_4 on Aerosil 200

Type	M_{tot}	M_{EO}	p_{cal} $\Theta \rightarrow 0$	p_{IR}
RPE 2510	2500	250	0.44	0.58
RPE 2520	2500	500	0.41	0.55
RPE 3110	3100	300	0.47	0.58
RPE 22R4	3350	1350	0.42	0.65
RPE 10R8	4550	3650	0.32	0.68
PE 3100	1100	200	0.57	0.63
PE 4300	1700	600	0.41	0.42
PE 6100	2000	250	0.38	0.50
PE 6200	2500	750	0.53	0.43
PE 6400	3000	1250	0.40	0.43
PE 6800	8500	6750	0.28	0.44
PE 8100	2600	300	0.52	0.43
PE 9200	3650	900	0.48	0.42
PE 9400	4600	1750	0.41	0.56
PE 10100	3550	300	0.37	0.54
PE 10500	6500	3250	0.25	0.49
ZN 2364	1800	1350	0.54	0.82
ZN 1832	1800	800	0.59	0.60
ZN 1833	3000	1250	0.47	0.53

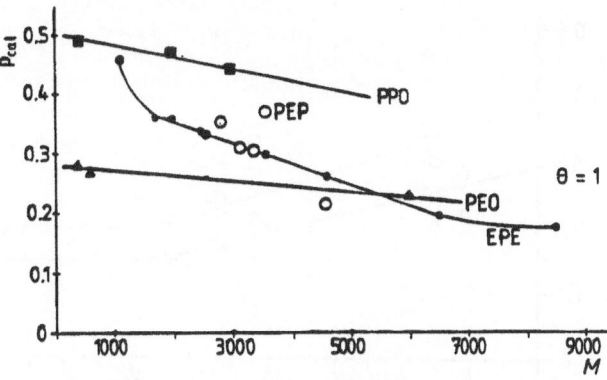

Fig. 12. Dependence of the fraction of adhered segments p_{cal} on the molar mass M at full coverage for PEO ▲, PPO ■, EPE ●, PEP ○ CCl_4; Aerosil 200; 25 °C

Discussion

Dependence of the adsorption enthalpies on solvent and molar mass

The different adsorption enthalpies of the polymers in CCl_4 and water are produced by the desorption of the solvent molecules from the silica interface and by the desolvation of the polymer chains. These processes compete with the intrinsic binding of the polymer segments. Because of the hydrogen bonds the interaction energies of the H_2O-molecules with the surface groups, as well as with the polymer segments are essentially higher than those of the CCl_4 molecules. The stronger interaction of H_2O with silica becomes apparent in the higher wetting enthalpy, -30.1 J/g_{Ae} compared to CCl_4, -23.4 J/g_{Ae}.

Microcalorimetric mixing enthalpies [11—13] and polymeranalytic measurements demonstrate the better solvency of H_2O for PEO and PPO compared to CCl_4. Therefore, the desolvation of the polymer occurring with the adsorption process needs a

smaller energy in CCl_4 than in H_2O. The smaller polymer solvent interaction, as well as the smaller adsorbent solvent interaction, favors the adsorption of the polymer from CCl_4 compared to H_2O. With this assumption, the adsorption theories predict higher adsorbed amounts and adsorption enthalpies from CCl_4 than from H_2O. Correspondingly, the measured adsorption enthalpies from H_2O are distinctly smaller than from CCl_4 (Table 3).

The decisive enthalpic contribution of the OH-endgroups is especially guaranteed by the adsorption of endgroup-modified, methylated PEO-samples. From H_2O the enthalpy amount increases with methylation of the OH-endgroups and with molar mass (Fig. 6). However, from CCl_4 with the homopolymers PEO and PPO a decrease of the adsorption enthalpies is measured with methylation and with increasing molar mass (Figs. 5 and 6).

This opposite behavior can be explained by the smaller enthalpic interaction of the methoxy group with H_2O compared to the hydroxyl group. Conversely, with CCl_4 the interaction enthalpy of the methoxy group is larger than that of the hydroxyl group.

At full coverage practically all SiOH-groups are involved in the adsorptive binding. Polymers of low molar mass are adsorbed with higher fractions of OH-endgroups compared to those of high molar mass. Beyond that, an additional exothermic enthalpy contribution can be created by interactions of OH-endgroups with polymer segments in the adsorbed layer. This is confirmed by mixing enthalpies measured for ethylene oxide oligomers with and without OH-endgroups [12, 13].

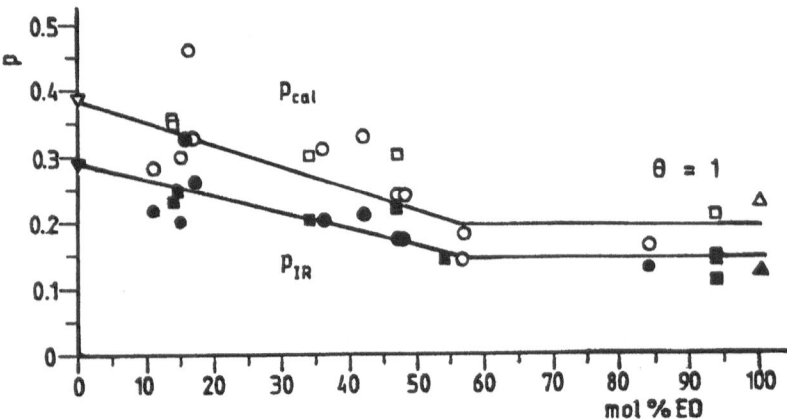

Fig. 13. Dependence of the fraction of adhered segments p_{cal} o□ and p_{IR} ●■ on the EO-fraction of EPE o● and PEP □■ at full coverage, $\Theta = 1$; CCl_4; 25°C; comparison with homopolymers PEO △▲, PPO ▽▼

The further decrease of the adsorption enthalpies with molar mass in CCl_4 to lower values than those of the dimethylized oligomers can be possibly attributed to an incomplete coverage of the surface SiOH-groups that are screened by loops. Unfortunately, with IR measurements small concentrations of free SiOH-groups at high but incomplete coverage cannot be detected and evaluated from the extinction values [10].

Dependence of the adsorption enthalpies on block composition

The adsorption enthalpies of the EPE and PEP block copolymers are dependent on the composition and on the molar mass of the blocks. From CCl_4, block copolymers with high EO-fractions show plateau adsorption enthalpies similar to those of polyethylene oxides whereas with high PO-fraction the plateau enthalpies approach the values of PPO-samples (Figs. 10, 11).

In the series Pluronic PE 6100, PE 8100, PE 10100 the adsorbed amounts in the plateau increase, whereas the adsorption enthalpy is practically constant (Table 3). The PO-fraction of the PO-middle block increases in this series at constant length of the EO-endblocks. The constant enthalpy values make clear that the EO-fraction is decisive for the height of the adsorption enthalpy, whereas the increase of the PO-fraction has no influence.

Because of the preferential interaction of the EO-segments with the SiOH-groups, established by IR measurements [10], the adsorption of the EPE block copolymers mainly takes place by the EO-segments in the endblocks.

PO-segments of the middle block are only partly adsorbed and project as loops into the solution. The comparison of PE 8100 with PE 4300 with comparable adsorbed amounts demonstrates that doubling of the EO-segment number results in a distinct increase of the adsorption enthalpy (Table 3).

The molar mass dependence of the plateau adsorption enthalpy, produced by the decreasing influence of the OH-endgroups with increasing molar mass is also recognized with EPE block copolymers of different molar mass. Despite the increased EO-fraction a smaller adsorption enthalpy is observed with PE 6800 (84% EO, $M = 8500$) compared to PE 4300 (42% EO, $M = 1700$). The reduction of the OH-endgroups with molar mass overcompensates the expected rise of ΔH by the increased EO-fraction. The competing effects are demonstrated clearly in Fig. 10. The adsorption enthalpies first increase with increasing molar mass of the EO-block to a maximum at $M_{EO} = 800—850$, followed by a decrease at higher molar masses.

In order to study the influence of the block position on the adsorption behavior, reverse block copolymers PEP with central EO-block are measured in comparison with the EPE block copolymer samples. The smaller plateau adsorption enthalpies of the PEP-samples increase only slightly with increasing molar mass of the EO-middle block and with decreasing PO-fraction (Fig. 11). No competing influence of the endgroups (as with EPE) is recognized.

From this behavior it can be concluded that the adsorption is determined by the preferentially adsorbing EO-segments of the central EO-block. If the PO-endblocks with secondary OH-endgroups would adsorb, the doubling of their length should lead to a decrease of the plateau adsorption enthalpy, as observed with EPE. Because no decrease is observed, the PO-endblocks could not be involved dominantly in the adsorption of the PEP-macromolecule and they will dangle into solution. According to mixing enthalpies [12], interactions of OH-endgroups with PO-segments in the layer would lead only to a very small enthalpy contribution.

The three statistic copolymers have nearly the same plateau adsorption enthalpies (Table 3). In comparison to this behavior doubling of the EO-fraction in EPE-copolymers leads to an increase of the adsorption enthalpy by 2 J/g_{Ae}. This demonstrates the clearly larger influence of the EO-fraction on the adsorption of block copolymers than on that of statistic copolymers.

Fraction of adhered segments

With all polymers at low coverage, higher fractions of adhered segments are measured than in the plateau region (Tables 1—4). At low coverage the macromolecules are able to adsorb undisturbed under displacement of solvent molecules with the highest possible number of segments. The obtained fractions $p > 0.5$ at very low coverage have to be explained by multiple interactions [4, 6, 10] between more than one SiOH-group and one polymer segment. With increasing coverage the macromolecules are not allowed to be adsorbed with the highest number of segments because a competition exists for the available surface sites with segments of other macromolecules. An equilibrium state with lower average number of adhered segments is adapted.

The IR-spectrometric measured p_{IR}-values [10] are smaller than the p_{cal}-values at full coverage (Fig. 13, Tables 1 and 3). With IR the specific hydrogen bridges are measured, whereas the microcalorimetric measurment considers the displacement of the solvent molecules [6]. Taking the p_{IR}-values, one can calculate with Eq. (2) the contribution $\Delta H_{L,A,M} - \Delta H_{L,A,P}$.

The number of solvent molecules that have to be additionally displaced by the adsorption of diethyl-ether- or ethanol-molecules results from Eq. (4) [6]

$$ n = \frac{\Delta H_{L,A,M} - \Delta H_{L,A,P}}{\Delta H_{L,A}(CCl_4)} . \tag{4} $$

In comparison to the EO- and PO-segments, both monomers have larger segmental areas. According to values calculated with Eq. (4), ethanol and diethylether displace 40—50% more solvent molecules compared with EO-segments, and 15—20% more compared with PO-segments.

At full coverage the fraction p for PEO is distinctly smaller than for PPO. PEO adsorbs in the plateau zone with each fourth to fifth segments, whereas PPO adsorbs with each second to third segment, on the average.

The p-values of the block copolymers EPE and PEP at full coverage are positioned between the values of both homopolymers. At low EO-fractions (PE 3100) they are similar to those of PPO, at high EO-fraction (PE 10500) they are similar to those of PEO (Table 3). The block copolymers demonstrate an essentially stronger decrease of the p-fractions with molar mass compared to the homopolymers, and they even fall under the p-values of PEO (Fig. 12). The reason for this behavior is the preferential participation of the EO-segments in the adsorptive binding, which has been demonstrated before [10]. With increasing molar mass at constant EO-PO-fraction of the EPE copolymers (PE 3100—6100) the chain length of the preferentially adsorbed EO-endblocks increases and a sufficient number of adhered EO-segments is able to the whole copolymer. The PO middle block can form a loop or, combined with one EO-endblock, form a tail that pro trudes into the solution. Also, with increasing EO-fraction in the series PE 6100, 6200, 6400, 6800 (Fig. 13) the p-fraction decreases down to the value of pure PEO; already at 50—60 mol% EO the p-values of PEO are achieved. Therefore, above this limit the PO-segments of EPE are participated in adsorption in less degree.

The p_{cal}-values of PEP block copolymers with central EO block decrease also with increasing molar masses M_{tot} and M_{EO}, respectively (Table 3, Fig. 12). Because no endgroup effect is observed from this behavior the conclusion is made that the PO-endblocks of this polymer are not adsorbed and protrude as tails into the solution.

With statistic copolymers only a small decrease of the p-values with increasing molar mass M_{EO} is observed. The p-values are essentially higher than those of the block copolymers of comparable molar mass and EO fraction. At full coverage each third segment of the SEP-polymers is adsorbed. Therefore, the adsorption behavior is similar to that of the homopolymer PPO that adsorbs with each second or third segment corresponding to a very flat conformation.

Adsorptive binding

Mainly from the behavior at low coverage and from the same $\Delta H_{P,A}$-values of EO- and PO-segments, it can be concluded that the nature and strength of the interaction with the surface groups is nearly the same for EO- and PO-segments (Fig. 4, Tables 1—3). IR-spectrometric measurements confirm that EO- and PO-segments form hydrogen bonds with the surface SiOH-groups and no essential enthalpic difference exists [2, 5]. Therefore, the different adsorption behavior of PEO and PPO, which is clearly documented by the competition and displacement experiments [10], has its origin in the different free energy interactions of the polymer with the solvent CCl_4. The larger adsorption amounts and the adsorption enthalpies in the plateau for PEO compared to PPO can be explained by the fact that CCl_4 is a worse solvent for PEO than for PPO. The different adsorption behavior of the block copolymers in dependence of block position, EO-PO-fraction, and molar mass compared to the homopolymers can also be understood by the different solvencies of the blocks. Because of the poorer interaction of the EO-segments with CCl_4 a preferential adsorption of this comonomer compared to PO-segments results. According to the position of the PO-block, a tendency to loop- or tail-formation occur depending on molar mass and EO-PO-fraction.

The binding enthalpies $\Delta H_{P,A}$ for all polymers measured lie between 40 and 50 kJ/mol. The comparison of the plateau adsorption enthalpies of PEO 600 with two OH-endgroups ($\Delta H_\infty = -15$ J/g_{Ae}), of PEO 550 with one OH- and one methoxy endgroup ($\Delta H_\infty = -12.3$ J/g_{Ae}) and of PEO 500 with two methoxy endgroups ($\Delta H_\infty = -10.4$ J/g_{Ae}) (Fig. 5) demonstrates that the adsorption by hydroxyl groups takes place with higher enthalpy than by the ether groups.

Structure of the adsorbed layer and conformational model

The fraction of adhered segments, the adsorption enthalpies, the binding enthalpies and their dependence on molar mass and polymer composition demonstrate a different adsorption behavior of the block copolymers of ethylene oxide and propylene oxide compared with the homopolymers and the statistic copolymers in CCl_4. With PEO in the plateau zone each fifth EO-segment and with PPO each third PO-segment is adsorbed in average. The OH-endgroups of the oligomers (M 500) participate in adsorption and, relative to their fraction, they contribute with a large adsorption enthalpy. The deciding indication that both OH-endgroups are concerned in the adsorption process is the two fold decrease of the plateau adsorption enthalpy with mono- and dimethylation of the endgroups. At higher molar mass the influence of the endgroups is of minor importance and the formation of tails that protrude into the solution becomes probable.

Block copolymers with low molar mass and low EO-fraction have similar p-values to PPO homopolymers that demand a comparable, very flat adsorbed conformation in the whole coverage range. Block copolymers EPE and PEP with high molar mass and high EO-fraction adsorb with lower fractions of adhered segments, like as do the homopolymer PEO. The adsorption of the EPE-samples with central PO-block is determined by the high number of preferentially adsorbing EO-segments in the endblocks. The central PO-block does not adsorb and protrudes as a loop into the solution. This is established mainly by the results that, at molar masses higher than $M = 3500$ and EO-fractions higher than 50% the p-values are constant and correspond to the p-values of PEO, and that a constant length of the EO-endblocks and increasing PO-block length the plateau adsorption enthalpy does not change, and that with increasing molar mass and fraction of EO-segments at first the plateau adsorption enthalpy grows distinctly, but at high molar mass decreases again, and the endgroup influence becomes increasingly insignificant.

PEP block copolymers are adsorbed by the central EO-block. Especially, the non existent endgroup effect and the fact that the plateau adsorption enthalpies approach a plateau value with increasing EO-fraction, demonstrate that the PO-endblocks are not bound to the silica surface and protrude into the solution.

Acknowledgement

We are grateful for the financial support of the Deutsche Forschungsgmeinschaft.

References

1. Killmann E, Eckart R (1971) Makromol Chem 144:45
2. Killmann E, Winter K (1975) Die Angewandte Makromolekulare Chemie 43:53
3. Cohen Stuart MA, Fleer GJ, Bijsterbosch BH (1982) J Colloid Interface Sci 90:321
4. Korn M, Killmann E, Eisenlauer J (1980) J Colloid Interface Sci 76:7
5. Killmann E, Korn M, Bergmann M, Adsorption from Solution Ottewill RH, Rochester CH, Smith AL (eds) (1983) Academic, London, p 259—272
6. Killmann E, Bergmann M (1985) Colloid and Polymer Sci 263:381
7. Cohen Stuart MA, Fleer GJ, Scheutjens JMHM (1984) J Colloid Interface Sci 97:515
8. Cohen Stuart MA, Scheutjens JMHM, Fleer GJ, Polymer Adsorption and Dispersion Stability, Goddard ED, Vinvent B (eds) (1984) ACS Symp Ser 240:53
9. Cohen Stuart MA, Cosgrove T, Vincent B (1986) Adv Colloid Surface Sci 24:143
10. Killmann E, Fulka C, Reiner M (1990) J Chem Soc Faraday Trans 98, 9:1323
11. Koller J, Killmann E (1981) Makromol Chem 182:3579
12. Cordt F (1985) Dissertation TU München, publication in press (1990) Makromol Chem 191
13. Moeller F (1989) Dissertation TU München, publication in preparation

Authors' address:

Prof. Dr. Erwin Killmann
Institut für Technische Chemie
Technische Universität München
Lichtenbergstr. 4
8046 Garching, FRG

Progress in Colloid & Polymer Science   Progr Colloid Polym Sci 83:96—103 (1990)

Self-mobility of flexible polymers adsorbed at a solid-liquid interface

R. Varoqui*) and E. Pefferkorn

Institut Charles Sadron (CRM-EAHP), CNRS-ULP Strasbourg, France

Abstract: In the present paper, we report on the rate of exchange of ^{3}H-labelled polymers in the adsorbed state with non-labelled polymers in the solution. The method consisted in carrying out preliminary adsorption with radioactive polymers, and subsequently exposing the surface to a solution of non-labelled polymers, and in continuously analyzing the radioactivity of the solution. In that way, chemical composition at thermodynamic equilibrium was maintained constant; only the isotope concentration ratios at the surface and in solution were different. — Investigations were carried out on two systems: — non hydrolyzed polyacrylamide adsorbed from aqueous solutions onto well defined alumino-silicate surfaces with $(SiO)_2AlOH$ and $SiOH$ surface groups statistically distributed; and — polystyrene adsorbed from carbon tetrachloride onto silica beads with $SiOH$ surface sites. — For both systems, we found that the rate of exchange of isotopes is very small, which means that polymers have a unusual long "lifetime" in the adsorbed state. The rate of exchange is also a function of polymer concentration and in a certain time interval the kinetics obeys a simple second-order rate reaction characterized by one relaxation time. This is consistent with interfacial equilibrium and kinetically indistinguishable molecules. However, the radioactive polymers desorbing first cannot be characterized by one relaxation time. For polystyrene, the isotherms display a plateau region followed by a region were adsorption increases. The kinetic properties are very different in both regions. Especially at large solution concentration the interfacial exchange cannot be described in terms of one rate constant; instead, a whole spectrum of relaxation times must be considered.

Key words: Polymer; self-mobility; solid-liquid; interface

Introduction

Surfaces coated by adsorbed polymers have a good stability in the sense that usually one finds that the layer does not redissolve at all when the surface is exposed during long periods to pure solvent. This has lead to the preparation of chromatographic supports with specific surface properties by incubation of porous silica with a polymer solution and subsequent washing [1—3]. The stability of adsorption was sometimes explained in terms of binding energies which are usually large [4] (the energy is proportional to the number of attached monomers, i.e., of order fN with $f \simeq 0.1$ to 0.8 and N the number of segments). The lack of desorption was also explained by polydispersity [5] and the ef-

fect of molecular weight was carried out quantitatively by gel permeation. Preferential adsorption of long chains was found [6]. Strong preferential adsorption was also sometimes objected [7—8].

In all experiments performed so far, no information on the self-mobility of a polymer in its adsorbed state was obtained. In order to discuss the thermodynamic state (reversibility, irreversibility, [9]), we need additional data on the mean residence time of a chain at the interface. In order to obtain this information, we devised a very simple method: a preliminary adsorption of radioactive-labelled polymers was first carried out, then the surface was exposed to a solution of non-labelled polymers. By using a continuously stirred tank reactor, the flux of labelled polymers from the surface to the solution was determined without change of chemical composition, and a relaxation time characteristic of the dynamics of polymers at interfaces was obtained [10]. We report here on the dynamics of: i) polyacrylamide adsorbed from aqueous media onto modified silica, and ii) polystyrene adsorbed from carbon tetrachloride onto silica.

Materials and methods

Polymer

Polyacrylamide and polystyrene were prepared according to conventional techniques [11]. Polyacrylamide was fractionated and characterized (see Table 1)

$$+CH_2-CH+_{m1} \quad +CH_2-CH+_{m2}$$
$$\begin{array}{cc} | & | \\ C=O & CHOH \\ | & | \\ NH_2 & {}^3H \end{array} \quad .$$

$m_1/m_2 = 10^3$; specific radioactivity is 1.12×10^5 cpm/mg; for molecular weight and polydispersity, cf. Table 1.

$$+CH_2-CH(C_6H_5)+_{m_1}+CH_2-CHCH^3HOH+_{m_2} \ .$$

$m_1/m_2 = 2 \times 10^3$; $M_w = 3.6 \times 10^5$; specific radioactivity is 2×10^5 cpm/mg; polydispersity ratio (GPC):1.4.

Inclusion of a tritium tracer was accomplished by copolymerization with minute amounts of acroleine and reaction with KB^3H_4.

Absorbent and solvent

The adsorbent was non-porous spherical glass beads of 34-μm average diameter with a specific surface area of 7.8 $\times 10^{-2}$ m^2 g^{-1}. A scanning electron micrograph of the sorbent is given in [12]. The glass beads have a very smooth surface without significant roughness on a scale of molecular dimensions. The glass beads were treated with HCl to bring every surface siloxane $(SiO)_2$ into silanol form (SiOH):

$$(SiO)_2 \xrightarrow{HCl} (SiOH)_2 \ . \tag{1}$$

Polystyrene adsorbs well from CCl_4 solutions onto silanols [1], whereas non-hydrolyzed polyacrylamide (PAM) does not. However, PAM adsorbs on surfaces partly covered with (AlOH) groups. Part of the silanol was therefore transformed into aluminol as described elsewhere [11]:

$$3 \ (SiOH) \xrightarrow{AlCl_3 + OH_2} (SiO)_2AlOH-(SiOH) \ . \tag{2}$$

PAM was adsorbed from aqueous media onto aluminol-grafted glass beads [2] with a surface composition [Al]/[Si] of 5%. For the viscosity and light-scattering measurements, we used monosized colloidal silica with diameter 1.6 μm (an electron micrograph is given in [13]). The beads were then treated with $AlCl_3$ with a final surface composition [Al]/[Si] of 33%.

Measurements of the rate of exchange of radioactive and non-radioactive polymers.

The technique is an adaption of a continuously stirred tank reactor schematized below and fully described elsewhere [14]: Glass beads in the reactor vessel were first equilibrated with a radioactive solution under controlled agitation. After adsorption equilibria was reached, the beads were allowed to settle and the supernatant solution was sucked through orifice (B) (orifices (A) and (B) in figure 1 were fitted with a millipore filter "Mitex LSWP" of pore radius 5 μm). After all the supernatant had been collected, the cell was filled instantaneously with a non-radioactive solution of concentration C, and a non-radioactive solution of the same composition was circulated through the cell with an automatically driven syringe connected to orifice (A). The flow rate was 0.6 ml/mn. This procedure was used for PAM.

For the more polydisperse polystyrene, the procedure was slightly different. An adsorption with radioactive species was first performed. At the same time a similar adsorption run was carried out with non-radioactive polymer under identical conditions. The inactive supernatant was separated from the sedimented beads, one part of the inactive supernatant was then quickly brought into contact with the sediment of beads covered with radioactive polymers, and at the same time, the other part of the non-labelled solution was introduced into the cell at (A). This insured that the molecular weight distribution was the same in the radioactive and non-radioactive supernatant. Polydispersity effects are then avoided [5]. The effluent was collected at the outlet (B) in a large number n of successive samples after each time interval Δt and each sample was analyzed for radioactivity. The residual radioactivity of the beads was determined from the conservation equation:

$$SA_{s,0} = VA_b(t) + SA_s(t) + J_v \int_0^t A_b(t)dt \ . \tag{1}$$

A is the specific radioactivity, subscripts s and b refer to surface and effluent, respectively, $A_{s,0}$ is the specific radioactivity (cpm/cm^2) at time zero of the surface. Since agitation was maintained throughout, $A_b(t)$ is also the specific radioactivity of the solution in the cell, and $VA_b(t)$ is the total radioactivity of the solution in the cell. The surface concentration C_s^* of radioactive polymers was determined, combining (1) with the relation (2) between activity and concentration C^*, and approximation (3) could also be used:

$$C^* = A/A_{sp} \tag{2}$$

$$\int_0^t A_b(t)dt \simeq \Delta t \sum_{i=1}^n A_{b,i} \ . \tag{3}$$

Table 1. Characteristics of polyacrylamide in adsorbed and in solution state

M_w (g/mole) $\times 10^{-6}$	M_w/M_n (GPC)	$\langle R_G^2 \rangle^{1/2}$ * (Å)	C_s (mg/m²)	L_H (Å, viscometry)	L_H Å (q.e.l.s.)	m_c (g/l)	m_s (g/l)
0.61	1.05	444	4.32	795	720	7.4	54
0.85	1.28	542	4.70	920	880	5.7	51
1.08	—	625	4.80	1030	—	4.8	47
1.14	—	646	4.85	1090	—	4.6	44
1.34	—	712	4.95	1160	—	4.1	43
1.45	1.20	—			1150		—

*) $\langle R_G^2 \rangle^{1/2}$ is the radius of gyration determined from the light scattered by the polymer solution.

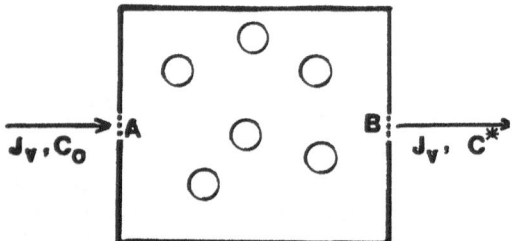

Fig. 1. Experimental device (schematic) for measurements of exchange rate of isotopes

Surface layer thickness determination

The layer thickness was determined for polyacrylamide adsorbed on silica particles of 1.6 μm with 33% AlOH grafts. Viscosity and inelastic light-scattering were used. The principle is the following: the viscosity η of beads covered with polymers is given by the Einstein formula:

$$\eta = \eta_0 (1 + 2.5 f\Phi) . \qquad (4)$$

Φ is the volume fraction of beads and the factor f takes into account the effect of the polymer layer. An apparent layer thickness L_H is defined from f [15]:

$$f = \left(1 + \frac{L_H}{R}\right)^3 , \qquad (5)$$

R being the radius of the uncoated beads. With inelastic light-scattering, the increase in the Stoke's radius is obtained from the Brownian diffusion coefficient D:

$$D = \frac{kT}{6\pi\eta(R + L_H)} . \qquad (6)$$

Viscosity measurements were performed with an Ubbelhode viscometer with a photoelectric detection system and inelastic light-scattering were made with a techniques set up by Duval [16].

Results and discussion

Adsorption isotherms

Adsorption isotherms for polyacrylamide of mass $M_w = 1.2 \times 10^6$ on a 5% AlOH modified silica are reported in Fig. 2. The pH dependence was previously explained [17]. It was found that the adsorption is proportional to the number of AlOH groups on the surface (PAM adsorbs through hydrogen-bonding between AlOH and the amide $CONH_2$ moiety). Since AlOH is amphoteric (in the form of $AlOH_2^+$ for pH < 4.6, and AlO^- for pH > 4.6), the number of residual AlOH is pH dependent.

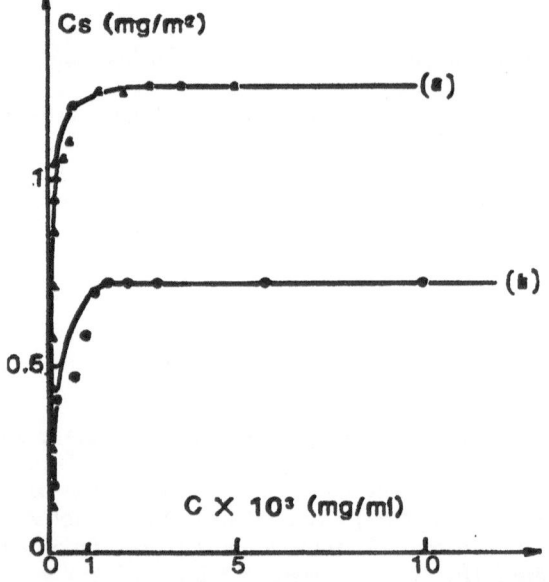

Fig. 2. Adsorption isotherms for polyacrylamide adsorbed at 25°C on aluminosilicate beads: curve (a): pH 4.5; curve (b) pH 4.0; $M_w = 1.2 \times 10^6$

In Table 1 are reported the L_H values determined at pH 4.5, $T = 25\,°C$ for PAM of different molecular weight. C_s in the second column is the adsorption at saturation (plateau value).

One infers from the L_H data (Table 1) that PAM develops in the adsorbed state large loops and a dense surface layer. The mass density of polymer at the interface (m_s), and in the free coil in solution (m_c) are reported in the last two columns:

$$m_s = C_s/L_H \qquad (7)$$

$$m_c = \frac{M}{4.18\langle R_G^2\rangle^{3/2}} \qquad (8)$$

$$\langle R_G^2\rangle = 2.1 \times 10^{-16}\ ([\eta]M)^{2/3}\ . \qquad (9)$$

Figure 3 represents the adsorption isotherm of polystyrened at the silica/carbon tetrachloride interface at 25 °C and 35 °C. The isotherm compared to the PAM behavior has a very different shape: i) no initial rapid rise is observed; and ii) at low solution concentration we observe a plateau ((AB) and (AD) regions) followed by regions (BC) and (DE) where the amount increases with the solution concentration. These isotherms do not resemble the adsorption isotherms found for polystyrene adsorbed from CCl_4 onto aearosil silica [18]. The aerosil which was most often used in adsorption studies was characterized by electron photomicrography [19]. It is a porous aggregate of 50 to 100 (≈ 150 Å) primary particles, irreversibly fused together. Usually, the specific adsorption is calculated on the basis of the surface of a primary particle. It must, however, be recognized that fused silica (aerosil or cabosil) is not a proper model for polymer adsorption studies; instead our sorbent can be truly considered as a non-porous, essentially infinite surface with respect to molecular dimensions (see [12]). Aerosil silica is also rather hydrophobic because the ratio of siloxane to silanols is large, whereas our silica is totally hydrated and is, therefore, chemically more homogeneous than fused silica.

The prominent feature in the polystyrene isotherm is the occurrence of a "kink" (points B and D in Fig. 3). This phenomenon has been found repeatedly in the study of the adsorption of proteins on solids [20]. It was interpreted in terms of structural transitions or by supposing the formation at the critical point of a two-dimensional (2-D) protein crystal [21]. For the polystyrene systems, it was admitted that in the plateau the monomer binds via a stoichiometric 1-1 association with surface hydroxyls, and the polymer in the plateau region lies flat on the surface. At higher concentrations, the formation of extended loops and/or tails was supposed [22]. It should be noted that the plateau value is temperature independent, so that adsorption is driven by a change in enthalpy, whereas in the ascending part the phenomenon is also related to an entropy change. Conformational reorganization therefore seems possible at large surface coverage. Layer-thickness measurements, not yet performed, should confirm the model.

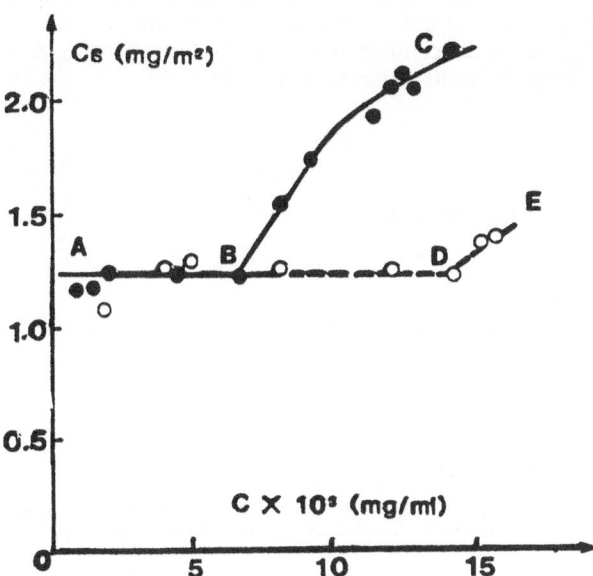

Fig. 3. Adsorption isotherm of polystyrene adsorbed on hydrated silica in CCl_4 at two temperatures: ($\bullet$) 25 °C; ($\circ$) 35 °C

Kinetics of exchange between adsorbed and free polymers

Polyacrylamide: The kinetics of desorption of labelled polymers is represented in Fig. 4 for the following parameter values: 5% grafted AlOH silica, pH = 4.0, $T = 25\,°C$, $C_{s,0}^* = 0.69$ mg/m² (this corresponds to the plateau value reported in Fig. 2). After adsorption equilibrium with labelled polymers, the solution concentration was $C^* = 3 \times 10^{-3}$ mg/ml. It was replaced in four different runs by a non-radioactive solution with concentration C reported in the lengend of Fig. 4. In all kinetic runs, apart from the case referring to pure solvent, the

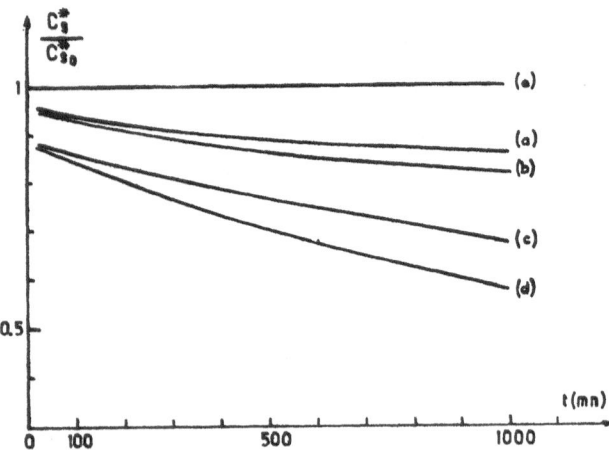

Fig. 4. Variation of the concentration of the radioactive polymers on the surface as a function of time for PAM in the presence of solvent (e) and in presence of polymer solutions at concentration C (mg/ml): 2×10^{-3} (a); 5×10^{-3} (b); 10×10^{-3} (c); 17×10^{-3} (d)

sum of labelled plus nonlabelled polymers did not change, neither on the surface nor in the solution during the kinetic run; only an asymmetric bulk/surface ratio in isotropic composition was established with the flow device.

Several observations arise from Figs. 4 and 5:

i) Desorption does not occur in the presence of pure solvent, but in the presence of a solution; labelled polymers, however, desorb at a very slow rate.

ii) The logarithmic derivative shown in Fig. 5 displays two distinct regions: up to 300—400 mns, which encompasses roughly 30% desorbed labelled polymer, the derivative decreases with time. Then it becomes constant and proportional to the solution concentration as seen in Fig. 6.

Assuming equilibrium, the kinetics of the desorption of labelled chains can be simply analyzed in terms of net fluxes [23].

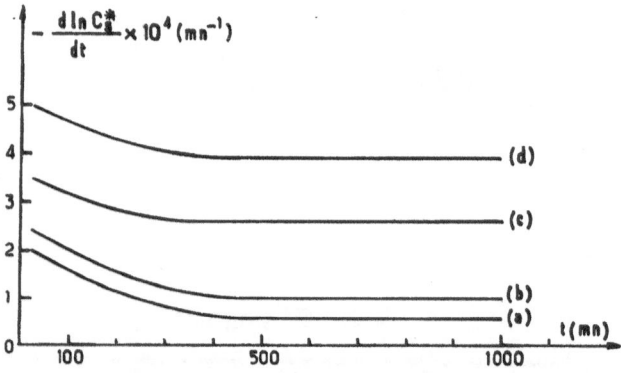

Fig. 5. Variation of the logarithmic derivative of the concentration of surface radioactive polymers with time for the four concentrations reported in Fig. 4

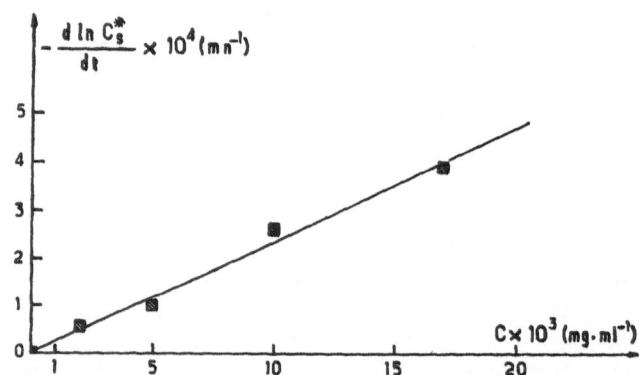

Fig. 6. Variation of the logarithmic derivative of the concentration of surface radioactive polymers as a function of the solution concentration. C_s^* is the value in the plateau region shown in Fig. 5

value of C was chosen in the region of maximum adsorption.

In Fig. 5 the decrease of $\ln C_s^*$ with time is shown. The experimental conditions (flow rate, concentrations, etc.) were always such that

$$\frac{C_s^*}{C_s} \gg \frac{C^*}{C} \; ; \tag{10}$$

C and C^* being, respectively, the concentration of nonlabelled and labelled species in solution. The

The ratio $J_{s \to b}/J_{s \to b}^*$ of the total number of molecules $J_{s \to b}$ — including, inactive and radioactive species — over the number of radioactive molecules $J_{s \to b}^*$, which leave the surface per unit time on account of Eq. (10) is given by:

$$\frac{J_{s \to b}}{J_{s \to b}^*} = \frac{C_s}{C_s^*} \; . \tag{11}$$

We consider now the total number of molecules coming from the solution, and those which hit the surface per unit time. A number of them are adsorbed, some are reflected. The adsorbed amount, which we call the backward flux $J_{b \to s}$, is proportional to the solution concentration C:

$$J_{b \to s} = KC . \tag{12}$$

Since K is completely defined by the structure and kinetics of the molecules populating the interface, it is natural to expect K to be a constant in all kinetics referring to the plateau region of the isotherms, which is actually our concern. If equilibrium holds, forward and backward fluxes are equal:

$$J_{s \to b} = -J_{b \to s} . \tag{13}$$

Equation (13) in conjunction with Eqs. (11)—(12) and, since $J_{s \to b}^*$ is dC_s^*/dt, gives

$$\frac{dC_s^*}{dt} = -KCC_s^*/C_s . \tag{14}$$

Since C_s is constant in the plateau, we replace K/C_s by a constant K_E, and we arrive at the result shown in Fig. 6:

$$\frac{dC_s^*}{dt} = -K_E CC_s^* . \tag{15}$$

Equation (15) with formally has the structure of a bimolecular reaction rate is merely a consequence of thermodynamic equilibrium, i.e., of equal forward and backward net fluxes.

The question that remains is the decrease of the logarithmic derivative at small time. From intuition we might imagine that at the end of the adsorption process, the last comers — which we believe do desorb first — are less tightly bound. This leads to an image of an interface populated by molecules with a spectrum of different "frozen" conformations (number of attachments might be very different) and consequently, different kinetic properties. Equation (14) transforms then into

$$\frac{d\ln C_s^*}{dt} = -\frac{\sum_i \tau_i C_s^{(i)} \exp(-\tau_i t)}{\sum_i C_s^{(i)} \exp(-\tau_i t)} \tag{16}$$

$$KC = -\sum_i \tau_i C_s^{(i)} , \tag{17}$$

where $C_s^{(i)}$ is the fraction of attached molecule with relaxation time τ_i^{-1}, and K represents the mean value (Eq. (17)). As the logarithmic derivative becomes a function of time, the simple phenomenology described by Eq. (15) no longer holds.

Polystyrene: In Figs. 7, 8 the rate of decrease of $\ln C_s^*$ with time for polystyrene adsorbed onto hydrated silica is reported. As observed in Fig. 7, the logarithmic derivative of C_s^* is remarkably constant for the kinetic run in the plateau: $C_s = 1.22$ mg/m^2, $C = 3.2 \times 10^{-5}$ mg/cm^2, $T = 25\,°C$. Again, the rate of desorption is very slow compared to what we know about molecular motions in solutions. However, for all kinetic runs with origin points located in the ascending part of the isotherm, the derivative decreases with time. In the plateau the molecules are characterized by one relaxation constant K_E in the sense of Eq. (15) and are, therefore, kinetically indistinguishable. In the ascending region of the isotherm, C_s is roughly proportional to c; K_E is, therefore, expected to vary as the inverse of C_s. The variation of K_E with C_s, seen in Fig. 9, is thus in line with Eq. (15), but K_E for a given C_s is also a function of C_s^*, i.e., a function of time as expected from Eq. (16) when a spectrum of relaxation times is considered. This recalls the behavior of C_s^* for PAM at small time, and

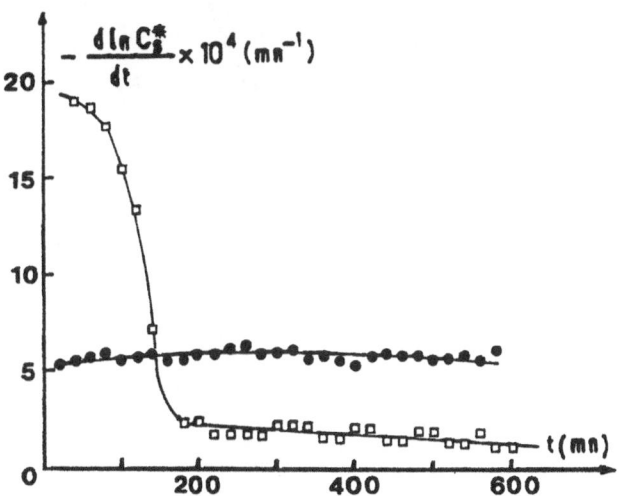

Fig. 7. Variation of the logarithmic derivative of the concentration of surface radioactive polymers with time for polystyrene adsorbed onto hydrated silica under the following conditions: (●), $C_s = 1.22$ mg/m^2, $C = 3.24 \times 10^{-3}$ mg/ml, $T = 25\,°C$, (□), $C_s = 1.5$ mg/m^2, $C = 18 \times 10^{-3}$, $T = 35\,°C$

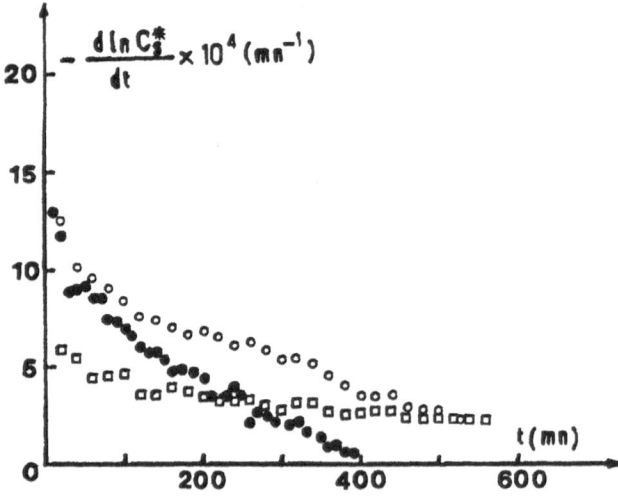

Fig. 8. Variation of the logarithmic derivative of the concentration of surface radioactive polymers with time for polystyrene adsorbed at 25 °C onto hydrated silica at solution concentrations beyond the plateau value: (○), = 2.17 mg/m², $C = 17.06 \times 10^{-3}$ mg/ml; (□), $C_s = 1.53$ mg/m², $C = 8.42 \times 10^{-3}$ mg/ml; (●), $C_s = 2.43$ mg/m², $C = 22.8 \times 10^{-3}$ mg/ml

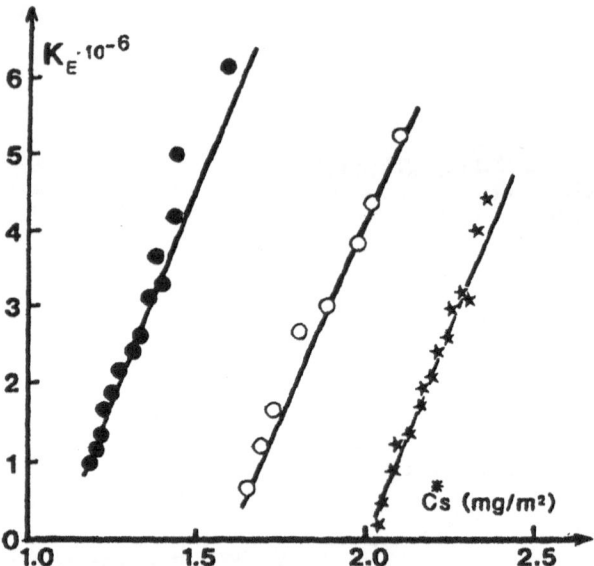

Fig. 9. K_E values defined by Eqs. (14)—(15) for polystyrene adsorbed at 25 °C onto hydrated silica as a function of the residual concentrations of adsorbed radioactive polymer: (●), $C_s = 1.53$ mg/cm², $C = 8.42 \times 10^{-3}$ mg/ml; (○), $C_s = 2.17$ mg/cm², $C = 17.06 \times 10^{-3}$ (mg/ml), (★), $C_s = 2.43$ mg/cm², $C = 22.80 \times 10^{-3}$ mg/ml

again in the region of overadsorption; the polymer layer cannot be considered as an homogeneous phase.

De Gennes theory: In a series of recent papers [23—25], de Gennes developed the theory of dynamics of polymers confined to a solid/liquid interface. He expressed the K coefficient (our Eq. (12)) in terms of the product of a tunneling amplitude T, — which is the statistical weight for a chain when starting at the limit of the diffuse layer to reach the surface — and a coefficient K_{si} which takes into account any slowing down factor (reptation effects, glassy states, etc.):

$$K = K_{si}T . \qquad (18)$$

T was expressed in terms of molecular parameters and chain length, $T \simeq N^{-3/10}$, however, K_{si} was not described at a molecular level. The theory is based on a homogeneous surface layer and on the assumption of the existence of a state of thermodynamic equilibrium. As we have seen, this must be amended — the kinetics we observe suggest that the first statement is restrictive; it probably applies to regimes of low adsorption.

Conclusion

Summarizing our observations, we arrive at following:

— Polymers in an adsorbed state are firmly "attached" at the solid/liquid interface.

— Forward and backward net fluxes at the interfaces, where found equal, and despite slow motion, guarantee a state of true thermodynamic equilibrium.

— In some situations, the isotope exchange cannot be characterized during the whole exchange process by one relaxation constant, and this suggests classes of polymers differing in their degree of "attachment". In view of the very slow motion at the interface, rapid kinetic turnover among conformations is not the rule, inhomogeneous layer structures can, therefore, be conjectured. Especially striking is the behavior of polystyrene on silica surfaces, for which the peculiar shape of the isotherm is remarkably reflected in the dynamic properties. Further investigations are needed to resolve some of these puzzling observations and especially, the role of chain length in the dynamics should be ascertained carefully.

Acknowledgement

We thank Dr. M. Duval of the Institut Charles Sadron for the inelastic light-scattering measurements.

References

1. Pefferkorn E, Tran QK, Varoqui R (1981) J de Chimie Physique 78:549
2. Letot L, Lesec J, Quivoron C (1981) J Liquid Chromatog 4:1311
3. Schmitt A (1984) In: Cazabat AM, Veyssié M (eds) Colloides et Interfaces. Les Editions de Physique, Paris, p 245
4. Benoît H (1988) Informal discussion
5. Cohen-Stuart MA, Scheutjens JMHM, Fleer GF (1980) J Polym Sci Polym Phys Ed 18:559
6. Vander Linden C, Van Leemput R (1978) J Colloid Interface Sci 67:48
7. Felter RE, Ray Jr LN (1970) J Colloid Interface Sci 32:349
8. Furusawa K, Yamamoto K (1983) Bull Chem Soc Japan 56:1960
9. McGlinn TC, Kuzmenka DJ, Granick S (1988) Phys Rev Lett 60:805
10. Brash JL, Uniyal S, Pusineri C, Schmitt A (1983) J Colloid Interface Sci 95:28
11. Pefferkorn E, Carroy A, Varoqui R (1985) Macromolecules 18:2252
12. Pefferkorn E, Haouam A, Varoqui R (1989) Macromolecules 22:2677
13. Stöber W, Fink A, Bohn E (1968) J Colloid Interface Sci 26:62
14. Pefferkorn E, Jean-Chronberg AC, Varoqui R Macromolecules (in press)
15. Varoqui R, déjardin Ph (1977) J Chem Phys 66:4395
16. Duval M (1982) Thesis, University Louis Pasteuer, Strasbourg, France
17. Pefferkorn E, Jean-Chronberg AC, Varoqui R J Colloid Interface Sci (in press)
18. Kawaguchi M, Meida K, Kato T, Takahashi A (1984) Macromolecules 17:1666
19. Eisenlauer J, Killmann E (1980) J Colloid Interface Sci 74:108
20. Norde W (1986) Adv Colloid Interface Sci 25:267
21. Fair BD, Jamieson AM (1980) J Colloid Interface Sci 77:525
22. Pefferkorn E, Haouam A, Varoqui R (1988) Macromolecules 21:211
23. de Gennes PG (1985) C. R. Acad Sci Paris, Serie 2 301:1399
24. de Gennes PG (1987) Adv Colloid Interface Sci 27:189
25. de Gennes PG (1988) In: Nagasawa M (ed) Molecular Conformations and Dynamics of Macromolecules in Condensed Systems, Elsevier Science Publishers, Amsterdam, p 315

Authors' address:

R. Varoqui
Institut Charles Sadron
6, rue Boussingault
67083 Strasbourg Cedex, France

Progress in Colloid & Polymer Science   Progr Colloid Polym Sci 83:104—109 (1990)

Correlation between adsorption and the effects of surfactants and polymers on hair

Th. Förster and M. J. Schwuger

Henkel KGaA, Düsseldorf, FRG

Abstract: The external horny layer of the hair, the cuticula, consists largely of amorphous keratin, which, in young undamaged hairs, contains only a few polar groups and is hydrophobic. External influences such as sunshine, permanent waving or bleaching lead to the formation of polar groups in the cuticula that hydrophilize the hair surface. Surfactants or polymers in hair care products are adsorbed on the hair and alter its wetting characteristics; this can be followed very sensitively by means of Wilhelmy's method. Cationic surfactants used to lubricate the hair are adsorbed in several layers on the hair surface. After the outer layers have been rinsed and desorbed in water, a single layer remains attached, whose hydrophobic hydrocarbon chains are oriented towards the aqueous phase. Cationic polymers are also adsorbed from aqueous solutions and solutions containing anionic surfactants, but in this case, however, the surface is hydrophilized. Parallel studies of combability were carried out, and showed that the reduced friction associated with cationic surfactants is a consequence of the presence of the hydrophobizing adsorption film.

Key words: Wettability; combability; lubricants; hydrophobic film

Introduction

Cationic surfactants and polymers are used in hair cosmetics as softening and conditioning additives. Their effectiveness depends on them being adsorbed on the hair surface from shampoos and rinses [1]. In an aqueous environment, hair keratin behaves as an ion exchanger that is capable of swelling [2—4], so that surfactants and polymers are adsorbed, not only at the hair surface, but can also penetrate into the inner hair. The adsorbed surfactants and polymers in the cuticula, which are responsible for lubrication effects, can therefore only be quantitatively determined by means of methods that exploit specific surface effects. The adsorption of surfactants or polymers brings about a change in the surface energy of the hairs, and this change can be sensitively measured by means of Wilhelmy's method [5]. The cuticula of undamaged hair is strongly hydrophobic [6]. After surfactant adsorption had occurred, it was found that the hair

surface had been hydrophilized; this effect disappeared when the treated hairs were repeatedly immersed in water [5, 7]. However, these observations obtained from wetting studies do not permit any conclusions to be drawn about whether all of the adsorbed substance was rinsed off or whether a monolayer film of adsorbed surfactant molecules remained attached with its hydrophobic hydrocarbon chains oriented to the outside. Physical studies of the influence of long-chain surfactants on the frictional properties of polymers [8—10] and hairs [11] suggest that a monomolecular adsorption film does actually remain attached to the surface of the hair, and that this is responsible for the reduction in friction. This effect of softening treatment is observed not only in undamaged hair, but also in bleached hair [12, 13], which, as a result of the oxidation of disulfide bridges to polar sulfonic acid groups is more hydrophilic than undamaged hair [6]. It was hoped that wetting studies of this oxidized, hydrophilic hair would enable the presence of a hydrophobizing adsorption film to be proved, just as on a hydrophilic metal surface [14]. Extensive wetting tests were therefore carried out on bleached hair, during which the adsorption and desorption characteristics of a variety of surfactants and polymers were observed; special attention was paid to proving the presence of a hydrophobic adsorption film. It was shown that cationic surfactants are adsorbed on the hair surface in several layers. After repeated immersion in water a hydrophobizing layer remains attached. This adsorption mechanism explains the lubrication effect of cationic surfactants that was observed in additional dry combing studies.

Experimental

Measuring wetting tension

Wetting tension was measured by means of a modified Wilhelmy method with a recording tensiometer. Single hairs were immersed in double distilled water at a rate of 0.2 mm/min and the force exerted on the hair was measured with an inductive force absorber (s. Fig. 1). This force is made up of the earth's gravitational attraction, buoyancy, and the wetting force.

The wetting tension j or the dynamic angle of contact θ can be calculated directly from the measured wetting force F, the known surface tension γ of the distilled water, and the circumference L of the elliptical hair, calculated from the optically determined semi-major and semi-minor axes:

$$j = F/L = \gamma \cdot \cos\theta .$$

An undamaged hair is strongly hydrophobic, resulting in an angle of contact of more than 90° and a wetting tension of less than 0 mN/m [6].

To study the desorption characteristics of the substances adsorbed on the hair surface, each hair was immersed 5 times in succession in double distilled water and the wetting tension was measured. The wetting tension was usually measured before and after product treatment as a pair comparison on the basis of four to five single hairs per product. The absolute error of the mean was approximately ±5 mN/m.

To simulate a treatment as given in practice, several strands of hair were treated instead of single hairs; single hairs were taken from the strands before and after treatment and their wettability was measured. In this case no pair comparison could be made.

Dry combability

For each product, 20 hair strands were combed 10 times each in a machine, before and after treatment, at 25 °C

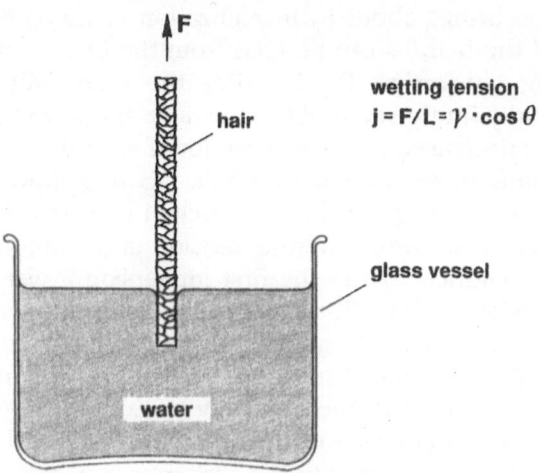

Fig. 1. Test set-up for measuring wetting tension by means of Wilhelmy's method

and 40% relative humidity, without any build-up of electrostatic charge. The comb force and work was measured with a load cell. The product effectiveness was determined by means of a comparison of the mean values of the combing force before and after treatment. The relative error of the mean of the maximum dry combing force was about 4%.

Oxidative hair damage: bleaching

Brown strands of hair (Alkinco, no. 6634) were each oxidized twice for 30 min in 6% hydrogen peroxide solution (pH 9.4) with a wash liquor ratio of 1:15, and subsequently rinsed thoroughly and dried.

Treatment with surfactants and polymers

The hairs were treated with the hair treatment agent for 5 min with a wash liquor ratio of 1:5 and were then rinsed and dried. The hair treatment agents were set to pH 6.5 or 4.5 with citric acid.

The cationic compounds used were the surfactants lauryl trimethyl ammonium chloride and cetyl trimethyl ammonium chloride, and the polymer guarhydroxypropyl trimethyl ammonium chloride; the anionic surfactant sodium lauryl diglycolether sulfate was used, as were the fatty alcohol mixture cetyl/stearyl alcohol. All of these compounds are commercial products of Henkel.

Results and discussion

Treatment of undamaged, hydrophobic hair with an aqueous cetyl trimethyl ammonium chloride solution brings about hydrophilization of the surface of the hair, as can be seen from the increased wetting tension in Fig. 2. After the hairs were repeatedly immersed in distilled water, the wetting tension decreased to values even lower than the initial value for the undamaged hair. Figure 3 shows the corresponding results for bleached hydrophilic hair, which showed a wetting tension of 45 mN/m before treatment. After the first immersion a wetting tension of about 30 mN/m was measured as for undamaged hair. A higher wetting tension is reported in the literature for a similar adsorption study [5]. This difference can only be explained by the fact that, in our study, the hair was rinsed with water after treatment with the surfactant to remove excess surfactant solution.

After the second immersion the surface was appreciably more hydrophobic; the wetting tension of the surface then remained unchanged after further repeated immersion.

These results point towards the following mechanism. The cationic surfactant cetyl trimethyl ammonium chloride is adsorbed from aqueous solutions, forming a number of layers on both undamaged and oxidized hair. The outermost adsorption layer, in which the surfactant molecules are arranged with their hydrophilic head groups oriented towards the aqueous solution, is desorbed by immersion in water, so that a monolayer of firmly adsorbed surfactant molecules remains on the hair, with its hydrophobic hydrocarbon chains oriented

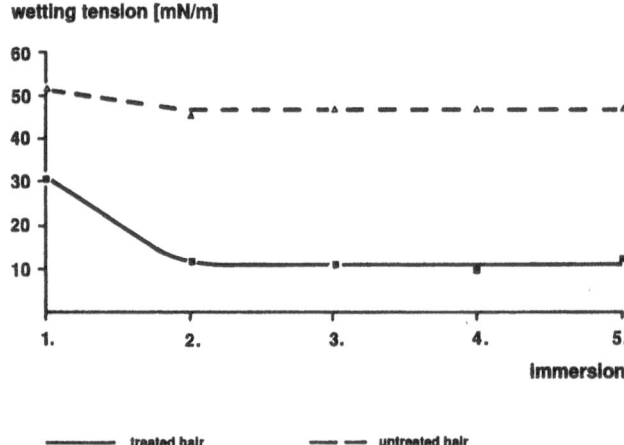

Fig. 3. Influence of the adsorption of cetyl trimethyl ammonium chloride from a 1% aqueous solution at pH 6.5 on the wetting tension of oxidized hairs

outwards. This hydrophobizing monolayer remains, even after repeated immersion in water, and is, therefore, effectively irreversibly adsorbed. A further factor indicating the veracity of this adsorption mechanism is the varying degrees to which the adsorption layers cover the hair surface. Whereas the inner, hydrophobic adsorption layer covers the hair surface evenly, the outermost adsorption layer is characterized by its irregular cover, revealed by the pronounced spikes in the locally resolved wetting curves in Fig. 4.

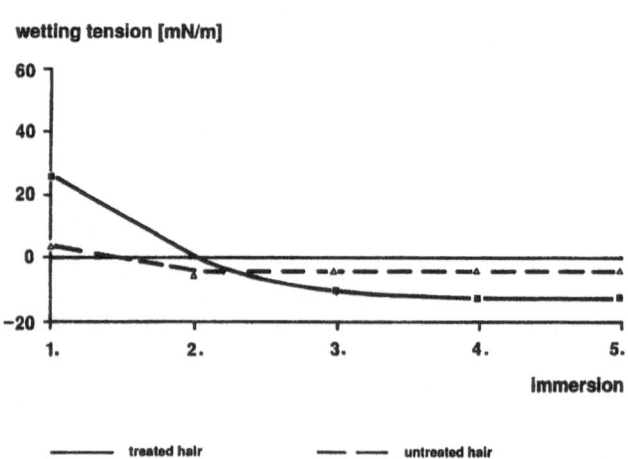

Fig. 2. Influence of the adsorption of cetyl trimethyl ammonium chloride from a 1% aqueous solution at pH 6.5 on the wetting tension of undamaged hairs

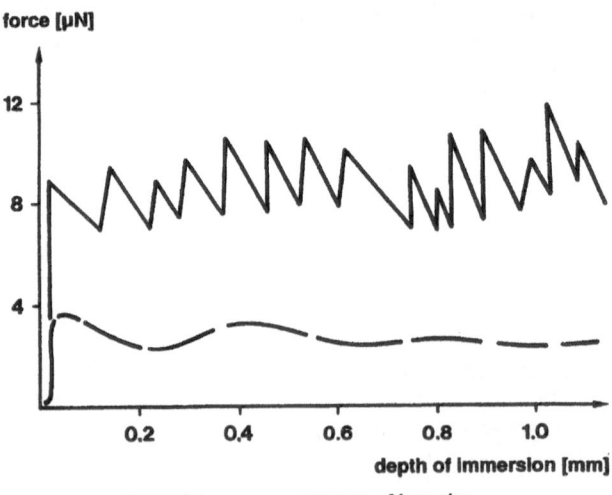

Fig. 4. Locally resolved wetting tension curves for an oxidized hair treated with a 1% aqueous cetyl trimethyl ammonium chloride solution at pH 6.5

The cuticula of the hair can be described in physicochemical terms as a hydrophobic surface which, in the pH range above the isoelectric point (approximately pH 3.5), contains a certain number of negatively charged ionic groups [15]. In this sense the hair surface resembles a weakly charged hydrophobic polycarbonate or polystyrene surface, on which surfactants are adsorbed in several layers [16—18]. In diluted surfactant solutions below the critical micelle concentration (CMC), individual surfactant molecules are adsorbed as a result of electrostatic, van der Waals, and hydrophobic interactions. Shortly before CMC is reached there are so many surfactant molecules adsorbed on the surface that interactions occur due to the forces of attraction between the hydrocarbon chains, and the surfactants are compressed into a densely packed monolayer. Above CMC a double layer may be formed.

This process is reflected in the adsorption isotherm in Fig. 5, in which the wetting tension of the treated hairs is shown as a function of cetyl trimethyl ammonium chloride concentration. Shortly before the CMC (shown to be 0.16 g/l by the Wilhelmy method) is reached, a hydrophobizing monolayer is adsorbed on the hair surface. Even after CMC has been exceeded, this layer gives a constant wetting tension of 10 mN/m, irrespective of the surfactant concentration.

Figure 6 shows the results obtained with the components of a softening hair rinse. The cationic surfactants lauryl trimethyl ammonium chloride and cetyl trimethyl ammonium chloride are both adsorbed on the hair surface in several layers. After the third immersion, the densely packed hydrophobic monolayers can be detected.

Cetyl/stearyl alcohol, which is insoluble in water, illustrates the importance of desorption characteristics in the detection of double-layer adsorption; its outermost, hydrophilic adsorption film cannot be removed, even by repeated immersion.

The use of a model hair rinse, consisting of a dispersion of fatty alcohol and a cationic surfactant in water, gives similar results to those obtained with a simple cationic surfactant solution. After hair is washed with an anionic surfactant shampoo (14% lauryl ether sulfate), the wetting tension increases almost to the initial value given by untreated hair. This shows that the hydrophobic monolayer containing cationic surfactant molecules is removed from the hair surface during the shampooing.

Because the external adsorption layers are easily removed, even in water, it may be assumed that

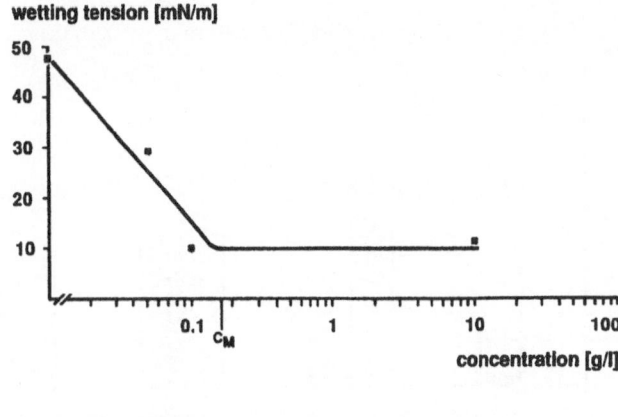

Fig. 5. Adsorption isotherm of cetyl trimethyl ammonium chloride adsorbed from an aqueous solution at pH 6.5 on an oxidized hair

an undesirable accumulation of cationic surfactants on the hair surface is unlikely to occur in practice.

Because the consequences of the wetting studies are of great practical importance with regard to hair rinsing, the same tests were carried out on strands of hair instead of single hairs to prove that the same results were obtained with this more practice-related method.

The wetting characteristics of the surface are affected, not only by softening hair rinses, but also by so-called conditioning shampoos, which usually contain cationic polymers as well as anionic surfactants.

Figure 7 shows the results obtained with a model formulation of a conditioning shampoo and the individual components. Whereas the anionic surfactant lauryl ether sulfate has no effect on the wetting characteristics of the cuticula, the cationic polymer guarhydroxypropyl trimethyl ammonium chloride is adsorbed on the hair. Surprisingly, polymer adsorption can bring about either hydrophilization or hydrophobization of the hair surface. This ambivalence explains the large error of the mean of ± 12 mN/m associated with the mean value obtained from six hairs after the first immersion. After repeated immersion a hydrophilic polymer layer, which is effectively irreversibly adsorbed, remains on the hair in all cases.

An empiric relationship between wettability and frictional characteristics, which can be explained in terms of the adhesion theory of friction [8, 9], has been found to exist for wool fibers [19]. This friction

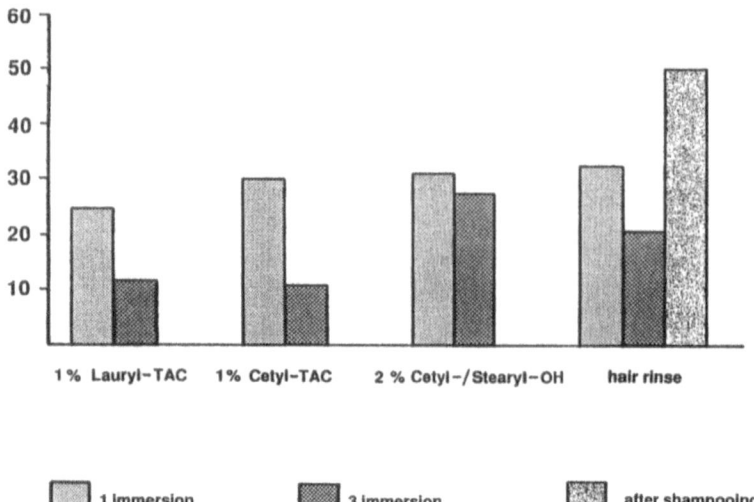

Fig. 6. Influence of cationic surfactants and fatty alcohol on the wetting tension of oxidized hairs. The cationic surfactants were in aqueous solution at pH 6.5, the cetyl/stearyl alcohol was in a 2% solution in ethanol. The hair rinse contained 1% cetyl trimethyl ammonium chloride and 3% cetyl/stearyl alcohol at pH 4.5

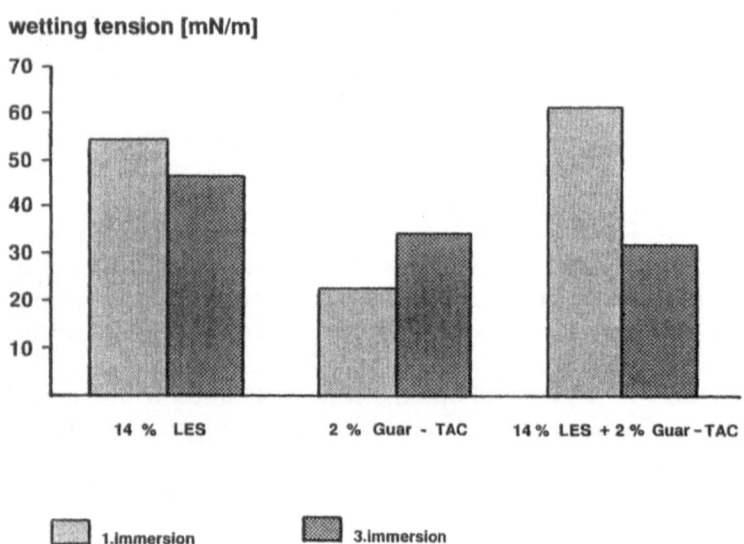

Fig. 7. Influence of sodium lauryl diglycolether sulfate and guarhydroxy-propyl trimethyl ammonium chloride on the wetting tension of oxidized hair. The anionic surfactant and the cationic polymer were in aqueous solution at pH 6.5

theory is based on the assumption that, when a smooth surface is lying against another smooth surface, they are only in contact at a few points on account of their microscopic roughness, and in effect, are welded together by the relatively large pressure exerted at these points. The surfaces cannot slide over each other until the shear strength of the points of contact is overcome.

Adsorbed surfactants acts as lubricants by forming densely packed, uniformly oriented adsorption films on the surface [8—10]. These films can easily slide over each other, because the forces of attraction between the hydrophobic terminal methyl groups of the hydrocarbon chains are only weak. This correlation between hydrophobization and reduction of friction was also found in dry combing studies. Figure 8 shows the reduction of the maximum dry combing force as a function of the concentration of aqueous cetyl trimethyl ammonium chloride solutions. A comparison with the adsorp-

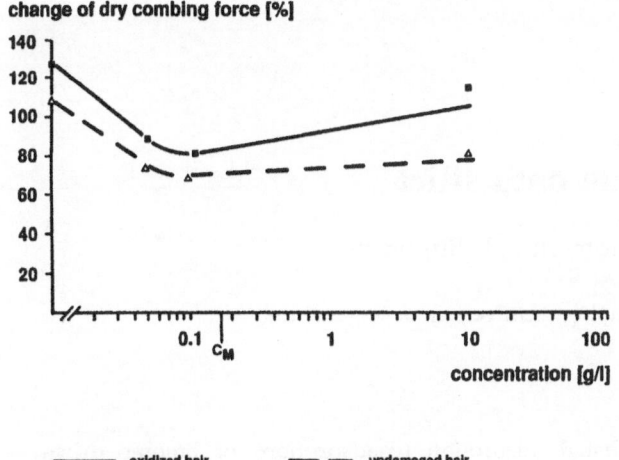

Fig. 8. Reduction of dry combing force as a function of cetyl trimethyl ammonium chloride concentration. Conditions as in Fig. 5

tion isotherm in Fig. 5 shows that the friction during dry combing decreases with increasing adsorption of the cationic surfactant. Wetting studies are thus not only an aid to explaining the adsorption mechanism, they also allow conclusions to be drawn with regard to practice-related parameters such as dry combability.

References

1. Tokiwa F, Hayashi S, Okumara T (1976) In: Kobori T, Montagna W (eds) Biology and Disease of the Hair, University of Tokyo Press, pp 631—640
2. Scott GV, Robbins CR, Barnhurst JD (1969) J Soc Cosmet Chem 20:135—152
3. Faucher JA, Goddard ED (1976) J Colloid Interface Sci 55:313—319
4. Faucher JA, Goddard ED, Hannan RB (1977) Text Res J 47:616—620
5. Kamath YK, Dansizer CJ, Weigmann HD (1984) J Appl Polym Sci 29:1011—1026
6. Kamath YK, Dansizer CJ, Weigmann HD (1977) J Soc Cosmet Chem 28:273—284
7. Kamath YK, Dansizer CJ, Weigmann HD (1985) J Appl Polym Sci 30:925—936
8. Schlatter C, Olney RA, Baer BN (1959) Text Res J 29:200-210
9. Fort T, Olsen JS (1961) Text Res J 31:1007—1011
10. Fort T (1962) J Phys Chem 66:1136—1143
11. Schwartz AM, Knowles DC (1963) J Soc Cosmet Chem 14:455—463
12. Newman W, Cohen GL, Hayes C (1973) J Soc Cosmet Chem 24:773—782
13. Garcia ML, Diaz J (1976) J Soc Cosmet Chem 27:379—398
14. Förster Th, von Rybinski W, Schwuger MJ (1988) Ber Bunsenges Phys Chem 92:1079—1083
15. Goddard ED, Leung PS (1987) Parfümerie und Kosmetik 68:546—561
16. Connor P, Ottewill RH (1971) J Colloid Interface Sci 37:642—651
17. Bisio PD, Cartledge JG, Keesom WH, Radke CJ (1980) J Colloid Interface Sci 78:225—234
18. Keesom WH, Zelenka RL, Radke CJ (1988) J Colloid Interface Sci 125:575—585
19. Lindberg J (1953) Text Res J 23:585—588

Authors' address:

Th. Förster
Henkel KGaA
Postfach 1100
4000 Düsseldorf, FRG

Progress in Colloid & Polymer Science   Progr Colloid Polym Sci 83:110—117 (1990)

The mediated adsorption (cosorption)
of anionic drugs by cationic surfactants onto silica

R. Bernard, E. Fuchs, M. Strnadova, J. Sigg, J. Vitzthum and H. Rupprecht

Institute of Pharmacy, Department of Pharmaceutical Technology, University of Regensburg, Regensburg, FRG

Abstract: The mediated adsorption (coadsorption) of anionic drugs — genetisic acid (GA) ethylene diaminotetraacetic acid (EDTA) — and cationic surfactants — alkylpyridinium chlorides (C_{10}—C_{16}), tetradecylpyridinium-salicylate and nitrate — on silicas with different structures and on *Escherichia coli* was studied. The anionic drugs are not adsorbed to silica from their pure aqueous solutions. In the presence of cationic surfactants, however, they are coadsorbed, accompanying the surfactant cations as counterions in their corresponding adsorbates. The establishment of these coadsorbates is hampered by narrow pores ($d^-_{pore} < 6$ nm) of the silicas. This mediated adsorption depends on the aggregation behavior of the surfactants and, consequently, on their hydrocarbon chain length. The coadsorbed drug anions can be rapidly exchanged by other ions, demonstrating that the surfactant adsorbates on silica act as anion exchanger. — From mixtures of cationic surfactants with EDTA the same mediated adsorption was confirmed on *E. coli* bacteria, explaining the enhanced antimicrobial action of these mixtures.

Key words: Coadsorption; anionic drug; cationic surfactant; silica; E. coli

Introduction

Adsorption phenomena determine to a great extent the quality of pharmaceutical solid/liquid preparations. This concerns, in particular, the physico-chemical stability and bioavailability of oral or parenteral administered suspensions, of dermal lotions and vesicle-preparations [1, 2]. Most of the research work dealing with adsorption in pharmaceutical systems is essentially focused on only one adsorptive, the "reactive" or "active" component. Pharmaceutical products are, however, multicomponent preparations. In these systems coadsorption of dissolved substances must be expected on solid components (including the container material). Attention should be especially drawn on the phenomenon of mediated adsorption. In this case an usually nonadsorbing species is adsorbed to a solid surface by means of another adsorbing agent in the liquid phase. This unexpected effect may significantly influence the pro-

duct quality, but is only sparsely described in the literature [3].

As an example, in this paper the adsorption of anionic drugs from aqueous solution onto the negatively charged surfaces of silicas and *Escherichia coli* (*E. coli*), mediated by cationic surfactants is described.

Materials

The adsorbents and adsorptives are described in Table 1.

Adsorptives

Surfactants: Decyl-, Dodecyl-, Tetradecyl-Hexadecylpyridiniumchlorides (analytical grade, Henkel Cie, Duesseldorf, FRG).

Table 1. Adsorbents and adsorptives: Physiochemical data and sources

Adsorbents

	Mean pore diameter [nm]	Mean particle size [µm]	Specific surface area (BET; N_2) [m^2 g^{-1}]	Source
Silicas				
Kr 36	15	200	510	(4)
KG 100	10	63—200	300 ⎫	
KG 60	6	63—200	500 ⎬	E. Merck,
KG 40	4	63—200	650 ⎬	Darmstadt
KG 20	2	63—200	750 ⎭	
Fractosil 500	60	63—125	80	
Aerosil 200	nonporous	8—10 nm	230 ⎫	Degussa,
Aerosil 0 × 50	nonporous	40 nm	69 ⎭	Frankfurt

E. coli: AT CC 11775 (Prof. E. Stetter, Universität Regensburg)

Tetradecylpyridinium-nitrate:
Obtained by crystallization from the corresponding chloride-salt in 25% HNO_3 and subsequent crystallization (twice) from ethanol, Content 99.2%.

Tetradecylpyridinium-salicylate:
Obtained from equimolar mixtures of tetradecylpyridinium chloride and sodium salicylate in ethanol, NaCl developed was separated by crystallization. The surfactant-salicylate was extracted by chloroform and purified by crystallization from acetone ethylacetate (1:1). Content 99.8%.

Cetyltrimethylammonium bromide: E. Merck, Darmstadt, FRG.

Drugs: Gentisic acid (2,5 dihydroxybenzoic acid) and disodium-ethylene diamintetra-acetate (EDTA-Na$_2$) (E. Merck, Darmstadt)

Reagents: (NaCl, NaNO$_3$, HCl) analytical grade (E. Merck, Darmstadt)

Water: Aqua purificata (double distilled) according to [5].

Methods

The adsorption experiments on silica and *E. coli* are described elsewhere [6]. The desorption experiments were performed according to the dissolution test of Pharm. Eur. II [5] using the paddle apparatus.

Production of drug-loaded silica

2 g SiO$_2$ Kr 36 were dispersed in 200 ml solution of an equimolar mixture of tetradecylpyridinium (TDP$^+$) chloride and gentisic acid (15 mM · l^{-1}). After gentle shaking (48 h at 21 ± 0.1 °C) the liquid phase was removed by filtration and the solid was dried at 40 ± 1 °C.
Drug content of the loaded silica: TDP$^+$: 1.34 mM g^{-1}; gentisic acidi 1.05 mM g^{-1}.

Dissolution (desorption) test

Drug-loaded silica 100 mg, dissolution fluid 500 ml, paddle stirrer at 80 UPM; temperature 21 ± 1 °C. Assay of the drugs: recording of drug concentration in the dissolution fluid by a monitor system: spectral photometer Ultraspec II (LKB, Freiburg, FRG), flow-through quartz cell 10 mm, peristaltic pump Miniplus 2 (Gilson, Villiers-le-Bel, France) with Iso-Versinic tubes, 2 mm diameter.
Wave lengths of gentisic acid: λ = 320 nm (H$_2$O, 0.1 N NaCl), λ = 330 nm (0.1 N HCl).
Tetradecylpyridinium cations: λ = 259 nm (in this case the adsorbance of gentisic acid at this wave length was considered).

Electrophoretic measurement

Apparatus:	Lazer Zee™, Model 501 (Penkem Inc., Bedford Hills, New York, USA)
Cell:	Fused silica, 0.15 × 1.5 × 10 cm, anode: molybdenum;
cathode:	platinum;
Samples:	*E. coli*-suspensions 4—10^8 bacteria/ml in water (double distilled) pH 5.6
Measurement:	21 ± 1 °C; voltage: 100 V; precision in measurement: ±2 mV

Results

Coadsorption of gentisic acid and alkylpyridinium chlorides on silica

Gentisic acid (2,5 dihydroxybenzoic acid) (GA) a type of hydrophilic anionic drug is only marginally adsorbed onto the surface of porous silica from aqueous solution (Fig. 1). The negative surface charge of the silica obviously prevents both the adsorption of the drug anions, and — considering both the pK_a-value of 2.9 for gentisic acid and the pH of dispersion medium (5—6.5) — of the unionized acid molecules [3].

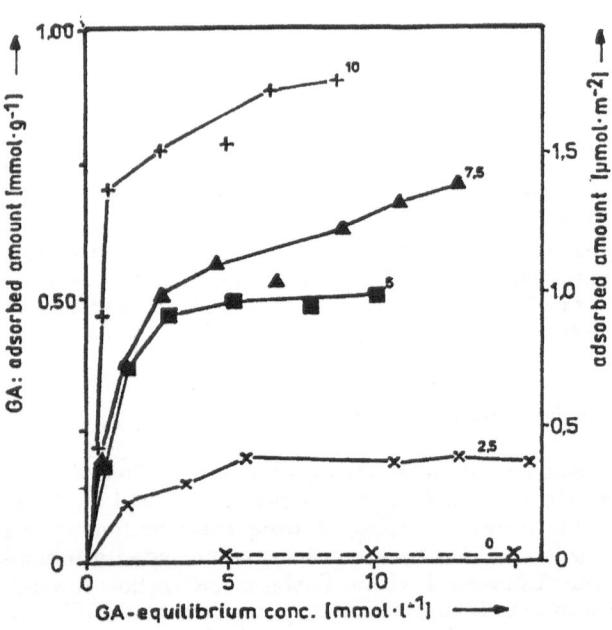

Fig. 1. Coadsorption (mediated) of gentisic acid on porous silica (Kr 36) in the presence of constant amounts of tetradecylpyridinium-chloride (conc. in mmol l^{-1}: numbers at the isotherms) at 20°C

In the presence of tetradecylpyridinium chloride (TDP-Cl), however, the anionic drug is bound to the silica surface: Adsorption isotherms of GA*), obtained in the presence of constant amounts of TDP-Cl clearly demonstrate the adsorption-mediating action of the cationic surfactant. An increasing

---

*) In the following text GA is used irrespective of the dissociation into anions of the acid molecules.

amount of TDP$^+$ results in a stronger adsorption, and higher saturation values are obtained for GA.

Correspondingly, the adsorption of TDP$^+$ cations onto silica is also significantly influenced by GA (Fig. 2):

The typical S-shape of the adsorption isotherm of the TDP$^+$ on silica [7, 8], due to electrostatic interactions in the first step and hydrophobic interaction in the second step, is replaced by a steep increase. The saturation values of adsorption of TDP$^+$ are also raised by GA, yet they level off at higher concentration of the anionic drug. The maxima in the adsorption isotherms are due to the fact that TDP-Cl was used. With increasing chloride concentration these anions (counterions) replace part of the GA in the coadsorbate. As a result, the saturation value decreases.

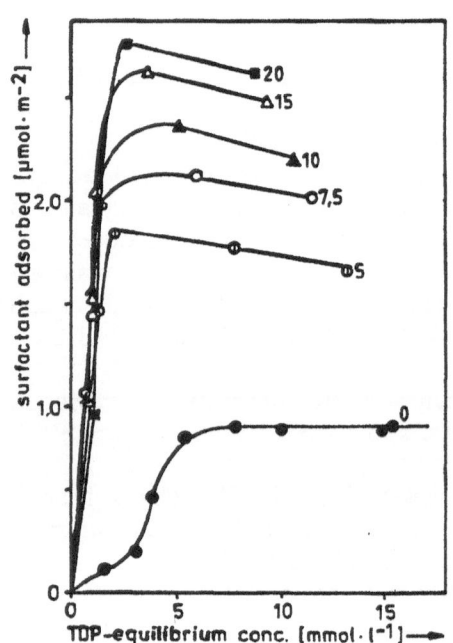

Fig. 2. Coadsorption of tetradecylpyridinium cations on porous silica (Kr 36) in the presence of constant amounts of gentisic acid (conc. in mmol l^{-1}: numbers at the isotherms) at 20°C

At equal molar concentrations of surfactant cations (5 mmol l^{-1}) the GA coadsorption appears to be strongly dependent on the alkyl-chain length of the surfactant (Fig. 3). With GA, the homologues

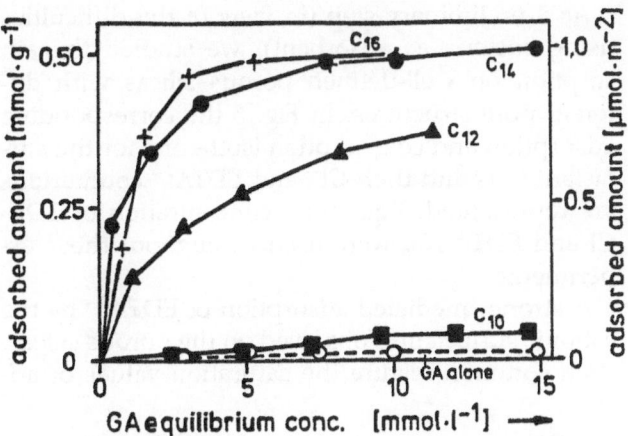

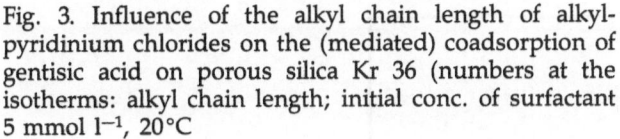

Fig. 3. Influence of the alkyl chain length of alkyl-pyridinium chlorides on the (mediated) coadsorption of gentisic acid on porous silica Kr 36 (numbers at the isotherms: alkyl chain length; initial conc. of surfactant 5 mmol l⁻¹, 20°C

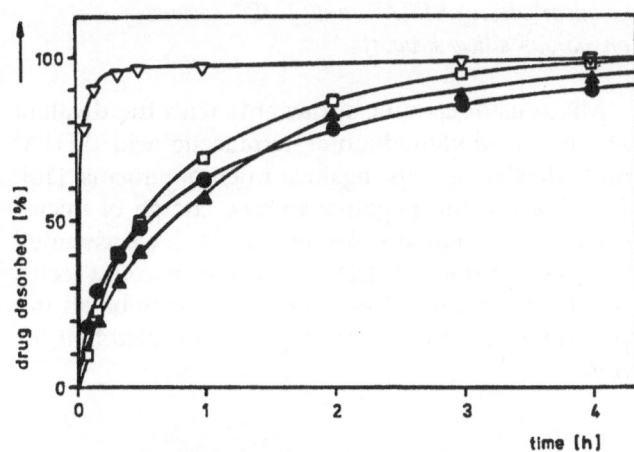

Fig. 4. Desorption kinetics of gentisic acid (GA) and tetradecylpyridinium ions (TDP⁺) from coadsorbates on porous silica (Kr 36) in H_2O and 0.1 NHCl; 100 mg drug-loades silica (corresponding to 0.69 mmol g⁻¹ GA and 0.83 mmol g⁻¹ TDP⁺; each = 100%). Volume of liquid phase 500 ml, 80 RPM, 21°C, paddle apparatus. ●—● TDP/H_2O; ▲—▲ TDP/HCl; □—□ GA/H_2O; ▽—▽ GA/HCl

show a significant mediating action, becoming stronger with the growing length of the alkyl chains. The effect obviously levels off between the C_{14} and the C_{16} homologue.

Coadsorbates of TDP-Cl with GA on porous silica were subjected to desorption experiments with respect to their possible use in controlled drug delivery. Drug-loaded silica was used in these experiments, produced by establishing adsorption:

After removal from the liquid and drying, the silica contained 1.05 mMol g⁻¹ GA and 1.34 mMol g⁻¹ surfactant ions. 100 mg of this drug-loaded silica was then subjected to desorption according to the "dissolution test" of Pharm. Eur. II [5]. 500 ml solvent was applied as to maintain "sink" conditions throughout the experiments (at 100% release the drug concentration was lower than 10% of the solubility of the drug).

In pure water as desorption liquid both the surfactant and gentisic acid show a sustained release pattern, typical for the release from hydrophilic porous matrices (Fig. 4) [9]. After an initial phase (<30 min) the rate of the drug desorption appears slightly higher than that of the surfactant cations. In 0.1 N HCl the dissolution pattern of TDP⁺ differs not essentially from that in water. The GA, however, desorbs rapidly in this acidic desorption liquid. More than 95% of the coadsorbed drug is desorbed within 20 min.

In 0.1 M NaCl solution the surfactant cations desorb more sustainedly than in water, while the GA desorption appears to be accelerated (Fig. 5). In contrast to 0.1 N HCl the desorption is slower and only 80% of the drug can be finally be desorbed.

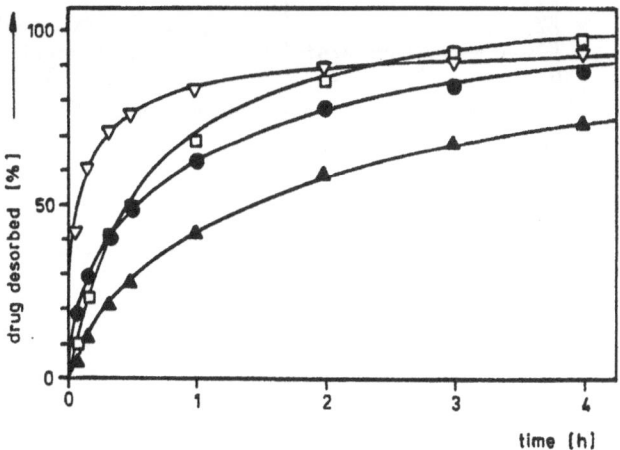

Fig. 5. Desorption kinetics of gentisic acid (GA) and tetradecylpyridinium ions (TDP⁺) from coadsorbates on porous silica Kr 36 in H_2O and 0.1 M NaCl; 100 mg drug loaded silica (corresponding to 0.69 mmol g⁻¹ GA and 0.83 mmol g⁻¹ TDP⁺; each = 100%). Volume of liquid phase 500 ml, 80 RPM, 21°C, paddle apparatus. ●—● TDP/H_2O; ▲—▲ TDP/NaCl; □—□ GA/H_2O; ▽—▽ GA/NaCl

*Coadsorbates of EDTA^{2-} and TDP$^+$ cations
on porous silica supports*

Mixtures of cationic surfactants with the divalent anions of ethylenediamine-tetraacetic acid (EDTA) are effective agents against microorganisms [10]. Considering the negative surface charge of microorganisms in aqueous dispersions [11], we assumed that adsorption of EDTA on the bacteria cells, mediated by the cationic surfactant, may be an important factor for the synergistic antimicrobial action.

As a preliminary step (in view of the difficulties using bacteria as adsorbent), we studied the adsorption on well-defined porous silicas with different pore structures. In Fig. 6 the corresponding adsorption and coadsorption isotherms for the surfactant ions and their Cl$^-$ and EDTA^{2-} counterions are represented. Equimolar concentration of TDP-Cl and EDTA-Na$_2$ were used throughout these experiments.

A strong, mediated adsorption of EDTA^{2-} by the cationic surfactant is observed on the porous silicas. As a common feature the saturation values of ad-

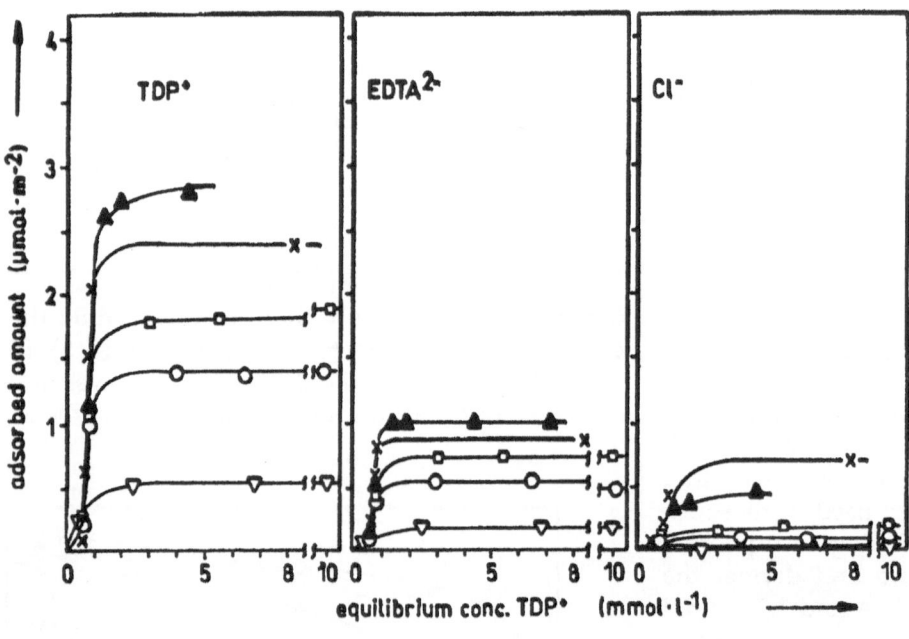

Fig. 6. Coadsorption of tetradecylpyridinium cations (TDP$^+$), EDTA^{2-} and chloride$^-$ on porous silicas with different mean pore diameters.
×—× Fractosil 500
□—□ KG 60
▽—▽ KG 20
▲—▲ KG 100
○—○ KG 40

Table 2. Composition of coadsorbates of TDP-Cl and Na$_2$-EDTA in the state of saturation on different porous silica adsorbents

Silicatype	Mean pore diameter [nm]	Saturation values of adsorption [µval g^{-1}]			Adsorbed EDTA^{2-} and Cl$^-$ [µval m^{-2}]	Ratio of equivalents: EDTA^{2-}/Cl$^-$	Ratio of equivalents: Σ anions/ TDP$^+$
		TDP$^+$	EDTA^{2-}	Cl$^-$			
Fractosil 500	60	2.36	1.64	0.69	2.33	2.40	0.98
KG 100	10	2.77	2.12	0.44	2.56	4.8	0.92
KG 60	6	1.84	1.50	0.19	1.69	7.9	0.92
KG 40	4	1.40	1.10	0.10	1.20	11.0	0.86
KG 20	2	0.56	0.40	0.05	0.45	8.0	0.80
A 200	nonporous	2.68	2.06	0.42	2.48	4.9	0.93

sorption of the surfactant cations (Fig. 6a), $EDTA^{-2}$ (Fig. 6b), and Cl^- (Fig. 6c) become smaller with a reduction of the mean pore size of the silica (adsorption values calculated on the basis of the specific surface areas of the silicas).

The influences of the pore structure upon the coadsorption processes can be clearly demonstrated by the composition of the adsorbates in the state of the saturation values of adsorption (Table 2). The sum of the coadsorbed counterions (in equivalents) declines to the same extent as the adsorbed amounts of the surfactant on Fractosil 500, KG 100, and KG 60. The ratio between the TDP^+ cations and equivalent amounts of the counterions is found to be between 0.96 and 0.92 for the pore sizes 60—6 nm. On KG 40 and KG 20 this ratio declines to 0.86 for KG 40 and 0.80 for KG 20, respectively. In the narrow pores obviously less counterions are adsorbed to the surfactant adsorbate.

The ratio between the coadsorbed $EDTA^{2-}$ and Cl^- counterions shows a remarkable shift, dependent on the pore size of the silica: the corresponding value increases from 2.5 on Fractosil 500 to a maximum value of 11 on KG 40. $EDTA^{2-}$ obviously becomes the preferred counterion with decreasing pore size of the adsorbent, until steric hindrance opposes this effect (on KG 20).

Saturation values of adsorption
of tetradecylpyridinium cations on nonporous silica
in the presence of different anions

In a further series of experiments the adsorption of TDP^+ was determined on a nonporous silica in the presence of different anions. From the corresponding saturation values of adsorption the specific effect of the counterions on the coadsorption process can be evaluated without the interference of any pore effects of the adsorbent (Table 3).

The saturation values of TDP^+ increase in the sequence $Cl^- < NO_3^- <$ salicylate$^-$, if pure surfactant salts of these anions are applied in the adsorption experiments. For comparison, the TDP^+ adsorption was determined for TDP-Cl in the presence of 10 mmol l^{-1} Na_2-EDTA, confirming the promoting action of organic counterions on the coadsorption on silica. In the coadsorbates the ratio between TDP^+ and the anions is significantly higher for organic cations (0.96—0.91) than for Cl^- (0.87).

Table 3. Saturation values of adsorption of tetradecylpyridinium cations and different counterions on nonporous silica Aerosil 0×50

TDP^+ salt	Adsorbed amount [μval m^{-2}]		Adsorbed anions/ TDP	Mean surface area [Å^2/ TDP-ion]
	TDP	Anions		
Cl^-	0.90	0.78	0.87	184
NO_3^-	2.58	2.34 (calc.)	0.91	64
Salicylate$^-$	4.95	4.75	0.96	33
Cl^- + $EDTA^{2-}$ 10 mmol l^{-1}	2.77	$EDTA^{2-}$ 2.12 Cl^- 0.44 ——— Σ 2.56	0.92	60

Coadsorption of hexadecylpyridinium cations and $EDTA^{2-}$ on E. coli

Adsorption isotherms of HDP^+ and $EDTA^{2-}$ were determined from equimolar mixtures of HDP-Cl and EDTA-Na_2 on *E. coli* (Fig. 7). Due to the variation of the adsorption data on this substrate the isotherms are given as hatched areas.

The most important feature is that coadsorption of the cationic surfactant and $EDTA^{2-}$ can be clearly seen on the surfaces of the *E. coli*: Indeed, there is a slow, but significant adsorption of $EDTA^{2-}$ on *E. coli*, even from the pure EDTA-Na_2 solution. In the presence of TDP^+, however, the $EDTA^{2-}$ adsorption appears considerably enhanced.

For the surfactant cations, both from HDP-Cl and HDP-Cl/EDTA-Na_2 mixtures, the typical S-shape of the adsorption isotherms is obtained. In the presence of EDTA the rise of the isotherm to the saturation level occurs at a lower equilibrium concentration, thus the adsorption value remains at the same level. From electrophoretic measurements ζ-potential values were evaluated for the *E. coli* bacteria. In the range of the first plateau of the HDP^+ adsorption isotherms there still exists a negative charge on the microorganisms. In the range of the saturation values of adsorption a positive charge is measured on *E. coli*, indicating charge reversal by a surplus of adsorbed surfactant cations. Without overemphasizing the absolute values of the ζ-potential the lower positive value of the coadsorbate is in good agreement with the

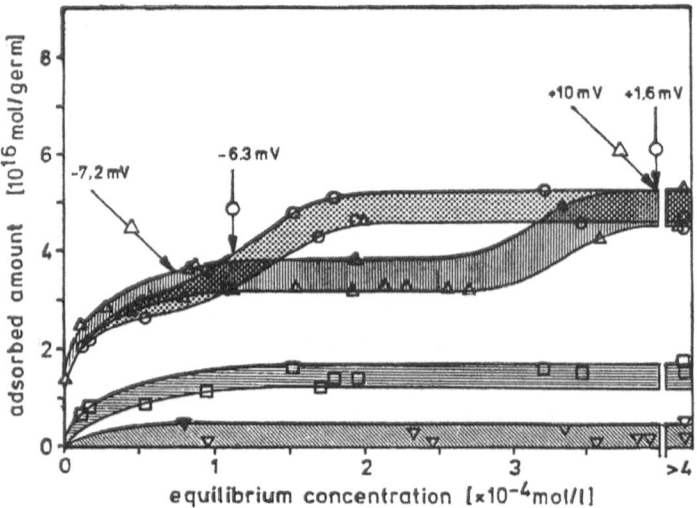

Fig. 7. Adsorption and coadsorption of EDTA-Na_2 and cetylpyridinium chloride (CP-a) on *E. coli*. Coadsorption from equimolar initial concentration.

△—△ CP$^+$
▽—▽ EDTA
○—○ CP$^+$ (EDTA^2)
□—□ EDTA (+ CP$^+$)

Arrows: vertical = ζ-pot. of CP$^+$/EDTA coadsorbates;
inclined = ζ-pot. of CP$^+$ adsorbates.
E. coli: 4.8 × 10^8 CPCl.
ζ-pot. *E. coli* suspension −32 mV; in the presence of EDTA −30 mV

higher ratio between counterion and HDP$^+$ in the presence of EDTA.

Discussion

Alkylpyridinium cations mediate the adsorption of drug anions onto negatively charged surfaces, both on silicas and on *E. coli*. This mediating effect is due to the formation of surface aggregates of the surfactant. They are considered to be similar in structure to micelles in the bulk phase and are therefore, called hemimicelles or admicelles [8, 12—14]. In these adsorbates a minor part of the surfactant ions are bound by electrostatic interaction to the solid surface (silica), compensating the negative surface charge there. (The first plateau in the adsorption isotherm represents the "saturation value" of this reaction). At higher concentrations, but still below the critical micelle concentration (CMC) in the bulk phase, the major part of the surfactant ions are bound to the primarily adsorbed species (the "anchors") by hydrophobic interactions. Most of these surfactant cations are accompanied by counterions that reduce the electrostatic repulsion of the surfactant head groups by charge compensation. As a consequence, most of the closely attached counterions are located within the plane of shear at the surface of the adsorbates. A positive, but weak ζ-potential, is therefore measured, according to the concept of the electrical double layer which is form-

ed between the silica surface and the bulk phase of the liquid [15].

On colloidal, nonporous silica about 90—97% of the surfactant cations are accompanied by their counterions in the state of saturation (Table 3). In silicas with small pores (mean pore diameter 2 nm) this ratio shifts to lower values of ≈80%. This reduction indicates that the structure of the adsorbates in narrow pores may be influenced by wall-to-wall interactions and pore exclusion effects.

The significant differences between univalent counterions in their influence upon saturated adsorption layers of the cationic surfactants is not sufficiently described in the adsorption models mentioned above. In the presence of Cl$^-$ ions one TDP$^+$ occupies a mean surface area of 184 Å^2 in saturated adsorption layers (Table 3), indicating a loose structure of the adsorbates. With NO_3^- as counterion this mean surface area is reduced to 64 Å^2, and in the presence of salicylate, to 33 Å^2. Closely packed adsorption layers must therefore be assumed at the silica surface in the presence of NO_3^- and salicylate$^-$. The differences between Cl$^-$ and NO_3^- were attributed to their different action on water structure [16]. In the case of the organic counterions their pK_a values, their bulkiness and conformation, as well as their hydrophilic-hydrophobic properties must be additionally considered. Nondissociated molecules of the weak acids (salicylic acid) may be additionally coadsorbed by a surface solubilization effect [17, 18]. In narrow pores of silica the coadsorption of organic counterions — EDTA^{2-} — out of a mixture with Cl$^-$ is obviously preferred (Table 2). This may be due to a wall-to-wall interaction of the divalent cation with the surfactant ions, but additional adsorption forces of the organic ion may also contribute in these mesopores to their preferred adsorption [19].

The kinetic experiments with gentisic acid demonstrate that the coadsorbed counterions are readily available, in particular, if exchangable ions are present in the desorption medium. The faster release in 0.1 N HCl compared with 0.1 N NaCl may be due to different influence of Na$^+$ and protons on both the ion exchange and the aggregation of the surfactant. (From the pH of the desorption medium of ≈ 4—5 in NaCl and ≈ 1 in HCl, respectively, one could expect a slower release of the weak acid. Only 6% of the gentisic acid is dissociated at this pH level (pK$_a$ = 2.9)).

The rapid release of the coadsorbed drug from the cationic surfactant coadsorbates distinguishes this adsorption principle from the coadsorption of drugs with nonionic surfactants [18]. From these adsorbates — coadsorbates — the drugs are slowly released, due to the solubilisate structure of the corresponding adsorbates.

Finally, the increase in the EDTA^{2-} adsorption onto *E. coli*, mediated by the cationic surfactant, confirms our hypothesis that coadsorption is an important preliminary step in the synergistic antimicrobial action of EDTA mixtures with cationic surfactants.

References

1. Gennaro AR (1985) Remington's Pharmaceutical Sciences, Mack Publ Comp, Easton Pa pp 301 and pp 1492
2. Harmia T, Speiser P, Kreuter J (1986) Int J Pharm 33:45
3. Rupprecht H (1988) in Enzyclopedia of Pharmaceutical Technology Vol 1, p 73
4. Rupprecht H, Unger K, Biersack MJ (1977) J Colloid & Polymer Sci 255:276
5. Pharmacopoea Europaea 2nd Ed. Part 1, Chapter V.5.4. Maisonneuve S.A., Sainte-Ruffine-France, 1986
6. Vitzthum J, Rupprecht H (1989) submitted to Acta Pharm Technol
7. Rupprecht H (1976) Acta Pharm Technol 22:37
8. Gao Y, Du J, Gu T (1987) J Chem Soc, Faraday Trans I, 83:2671
9. Higuchi T (1963) J Pharm Sci 52:1145
10. Denyer SP, Hugo WB, Harding VD (1985) Int J Pharm 25:245
11. Hugo WB (1967) J appl Bact 30:17
12. Harwell JH, Schechter R, Wade WH (1985) Collect Colloqu Sem Inst France Petrol 42:371
13. Harwell JH, Roberts BL, Scamehorn JF (1988) Colloid & Surfaces 31:1
14. Gu T, Gao Y, He L (1988) J Chem Soc, Faraday Trans I 84:4471
15. Myers D (1988) Surfactant Sciences and Technology, VCH, New York pp 286
16. Rupprecht H (1981) in Topics in Pharmaceutical Sciences (Breimer DD, Speiser P ed) Elsevier, Amsterdam p 443
17. Harwell JH, O'Rear EA (1988) in Surfactant Science Series Vol 33 (Scamehorn JF ed), M Dekker, New York p 162
18. Rupprecht H, Daniels R (1984) Progr Colloid & Polymer Sci 69:159
19. Schneider R (1988) Dissertation, Regensburg

Authors' address:

Prof. Dr. H. Rupprecht
L. S. Pharm. Technology
Universität Regensburg
Universitätsstrasse 31
8400 Regensburg, FRG

Progress in Colloid & Polymer Science   Progr Colloid Polym Sci 83:118—126 (1990)

Adsorption of paraquat ions on clay minerals.
Electrophoresis of clay particles

A. de Keizer

Wageningen Agricultural University, Dep. of Physical and Colloid Chemistry, Wageningen, The Netherlands

Abstract: The adsorption of paraquat (1,1'-dimethyl-4,4'-dipyridinium chloride) on montmorillonite, illite, and kaolinite was investigated. The organic cation was strongly adsorbed up to the cation exchange capacity (CEC) of montmorillonite and illite. On kaolinite, adsorption took place only up to 57% of the CEC. From the low concentration parts of the adsorption isotherms it was concluded that paraquat has a preference for the interlayer spacings over the exterior surface of the clay particles. Paraquat adsorption is reversible with respect to variation of the indifferent electrolyte concentration. The effect of electrolyte on the ζ-potential of the different clay particles is small. The effect of pH on the ζ-potentials is small for illite and montmorillonite, but large for kaolinite, corresponding with a high fraction of variable charge. The overall iso-electrical point of kaolinite is around pH = 2.5. It appears that paraquat adsorption has a minor effect on the ζ-potential until approximately 80% coverage is obtained. This typical effect is explained by the difference in affinity of paraquat for the interlayer spacings and the exterior surface. No charge reversal takes place. The behavior of kaolinite differs in some respects from montmorillonite and illite due to the large fraction of variable charge and its high charge density.

Key words: Adsorption; electrophoresis; paraquat; montmorillonite; illite; kaolinite

Introduction

The interaction of the herbicide paraquat (1,1'-dimethyl-4,4' dipyridinium chloride) with soils and soil-constituting minerals has been extensively studied during the last decade [1—5]. An important property of paraquat is the strong affinity for soils. Therefore, on contact, the herbicidal activity is essentially lost. It is likely that paraquat is practically immobile in soils, leading to some accumulation in the top layer. A detailed review on the properties of the bipyridinium herbicides has been given by Summers [6]. Paraquat, also known under the name of methyl viologen is a.o. frequently used as an efficient electron transfer-reagent in electrochemistry and bioelectrochemistry [7]. It is easily reduced to a stable, highly colored $PQ^{+\cdot}$ radical. The herbicidal activity is, in effect, also a consequence of the formation of $PQ^{+\cdot}$ radicals.

The adsorption properties of paraquat ions vary largely with the nature of the adsorbent. Paraquat ions adsorb strongly on clay minerals, somewhat less on activated carbon [1], silver iodide [8], and humic acids [6], whereas the adsorption on inorganic oxides [9, 10] is only weak.

The high affinity of paraquat for clay minerals is related to its highly polar nature. The affinity is particularly high for an expanding lattice clay, like montmorillonite. It has been shown that in the latter case paraquat adsorbs flat in the interlayer spacings [4, 5]. For a non-expanding clay like kaolinite the affinity is much lower. Coulombic forces play an important role in the adsorption of paraquat on clay minerals. However, additional forces seem to be involved. It was suggested that Van der Waals forces [2, 6] and charge-transfer complexes [11, 12] also contribute to the binding in the interlayer spacings. The adsorption is usually limited to amounts corresponding to the cation exchange capacity (CEC). Therefore, charge reversal is not likely.

Although the affinity of paraquat to silver iodide is much lower, charge reversal is readily obtained [8]. It follows that here specific interactions independent of available surface charges are present. A completely different behavior has been observed for the adsorption on inorganic oxides. For rutile (TiO_2) the affinity is much lower than on AgI, and paraquat ions are easily exchanged by simple inorganic cations [10]. It was concluded that the adsorption of paraquat is predominantly in the diffuse layer; no charge reversal has been observed.

An intriguing aspect in the adsorption of paraquat on soils and clay minerals is the phenomenon of a so-called strong adsorption capacity (SAC). This means that, to some degree of coverage, no paraquat could be analytically detected in solution [2]. This SAC was attributed to binding in the interlayer of clay minerals.

It is our aim to study the difference in the preference of the paraquat ion for adsorption in the interlayer spacings and on the exterior surface of some clay mineral particles. This can lead to more insight into the nature of the assumed strong adsorption capacity, which is important both from a practical, as well as a theoretical point of view. Besides direct adsorption measurements, also determination of the electrophoretic mobilities of clay mineral particles gives information to solve our problem. Electrophoresis basically gives information on the exterior surface of clay particles or aggregates. As far as we are aware, this method has not yet been applied to study the behavior of paraquat adsorption on clay particles until now.

In this study, we investigated the adsorption of paraquat on montmorillonite, illite, and kaolinite. Also, the electrophoretic mobilities of colloidal solutions of these clays, either or not in the presence of paraquat were determined. The effect of electrolyte and pH will be given in some detail.

Materials and methods

General

All chemicals used in this study were of pro-analysi quality, and used without further purification. Paraquat ($C_{12}H_{14}Cl_2N_2 \cdot 3 H_2O$) was obtained from Fluka. Water was pre-purified by reverse osmosis and subsequently

passed through a Millipore Super-Q system. All glassware used was made of borosilicate glass. For low concentrations of paraquat, polyethylene bottles were used.

All experiments were performed at 20°C.

Clay minerals

In our study, we used a Wyoming (Clay Spur) Montmorillonite [13]. A 50-gr sample was wetted carefully for 2 days. A suspension was prepared of the expanded clay in 2 l of water. A fraction <2 μm was obtained by sedimentation. The clay is saturated with sodium ions by shaking overnight in 1 M NaCl. After sedimentation the saturation is twice repeated. The suspension is then washed with demineralized water until clear supernatant is obtained after centrifugation for 10 min at 2000 rpm. The clay is then dialyzed against demineralized water until an electrolyte concentration of 1.10^{-6} M NaCl. A clay concentration of 3 g per litre was ultimately obtained. The cation exchange capacity as determined by the silver thiourea method [14] amounted to 1000 μeq/g. The surface area, as determined by methylene blue adsorption ($a_0 = 102$ Å^2), amounted to 850 m^2/g.

Illite was a natural sample obtained from Buresse (Belgium). After sieving the sample was wetted and fractionated (<2 μm) as described above. The clay was saturated with 1 M KCl and stirred overnight in order to close the wedges. Afterwards, the clay was saturated with 1 M NaCl/0.01 M KCl. The sample was finally dialyzed and the KCl concentration increased to 10^{-5} M. The clay concentration was 3.9 g/litre. By x-ray diffraction a c-spacing of 9.9391 Å, was found. Small peaks at 7.1623 Å and 3.5687 Å indicate the presence of a small amount of kaolinite (ca. 4%). Also a few percent of quartz may be present. X-ray fluorescence gave the following fraction for the different constituents: SiO_2 53.22%, Al_2O_3 26.53%, Fe_2O_3 1.35%, MgO 3.97% and K_2O 8.08%. The cation exchange capacity was 82 μeq/g. The specific surface area of the dried sol sample was 28 m^2/g.

A sodium-saturated kaolinite was obtained from Sigma. The powder is directly used in the adsorption measurements. A stable sol was obtained by the procedure described for montmorillonite except that a final electrolyte concentration of 7.10^{-5} M NaCl and a sol concentration of 13.6 gram per litre was obtained. By x-ray diffraction the clay appeared to be nearly 100% pure, with a high crystallinity. The specific surface area (BET) as determined by N_2-adsorption was 13.0 m^2/g; the CEC amounted to 57 μeq/g.

Determination of adsorption isotherms

Adsorption isotherms were determined by depletion from solution. Bottles with clay particles dispersed in 50 ml of paraquat solution were shaken overnight. The sol/suspension was precipitated by centrifugation. The equilibrium concentration was determined by the method of Calderbank and Yuen [15]. According to this procedure, paraquat is reduced in an alkaline sodium dithionite reagent to its blue-violet radical. The absorption was measured at 397 nm and the adsorption calculated after correction for the blank.

Paraquat concentrations lower than 1 μmol/litre were measured by concentrating the solution according to Tucker et al. [16]. In this method Na-cation exchange resin (Dowex AG50W-X4) was brought into a column after which a solution of 250 ml with low concentration of paraquat was percolated. Paraquat was displaced from the column with 25 ml saturated NH_4Cl. The paraquat concentration was ultimately determined according to the method described above.

Determination of electrophoretic mobilities

Electrophoretic mobilities of the clay particles were determined with a Malvern Zetasizer II micro-electrophoresis apparatus. For our measurements the cell was equipped with a quartz capillary and the mobility was determined at 1/7 and 6/7 of the diameter of the cell. The clay was diluted in polyethylene bottles to 0.045, 0.052, and 0.014% in weight for kaolinite, illite, and montmorillonite, respectively. After addition of NaCl, KCl and/or paraquat, the bottles were shaken for 2 h, after which the electrophoretic mobility was measured. At higher concentrations of paraquat, flocculation occurred with the montmorillonite and the illite samples. The flocs were redispersed by placing the bottles in a ultrasonic bath for 10 min. We have not observed problems with settling on the capillary wall, as observed earlier [17], probably because in the present case the measuring time was shorter and the diameter of the capillary was larger.

Results and discussion

Surface charge of clay minerals

The interpretation of the adsorption properties and electrophoretic mobilities of clay minerals is complicated because of the nature of the surface charge of clays. Different types of surface charge can be distinguished. First, isomorphic substitution of Si- and/or Al atoms in the lattice is the origin of the permanent charge on the faces. With respect to the three-layer clays, we note that for montmorillonite adsorption is mainly in the octahedral Al-sheet, while for illite substitution in the tetrahedral Si-sheet is dominant. This is reflected in the clay properties: montmorillonite is a swelling clay, illite is a non-swelling clay showing potassium ion-fixation. Kaolinite, a two-layer mineral, is also a non-swell-

ing clay. Montmorillonite and illite have a relatively large permanent charge, the permanent charge on kaolinite is small. This is reflected in the values of the cation exchange capacity (CEC). The second origin of the surface charge is due to dissociation of hydroxyls at the edges. The contribution of the edges in the total surface charge is both dependent on the fraction of edge surface as well as the pH of the solution and is very different for our minerals. For montmorillonite the variable charge is practically negligible, whereas for kaolinite the edges constitute a large fraction of the total surface. Illite has an intermediate position. Williams et al. [17] propose a primitive model in which they assume that for kaolinite the variable charge is a linear combination of the charges on quartz and alumina. In this picture for a three-layer clay, quartz and alumina contribute in a ratio of 2:1, probably leading to a lower point of zero charge (pzc). It is generally assumed [17] that the edges have a point of zero charge between 5 and 8. This means that the contribution of the edge charge will be relatively small for the pH-values chosen in our experiments.

Both the CEC-values, as well as the specific surface areas, differ largely for our clay minerals as is shown in Table 1. The consequence is that the surface charge density for kaolinite is much higher than for montmorillonite. This aspect proves to be important in the interpretation of the adsorption and electrophoretic results.

Table 1. Properties of our clay minerals

	Mont-mor-ilonite	Illite	Kaolinite
Swelling	+	—	—
Specific Area (m²/g)	850	28	13
CEC (μeq/g)	1000	82	57
Surface charge (C/m²)	0.11	0.28	0.42

Adsorption isotherms

The adsorption of paraquat on Na^+-montmorillonite at two electrolyte concentrations is given in Fig. 1. Typical high-affinity isotherms are obtained, both at 10^{-2} M, as well as at 10^{-3} M NaCl. At these concentration levels the effect of

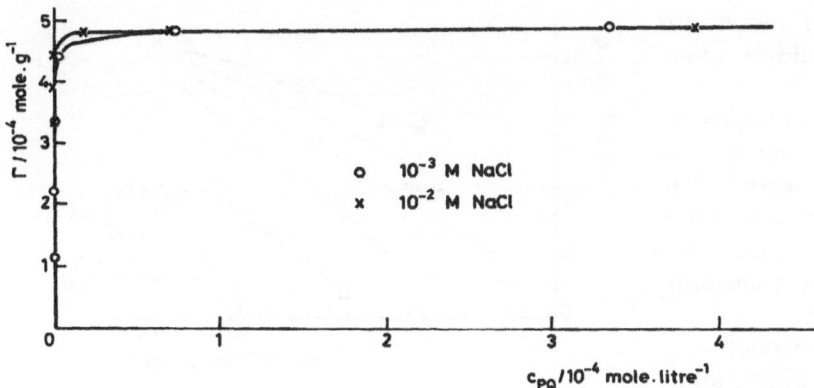

Fig. 1. Adsorption isotherms of paraquat on montmorillonite at 0.001 and 0.01 M NaCl

electrolyte on adsorption is almost negligible. This is due to the fact that adsorption takes place in the interlayer. A well-defined plateau value of 985 µeq/g is obtained. If we assume that both sides of a paraquat molecule occupy an area of 1.0 nm^2 [4] and, realizing that the specific surface area of (swollen) montmorillonite is 850 m^2/g, it can be calculated that about 70% of the internal surface area of montmorillonite is occupied, in agreement with Hayes et al. [4]. The plateau value corresponds with 99% of the CEC. It follows that the surface charge is practically fully compensated by the organic cations and no charge reversal is expected. As is shown later, this is verified by electrophoresis. It is, therefore, unlikely that forces independent of the paraquat charges, like Van der Waals forces, are significant.

The adsorption isotherms in Fig. 1 show the presence of a strong adsorption capacity for paraquat on montmorillonite. However, the SAC essentially depends on the analytical detection limit. By exchange of paraquat by ^{14}C-labelled paraquat it was shown [18] that the adsorption is reversible and probably is essentially an equilibrium property. This means that finite, but probably very low equilibrium concentrations must be present in solution. Experimental evidence will be given below.

Although the observed high affinity is probably the consequence of adsorption in the interlayer, we have no information on the difference between adsorption in this interlayer and that on an isolated surface of a completely expanded clay. To that order, we performed adsorption measurements at extremely low concentrations on the Wyoming montmorillonite sol. The results are given in Fig. 2. A very unusual behavior is observed: at some level of adsorption the equilibrium concentration strongly

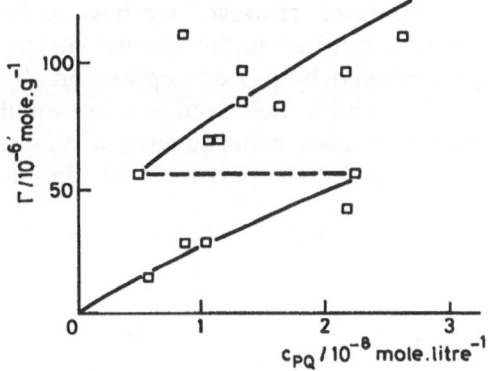

Fig. 2. Adsorption isotherm of paraquat on montmorillonite at low paraquat concentrations; c_{NaCl} = 0.001 M

reduces, after which it increases again. This phenomenon has been repeated in many independent experiments. Our explanation is as follows. At very low concentration of completely expanded montmorillonite particles, adsorption takes place exclusively at the exterior faces, because interlayers are absent. However, after some degree of occupation has been reached, partially covered plates start to flocculate so that paraquat ions can now also adsorb in the interlayer between the clay platelets. Because the affinity for the interlayer is much higher than for the isolated surface the equilibrium concentration reduces almost to zero. The idea of a strong difference in affinity between interlayer and exterior surface follows also from the electrophoretic mobility measurements discussed below. For soil systems montmorillonite is not in the expanded state and all paraquat will be present in the interlayer. Thus, below 50 µmol/gram, i.e., 10%

of the plateau value, equilibrium concentrations are below $0.5-1 \cdot 10^{-8}$ mole/l, which could be alternatively defined as a SAC-value.

In Fig. 3 the adsorption isotherms of paraquat on *illite* are given at pH values of about 4 and 6. Also for illite, a relatively high affinity is observed. It is obvious that decreasing the pH from ca. 6 to 4 has only a minor effect on the adsorption. Apparently, the variable charge of illite does not contribute significantly to the adsorption process.

According to Fig. 3, a maximum adsorption Γ_{max} can be estimated to be 41 μmol/g or 82 μeq/g, identical (within experimental error) to the CEC value. The coverage is much higher than for montmorillonite and amounted to 88% if adsorption only takes place on the faces. However, we have to be aware that for illite (and kaolinite) specific surface areas were determined by gas adsorption on the dried sol particles, which could lead to a too small specific surface area and, consequently, a higher degree of coverage in comparison with montmorillonite.

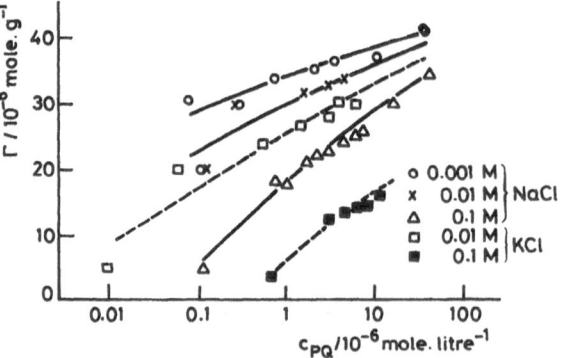

Fig. 4. Adsorption isotherms of paraquat on illite at different NaCl and KCl concentrations; pH = 5.5—6.5

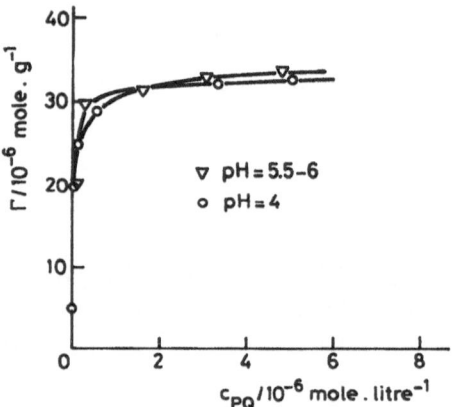

Fig. 3. The effect of pH on the adsorption of paraquat on illite; $c_{NaCl} = 0.001$ M

In Fig. 4 the adsorption isotherms of paraquat on illite are given at different NaCl and KCl concentrations. A significant decrease in adsorption is obtained with increasing electrolyte concentration. Potassium ions are able to displace paraquat ions more effectively than sodium ions, due to a stronger specific interaction of the potassium ions with the illite surface. The slope of the curves increases with increasing electrolyte concentration. According to

the Frumkin-Fowler-Guggenheim model, the slope (at $\theta = 0.5$) corresponds with the lateral interaction between the adsorbate molecules [19]. Due to screening of the adsorbing charges the lateral interaction decreases with increasing electrolyte concentration. There are no indications that illite possesses a strong adsorption capacity (SAC).

An important issue with respect to adsorption of paraquat is whether adsorption is reversible or not. As it is difficult to desorb paraquat by dilution of the equilibrium solution due to the relatively high affinity, an alternative procedure was followed. We adsorbed paraquat at 10^{-3} M, according to the usual procedure, and added additional NaCl in order to increase the electrolyte concentration to 0.1 M NaCl. The adsorption isotherm obtained in this way coincided perfectly with an isotherm measured directly at 0.1 M NaCl. Thus, no irreversibility could be detected with respect to the electrolyte effect.

In Fig. 5 the adsorption isotherm is given for *kaolinite* at 0.001 M NaCl and pH = 5. The affinity of the adsorption is not very different from that of illite. The dependence on indifferent electrolyte (not shown here) is also analogous. However, the most remarkable difference is that only 57% of the CEC is compensated. This may be due to the dimensions of the paraquat molecule with respect to the positions of the surface charges, as follows from the surface charge density. Apparently, paraquat ions are not able to fully compensate the surface charge up to the CEC-value and, accordingly, the surface charge remains negative at full coverage. This conclusion is corroborated by electrophoresis. It seems that, due to the high surface charge density, para-

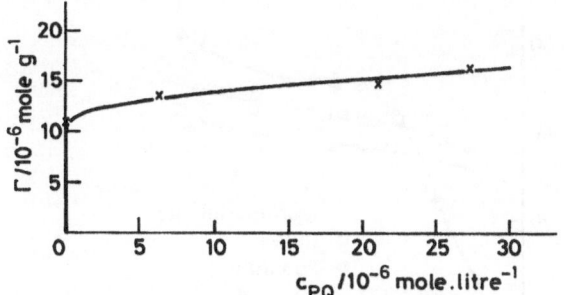

Fig. 5. Adsorption of paraquat on kaolinite; $c_{NaCl} = 0.001M$; pH = 5

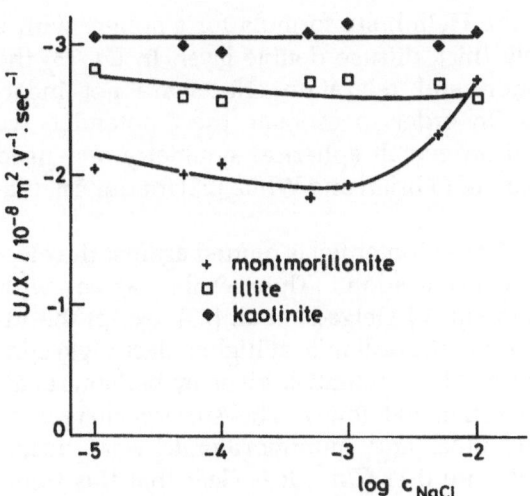

Fig. 6. The effect of electrolyte concentration on the electrophoretic mobility (U/X) of clay particles (c in mole/litre); pH = 6

quat adsorbs flat on the surface. It follows from the specific surface area and the assumed molecular area that in the plateau 75% of the surface (edges and faces) is occupied. A second reason for the smaller coverage could also be the large contribution of variable edge charges to the CEC of kaolinite. If paraquat is only weakly adsorbed at the edges, as is the case for inorganic oxides, it is reasonable to assume that in the plateau only the faces are fully occupied. Moreover, the edges are probably near their zero point of charge.

Electrophoresis

In Fig. 6 the effect of indifferent electrolyte (NaCl) on the electrophoretic mobility of montmorillonite, illite, and kaolinite is given. It appears that electrolyte has only a minor effect on the mobility. For illite and kaolinite, which are non-expanding clays, a small decrease with increasing electrolyte concentration is obtained; for montmorillonite, which is a expanding clay, a minimum seems to be present.

Interpretations of electrophoretic mobilities and calculations of the ζ-potential are thwarted by a number of complications. First, the surface charge is different for the edges and the faces so that the measured mobility involves an unknown averaging over a heterogeneous surface. A second problem is the symmetry of the particles. In the ideal case, a flat plate would be the most likely shape. In the undisturbed solution the platelets are randomly oriented. It is not known to what extent this remains so during migration, but we have found no influence of the value of the electrical field strength on the electrophoretic mobility of the clay particles. This is a strong indication that a random distribution is maintained during the electrophoretic pro-

cess. On aggregation a more spherical shape is obtained.

A primitive calculation of ζ-potential from the electrophoretic mobility of *randomly oriented* charged plates can be carried out according to the procedure described earlier for cylinders [20]. The average electrophoretic mobility (velocity of the particles per unit of electrical field strength) of a randomly orientated plate (U/X) can be obtained from the mobility of a plate parallel to the electrical field $(U/X)_\parallel$ and perpendicular to the field $(U/X)_\perp$. It can be derived [20] that the electrophoretic mobility of a randomly oriented plate is obtained by

$$U/X = \frac{1}{3} [U/X]_\perp + \frac{2}{3} [U/X]_\parallel . \qquad (1)$$

The mobility of a plate parallel to the field follows from Smoluchowsky's equation [21]

$$[U/X]_\perp = \frac{\varepsilon \zeta}{\eta} . \qquad (2)$$

The mobility of a plate perpendicular to the field will be extremely low because of strong hydrodynamic resistance. We assume $(U/X)_\perp = 0$. Hence,

$$U/X = \frac{2}{3} \frac{\varepsilon \zeta}{\eta} . \qquad (3)$$

This is the Helmholtz formula for a sphere with a relatively thick diffuse double layer. In Eq. (3) the retardation and relaxation effects are not incorporated. In order to estimate the ζ-potential, we assumed an overall spherical symmetry and used the theory of O'Brien and White [22] for our calculations.

In Fig. 7 the ζ-potential is plotted against the electrolyte concentration. The results agree with measurements of Delgado et al. [23], except the increase for montmorillonite at higher electrolyte concentrations. The ζ-potentials given by Williams et al. [17] are somewhat lower. The surface charges of kaolinite, illite, and montmorillonite were, resp., 0.42, 0.28, and 0.11 C/m^2. It is clear that this trend is reflected in the ζ-potential. The surface potentials as calculated from Gouy-Chapman theory [24] based on the surface charge are much higher (—100 to —200 mV). This means that the surface of shear must be at some distance from the surface and/or compensating charges must be present between the surface and the shear plane. Moreover, simple Gouy-Chapman theory predicts a substantial decrease of the ζ-potential with increasing electrolyte concentration. Zukovski et al. [25] introduced a strong co-ion adsorption in his dynamic Stern layer model to interpret the electrolyte dependence of the mobility of polystyrene latices. Although this contribution cannot be straightforwardly excluded, it is not likely that for clay minerals co-ion adsorption plays such an important role. Therefore, we explain constancy of the ζ-potential by assuming that the surface of shear approaches the clay-surface with increasing electrolyte concentration.

In Fig. 8 the ζ-potential is given as a function of pH. For illite and montmorillonite the pH has only

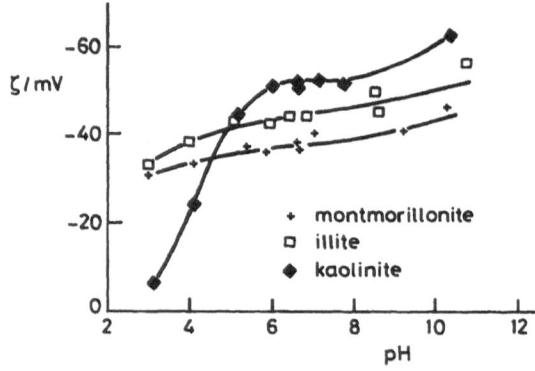

Fig. 8. The effect of pH on the ζ-potential of clay particles; $c_{NaCl} = 0.001$ M

a minor effect. Kaolinite contains a much higher amount of variable charge, which is reflected in the strong decrease in ζ-potential below pH = 6. The same trend has been observed by Williams et al. [17] for a Cornwall kaolinite. By extrapolation, it can be concluded that the isoelectric point of kaolinite (sum of the isomorphic substitution charge and the variable charge is zero) is around pH = 2.5.

Paraquat has a high affinity for clay minerals, particularly for expanding clays. Usually strong specific adsorption of countercharges, e.g., in the case of cationic surfactants, leads to a sharp decrease in ζ-potential. However, as shown in Fig. 9, the experimental results show a different behavior. For illite and montmorillonite the behavior seems quite similar. Instead of decreasing, the ζ-potential even has a tendency to increase somewhat; the reason

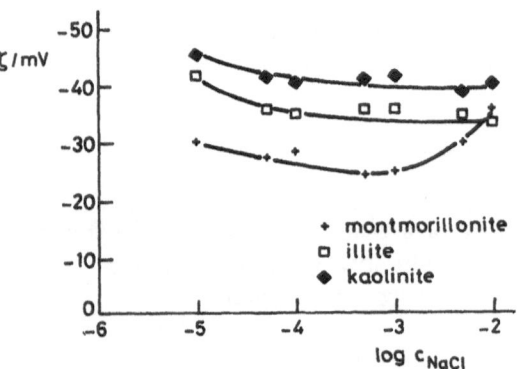

Fig. 7. The effect of electrolyte concentration on the ζ-potential of clay particles (c in mole/l); pH = 6

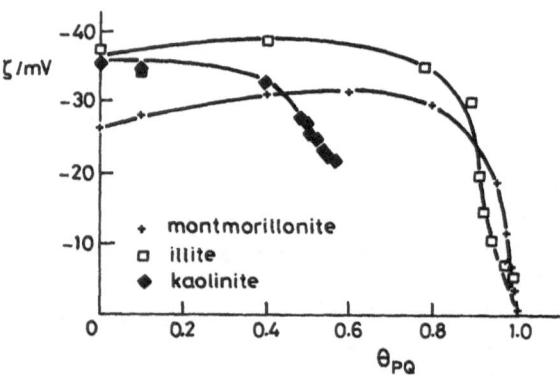

Fig. 9. The effect of paraquat adsorption on the ζ-potential of clay particles; θ_{PQ} is the degree of coverage with respect to the CEC; $c_{NaCl} = 0.001$ M

may be found in a shift of the position of the surface of shear perhaps due to the degree and the nature of aggregation of the particles. Apparently, for the exterior of the particles the electrophoretic charge is not influenced. We conclude that, before the CEC, adsorption must take place in the interior of the particles in the interlayer region. Near the CEC, adsorption also takes place on the outer surface and the electrophoretic mobility and the ζ-potential decrease sharply to zero. However, charge reversal does not take place. Apparently the affinity for the interlayer is much higher than for the exterior surface in agreement with earlier conclusions based on the low concentration isotherms for montmorillonite. For kaolinite the ζ-potential also decreases near the plateau value. However, the behavior for kaolinite differs in some respects. First, the increase of the ζ-potential is absent, but the main difference is that the ζ-potential does not decrease to zero. Apparently, the surface charge is only partially compensated by paraquat ions as discussed earlier. Also, the variable charge could contribute to the non-zero ζ-potentials at the maximum coverage with paraquat of the kaolinite particles if adsorption on the edges is only in the diffuse layer. Finally, it may seem somewhat unexpected that the ζ-potential of kaolinite does not decrease with coverage in the first part of the isotherm as adsorption in the interlayer is only minimal. Perhaps, probably due to some aggregation, paraquat is initially adsorbed inside the aggregate.

Conclusions

From adsorption and electrophoresis measurements it can be concluded that the bond strength of paraquat with respect to an exterior surface is much smaller than to the interlayer. There are indications that adsorption on variable charge sites is very weak and is similar to that on inorganic oxides. In our experiments, always finite, but very low equilibrium concentrations in solution could be detected. From our experiments it was concluded that paraquat adsorption is reversible and that it is an equilibrium process.

It appeared that interpretation of electrophoretic mobilities of clay particles is rather complex and up to now, has not been fully developed. Nevertheless, determination of electrophoresis is a useful method to study the behavior of clay minerals, particularly in the presence of cationic adsorbates.

We want to stress that, from studying the adsorption and electrophoretic properties of paraquat (and other bipyridinium ions) on clays, useful information on the general properties of clay minerals can be derived.

Acknowledgements

The experimental contributions of W. Bussink, P. van Erp, H. J. Gerrits, H. Rienks, and H. G. M. van de Steeg are gratefully acknowledged. Discussions with Dr. J. J. T. I. Boesten are appreciated. Thanks are due to Mr. J. van Doesburg for the characterization of our clay minerals and placing the illite sample at our disposal.

References

1. Weber JB, Perry PW, Upchurch RP (1965) Soil Sci Soc Proc 29:678
2. Knight BAG, Tomlinson TE (1967) J Soil Sci 18:233
3. Hayes MHB, Pick ME, Toms BA (1975) Res Rev 57:1
4. Hayes MHB, Pick ME, Toms BA (1978) J Colloid Interf Sci 65:254
5. Raupach M, Emerson WW, Slade PG (1979) J Colloid Interf Sci 69:398
6. Summers LA (1980) The Bipyridinium Herbicides, Academic Press, London
7. Kleijn JM, Rouwendal E, van Leeuwen HP, Lyklema J (1988) J Photochem Photobiol A: Chem 44:29
8. de Keizer A, Fokkink LGJ (1990) Colloids and Surfaces 51
9. Furlong DN, Sasse WHF (1983) Colloids and Surfaces 7:29
10. de Keizer A (1991) Colloids and Surfaces to be published
11. Haque R, Lilley S, Coshow WR (1970) J Colloid Interf Sci 3:185
12. Burdon J, Hayes MHB, Pick ME (1977) J Environ Sci Health B12:37
13. Weaver CE, Pollard LD (1973) The Chemistry of Clay Minerals (Developments in Sedimentology, Part 15), Chapter 5, Elsevier, New York
14. Chabra R, Pleysier J, Cremers A (1975) Proc Intern Clay Conf 439
15. Calderbank A, Yuen SH (1965) Analyst 90:99
16. Tucker BV, Pack DE, Ospenson JN (1967) J Agr Food Chemistry 15:1005
17. Williams DAJ, Williams KP (1978) J Colloid Interf Sci 65:79
18. Tomlinson TE, Knight BAG, Bastow AW, Heaver AA (1968) Monogr Soc Chem Ind 29:317
19. de Keizer A, Lyklema J (1980) J Colloid Interf Sci 75:171
20. de Keizer A, van der Drift JT, Overbeek JTG (1975) Biophysical Chemistry 3:107

21. Overbeek JTG, Wiersema PH (1967) In: Bier M (ed) Electrophoresis II: Theory, methods and applications. Academic Press, New York
22. O'Brien RW, White LR (1978) Faraday Trans II 74:1607
23. Delgado A, Gonzalez-Caballero F, Bruque JM (1986) J Colloid Interf Sci 113:203
24. Gouy G (1917) Ann phys 7:129, Chapman DL (1913) Phil Mag 25:475
25. Zukovski IV CF, Saville DA (1986) J Colloid Interf Sci 114:32

Authors' address:

Dr. A. de Keizer
Dept. of Physical and Colloid Chemistry
Wageningen Agricultural University
P.O. Box 8038
6700 EK Wageningen, The Netherlands

Progress in Colloid & Polymer Science   Progr Colloid Polym Sci 83:127—135 (1990)

On the adsorption of surfactants on non-sulphide minerals

B. Burg, M. Liphard, B. Schreck and H.-D. Speckmann

Henkel KGaA, Düsseldorf, FRG

Abstract: This work is concerned with basic studies on the flotative separation of three minerals: apatite, hornblende, and magnetite. Two anionic surfactants that differed in the solubility of their calcium salts were chosen as collectors (oleate and alkylsulphosuccinate). The flotation behavior of the minerals was characterized by determing adsorption isotherms and zeta-potentials, and by carrying out microflotation tests. It was shown that both collectors are adsorbed on all mineral surfaces, but at different equilibrium concentrations. As the concentration range of the onset of adsorption on apatite is lower than that on hornblende and magnetite, apatite can be separated from a mineral mixture or a respective ore. Calcium ions are often present in the flotation pulp and influence the adsorption behavior of the collectors. Due to the slight solubility of the calcium-containing minerals, apatite and hornblende, precipitation of calcium oleate occurs, as well as adsorption of oleate when a certain concentration is reached. Higher calcium ion concentrations induce more precipitation and make oleate less effective as a collector. Because of the higher solubility of the calcium alkylsulphosuccinate, precipitation does not occur in the considered concentration range. In contrast to oleate, the adsorption of alkylsulphsuccinate is shifted to lower equilibrium concentrations by calcium ions adsorbed on the mineral surface. This makes alkylsulphosuccinate a more effective collector in water with high calcium ion concentrations. The sequence of adsorption and hence, of flotation is not influenced by calcium ions, i.e., apatite, but not hornblende, can be separated from magnetitde.

Key words: Flotation; adsorption of surfactants; iron oxides; hornblende (silicate); apatite

1. Introduction

One of the techniques used to process non-sulphide iron ores is inverse flotation. If the gangue to be floated consists of salt-type minerals, fatty acid soaps and other anionic surfactants are generally used as collectors. They are thought to adsorb on the mineral surface and hydrophobize it, thus making it accessible to flotation. Different mechanisms have been suggested by various authors; a review is given by Hanna and Somasundaran [1].

There is much evidence that the adsorption of anionic surfactants involves their head groups being attracted to surface cations. For a number of salt-type minerals, the deposition of sparingly soluble metal soaps has also been proposed ("surface precipitation") [2—4]. Many authors have studies different regions in the adsorption isotherms, attributing the effects observed to different subsequent adsorption mechanisms [5—7]. However, calcium ions play an important role in solution, as well as on the mineral surface in most of the mechanisms proposed.

Though much work has been devoted to these mechanisms and their relevance to flotation recovery, the problem of separating different salt-type minerals from each other and from iron oxides has not yet been examined in detail. Thus, this study deals with the selectivity of anionic collectors towards certain minerals, as well as the influence of water hardness on the separation process. The work is primarily concerned with the differences in the adsorption behavior of collectors with respect to minerals and the influence of water hardness on adsorption. Three pure minerals characteristic for iron oxide and gangue (magnetite, apatite, and hornblende), together with two anionic surfactants differing in the solubility of their calcium salts were selected. Direct adsorption studies were carried out, as well as collector concentration-dependent measurements of the zeta-potential. The results were correlated with corresponding microflotation tests of the minerals.

2. Experimental

Materials

A magnetically separated magnetite that occurs naturally was selected as model for the iron oxide. It has a purity of better than 96% Fe_3O_4. For the adsorption measurements, synthetic Fe_3O_4 with a high specific surface area was supplied by Bayer AG (Leverkusen), type Bayferrox 306.

For the phosphatic gangue, pue crystals of a fluoro-apatite from Durango (Mexico) were used. Chemical analysis showed that it had approximately the theoretic composition of $Ca_5[F(PO_4)_3]$. A hornblende with a typical composition: $(Ca, Na, K)_3(Mg, Fe, Al, Ti)_5Si_8O_{22}(OH)_2$, from Kragero (Norway) served as a model for silicates. According to x-ray powder pattern the mineral contains 95% hornblende of the above composition and 5% feldspar. The two minerals apatite and hornblende were supplied by Kranz (Aachen). They were broken into small pieces in a mortar mill (type: RMO, Retsch, Haan). Fractions of 50—100 µm were collected for the microflotation tests. For the adsorption measurements the fractions of smaller particle sizes (<25 µm) were used. The magnetic contained only particles below 100 µm. Fractions of 50—100 µm were separated by wet sieving for the microflotation tests and the fraction below 25 µm was used for adsorption measurements. The specific surface areas of the fractions used for adsorption were determined by the BET method (Sorptomatik 1800, Carlo Erba Strumentazione, Italy). They were of the order of 1.2 m^2/g for magnetite, 0.9 m^2/g to 1.9 m^2/g for apatite, 1.6 m^2/g to 3.2 m^2/g for hornblende, and 9 m^2/g for synthetic magnetite.

Two anionic surfactants were used as collectors. Sodium oleate of analytical grade (Riedel-de-Haen, Seelze) was employed throughout the experiments. The second surfactant was technical grade alkylsulphosuccinate-di-Na-salt (alkyl = tallow). Fresh solutions of surfactants were prepared daily and the pH was adjusted with NaOH.

Adsorption measurements

The amounts of oleate and alkylsulphosuccinate adsorbed on apatite, magnetite, and hornblende were deter-

mined by the depletion method. Hence, 5 g of the mineral was shaken with 50 ml of surfactant solution of known initial concentration at pH 10 in a glass tube for 60 min. Preliminary kinetic studies showed that equilibrium was reached within even shorter times. The suspension was subsequently separated by centrifugation (labcentrifuge UJ II E Heraeus-Christ, Osterode) at 4000 rpm for 20 min.

The concentrations of the surfactants were determined by titration with cetylbenzyldimethylammonium chloride, which foams water-insoluble complexes with the anionic surfactants. Potentiometric titrations were carried out by Titroprocessor 672 (Metrohm, Switzerland) using a surfactant-sensitive electrode [8].

Flotation test

The flotation tests were conducted in a Hallimond tube, using 2 g of mineral and 165 ml of solution during conditioning with surfactants (10 min), and 210 ml during flotation. In special cases, preconditioning with calcium ions (5 min) was carried out to test the influence of hard water on the flotation results.

The total flotation time chosen was 12 min at a gas flow of 240 ml N_2/h. Because of the magnetic properties of the magnetite a special PE-stirrer was used at a stirring rate of 400 rpm. The flotation products were filtered and dried at 150 °C for 1 h before being weighed.

Electrokinetic measurements

The electrophoretic mobilities of the minerals were measured with a Zeta-Meter (Pen Kem Lazer Zee Meter, type 501, USA), with the mobilities being converted into zeta-potentials by use of the Helmholtz-Smoluchowski equation. The concentration of the suspension was 10 mg/l mineral; in the case of magnetite it was 20 mg/l. After addition of the desired surfactant solution, the pH was adjusted to 10. The ionic strength was maintained constant by addition of $1 \cdot 10^{-3}$ mol/l NaCl. Measurements were carried out after a conditioning time of 30 min.

3. Results and discussion

3.1. General considerations

All measurements were carried out at pH 10 in order to avoid complications due to decreasing dissociation of the fatty acid used [9—11] and increasing solubility of apatite and hornblende. But, also at this pH, the minerals provide a certain calcium concentration when dispersed in deionized water. Due to this ion concentration in solution, the interpretation of the results obtained from adsorption measurements by the depletion method is complex. Besides adsorption, precipitation of calcium salts of the surfactants may occur during the experiment. Both effects lead to a decrease in the concentration of the surfactant in the solution, which is normally considered to be due to adsorption only. A sensitive method of distinguishing between both effects is to carry out the experiments at different solid/liquid ratios. If only adsorption occurs, the adsorbed amount per m^2 of solid calculated from the concentration difference in solution should not depend on the solid concentration used. It should only be a function of the equilibrium concentration of surfactant. This does not hold if precipitation occurs. For slightly soluble solids, the total amount of precipitated calcium salts does not depend on the amount of solid (mineral) used, but only on the calcium ion and surfactant anion concentration in the pulp (solubility product). When, however, this effect is related to the surface area involved, the result should be strongly dependent on the solid/liquid ratio used.

Corresponding measurements have been carried out with the apatite/Na-oleate system by Rao et al. [12]. In the present work the hornblende/Na-oleate system (Fig. 1) has been investigated. Both isotherms show similar features. In the dilute concentration range, the observed depletion per square meter of mineral surface does not depend on the solid/liquid ratio used and, hence, can be considered to be due to pure adsorption. At higher concentrations, a steep increase in the isotherms is observed. The onset of these slopes is strongly dependent on the mineral concentration used. This suggests that the calcium salt of the surfactant precipitates. This may occur at the surface [9], but also in solution. In the latter case, the effect would only result in a loss of reagent. The ratio of lost (precipitated) to used (adsorbed) amount should increase with decreasing solid concentration. Indeed, for apatite [12] and hornblende (Fig. 1) it is found that in the precipitation region the onset of the steep increase of the isotherm is shifted to lower equilibrium concentration with decreasing solid/liquid ratio. Hence, the ratio of adsorbed to precipitated oleate simultaneously increases. The exact onset and amount of precipitation and the pure adsorption isotherm might be calculated from the solubility products of the calcium surfactant salts and the mineral. However, as natural minerals are used, a correct solubility product is difficult to determine. Also the solubility products of calcium

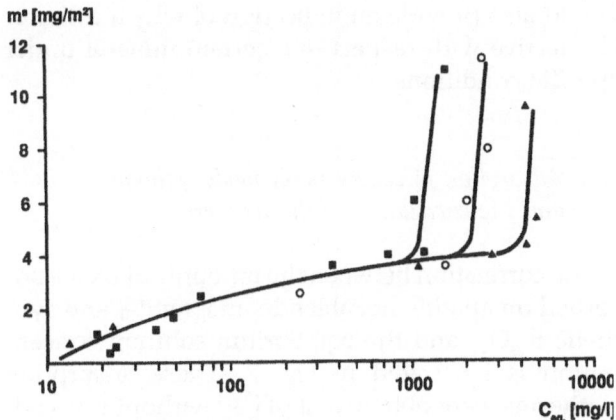

Fig. 1. Adsorption of Na-oleate onto hornblende from aqueous solution, deionized water, pH 10, at different solid/liquid ratios: ■ 10 wt % hornblende; o 20 wt % hornblende; ▲ 30 wt % hornblende

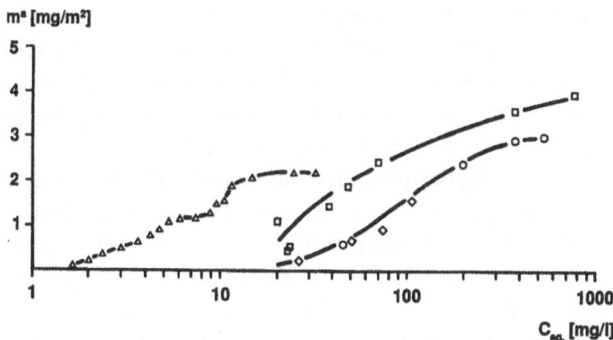

Fig. 2. Adsorption of Na-oleate onto different minerals from aqueous solution, deionized water, pH 10: △ apatite (from [12]); □ hornblende; o magnetite (Fe_3O_4); ◇ synthetic Fe_3O_4

oleate cited in the literature vary widely ($8.5 \cdot 10^{-8}$ $mol^3 \, l^{-3}$ to $2.5 \, 10^{-16}$ $mol^3 \, l^{-3}$ [4, 13—16]). This is probably due to the fact that the precipitation [13—16] of anionic surfactants is not only determined by the solubility product, but also by micelle formation [17, 18].

However, using deionized water, adsorption of the collectors used on both salt-type minerals was shown to occur at lower concentrations than precipitation. The main purpose of this paper was to look at the dilute concentration range. If adsorption starts at clearly distinct concentration regimes for the different minerals, separation should be possible if the dosage is chosen correctly. This would also provide an indication of why a collector is selective with respect to a certain mineral under specific conditions.

3.2. Adsorption of collectors on model minerals and the correlation to the recovery

The correlation between the amounts of oleate adsorbed on apatite, hornblende, magnetite, and synthetic Fe_3O_4, and the equilibrium solution concentration is illustrated in Fig. 2. These adsorption isotherms were obtained at pH 10 without extra addition of calcium ions. First of all, it can be seen that oleate adsorbs on all minerals used in this study, not only on the minerals subjected to flotation. However, the concentration at the onset of oleate adsorption differs, depending on the kind of mineral. Thus, on apatite adsorption occurs in the range from 2 to 10 mg/l, whereas the same adsorption density on hornblende and magnetite requires 10 times higher equilibrium concentrations of the collector. If the experimental error range is taken into account, the isotherms of oleate on magnetite and on synthetic Fe_3O_4 are identical.

Using alkylsulphosuccinate instead of oleate, again the adsorption on apatite starts at lower equilibrium concentrations than on magnetite, hornblende, and the synthetic Fe_3O_4 (Fig. 3). Evidently, the onsets of the adsorption isotherms of magnetite and hornblende are in the same concentration range. Because the collectors are adsorbed on apatite at lower equilibrium concentrations, a separation of apatite from the other minerals should be possible with both of the collectors investigated.

Corresponding experiments were performed in a microflotation cell (Hallimond tube) in order to show the relationship between the adsorbed amount of the collector and the flotation activity of the minerals. In this case, it is necessary to compare data obtained at the same solution (equilibrium) concentrations. While in the case of adsorption studies the equilibrium concentration is known from the experiment, a very low solid/liquid ratio has to be chosen in zeta-potential experiments and in flotation tests. Under these conditions the decrease in concentration due to adsorption can be neglected and the equilibrium concentration is approximately equal to the initial one. Thus, in this work all concentrations are denoted as c_{eq} (equilibrium).

Figure 4 presents the isotherms of oleate and alkylsulphosuccinate on apatite in comparison to the recovery in the flotation tests. The same plots are given for hornblende (Fig. 5) and magnetite (Fig. 6). From these results it can be concluded that the steep increase in the isotherms and in recovery occurs in the same concentration range. Obviously, relatively small adsorption densities of the collector are sufficient to float the minerals.

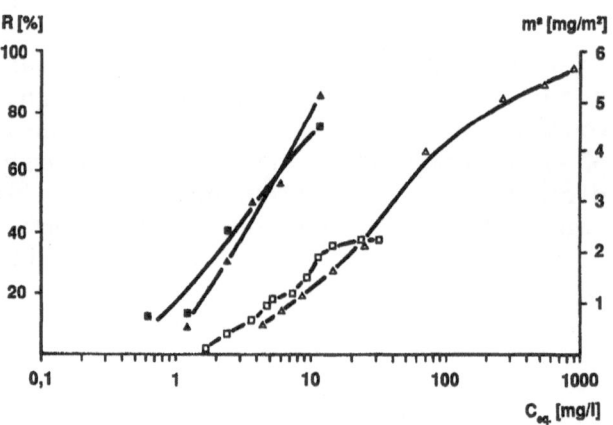

Fig. 4. Correlation between adsorption □, △ and recovery ■, ▲, for the systems apatite/Na-oleate □, ■, and apatite/Na-alkylsulphosuccinate △, ▲, deionized water, pH 10

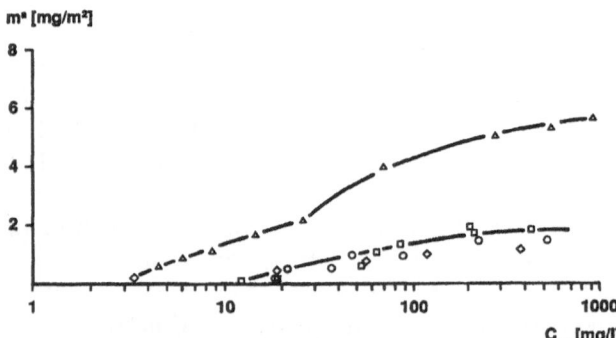

Fig. 3. Adsorption of Na-alkylsulphosuccinate onto different minerals from aqueous solution, deionized water, pH 10: △ apatite; □ hornblende; ○ magnetite (Fe_3O_4); ◇ synthetic Fe_3O_4

Information on the adsorption of ionic collectors can be obtained by direct adsorption experiments, as well as by zeta-potential studies. Figure 7 shows the values of the zeta-potential of apatite as a function of the concentration of oleate and alkylsulpho-

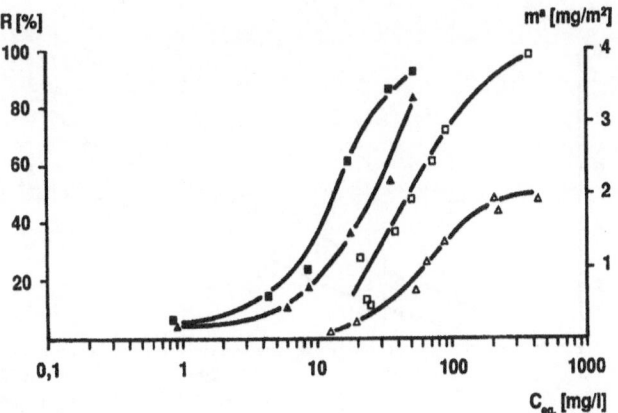

Fig. 5. Correlation between adsorption □, △ and recovery ■, ▲ for the systems hornblende/Na-oleate □, ■, and hornblende/Na-alkylsulphosuccinate △, ▲, deionized water, pH 10

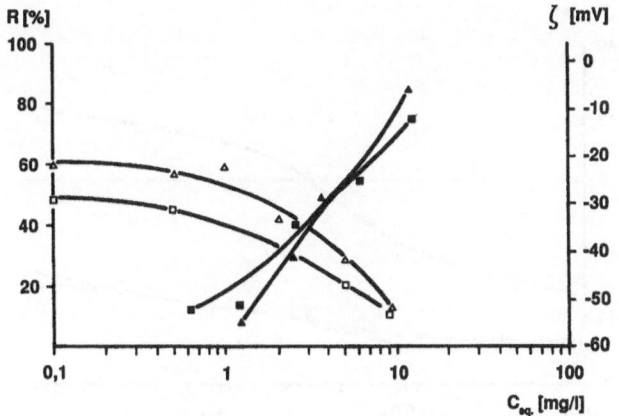

Fig. 7. Correlation between zeta-potential □, △ and recovery ■, ▲ of the systems apatite/Na-oleate □, ■, and apatite/Na-alkylsulphosuccinate △, ▲, deionized water, pH 10

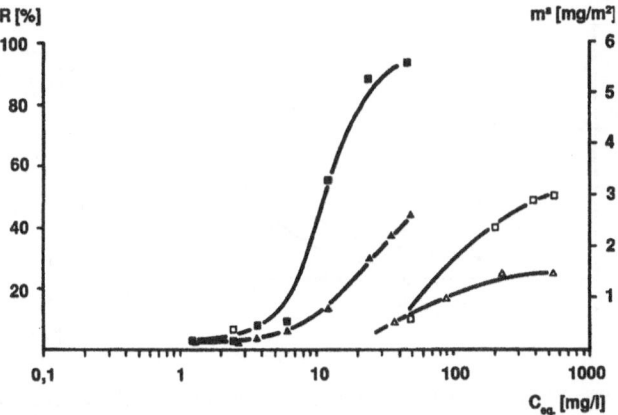

Fig. 6. Correlation between adsorption □, △ and recovery ■, ▲ for the systems magnetide/Na-oleate □, ■, and magnetite/Na-alkylsulphosuccinate △, ▲, deionized water, pH 10

succinate. As the collector amounts rise, the surface charge of the mineral becomes more and more negative. Increasing negative charge with an increase in anionic collector concentration indicates high affinity of oleate and alkylsulphosuccinate with surface calcium-sites [9, 19] and, hence, the successive collector adsorption. In Fig. 7 the recovery of apatite is once again presented along with the zeta-potential values. As in the case of direct adsorption experiments (Fig. 4) the increase in the negative potential is correlated with the increasing recovery of apatite.

Without extra addition of calcium ions, both oleate and alkylsulphosuccinate are effective collectors for apatite in this separation process. But, according to the flotation tests, all three minerals — apatite, hornblende, and even magnetite — can be floated by use of sufficiently high concentrations of the collectors. However, since the flotation of apatite requires only a small amount of the collectors, apatite can be separated from hornblende and magnetite. Therefore, both collectors are selective for apatite with respect to the other minerals investigated and the collector concentration applied. The entire separation of hornblende from magnetite should not be possible with any of the collectors. This result follows from the microflotation tests, as well as from the adsorption isotherms.

3.3. Influence of excess calcium ions on the adsorption and the selectivity of collectors

Water used for the processing of iron ores often contains large amounts of calcium salts. As mentioned in section 1, calcium ions affect the adsorption mechanism of anionic surfactants on salt-like minerals. For these reasons, the influence of calcium ions was investigated by zeta-potential and adsorption experiments as well as by flotation tests.

Figure 8 shows the zeta-potential of apatite, hornblende, and magnetite in water at different concentrations of calcium ions at pH 10 and constant ionic strength. In pure water and at calcium concentrations below $2 \cdot 10^{-5}$ mol/l all three minerals are

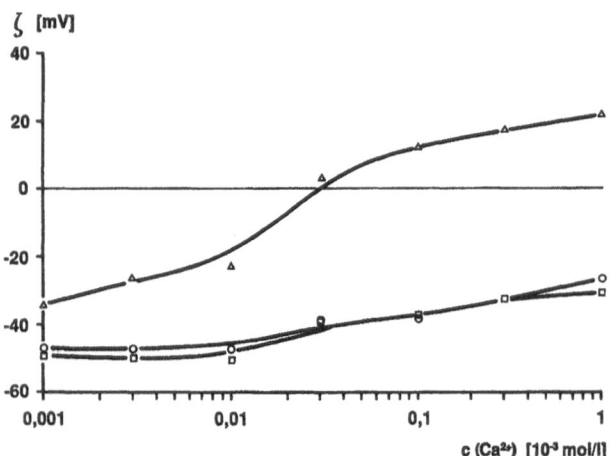

Fig. 8. Zeta-potential of apatite △, hornblende □, and magnetite ○ as a function of Ca^{2+}-concentration, pH 10

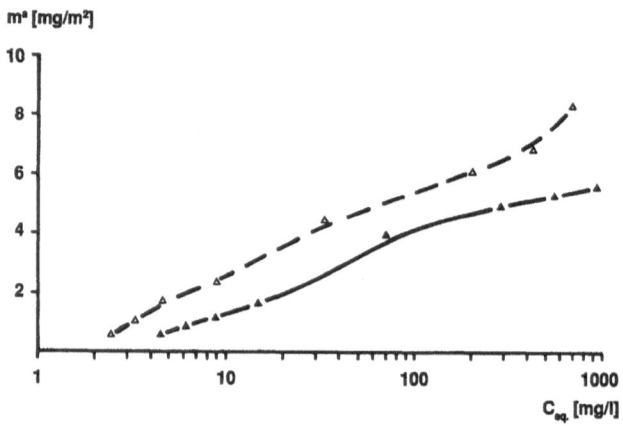

Fig. 9. Influence of water hardness on the adsorption of Na-alkylsulphosuccinate onto apatite, ■ deionized water, □ 5.36 · 10^{-4} mol/l Ca^{2+}, both pH 10

negatively charged. The further addition of calcium ions causes a shift of the potentials to less negative values and, in the case of apatite, the potential becomes positive at a calcium concentration of about $3 \cdot 10^{-5}$ mol/l. Decreasing negative charge with an increase in calcium concentration may be due to specific adsorption of the cations on the mineral surface.

The adsorbed amount of collectors at the same equilibrium concentration should increase with higher surface concentrations of multivalent cations independent of the different adsorption mechanisms of anionic collectors on salt-like minerals, as discussed in Section 1. Indeed, a look at the adsorption isotherms of alkylsulphosuccinate on apatite, with or without an addition of relatively low amounts of calcium ions ($c = 5.4 \cdot 10^{-4}$ mol/l, Fig. 9), shows that the adsorption density increases remarkably. Furthermore, it is more important that the onset of the isotherms shifts to lower equilibrium concentrations. Similar results are obtained on hornblende (Fig. 10) and on magnetite (Fig. 11). The behavior of the synthetic Fe_3O_4 differs considerably from the behavior of the natural one. Thus, neither a shift of the isotherm nor an increase in the adsorption density can be observed with synthetic Fe_3O_4 (Fig. 11).

The main difference between synthetic and natural magnetite is to be found in the chemical composition of the minerals. Chemical analysis shows that the natural magnetite ($Fe_xO_y = 96\%$) contains small amounts of silicates and phosphates (0.22% P, 1.11% SiO_2), whereas the synthetic Fe_3O_4

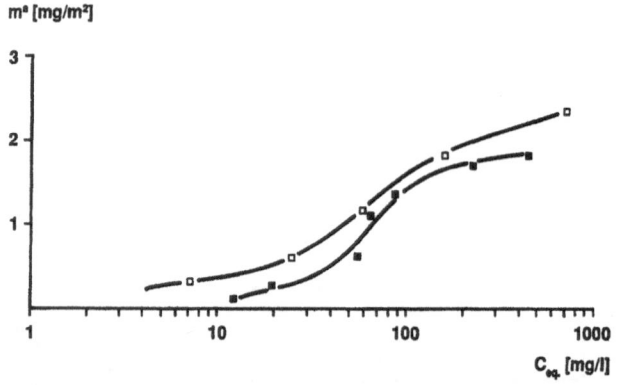

Fig. 10. Influence of water hardness on the adsorption of Na-alkylsulphosuccinate onto hornblende, ■ deionized water, □ 5.36 10^{-4} mol/l Ca^{2+}, both pH 10

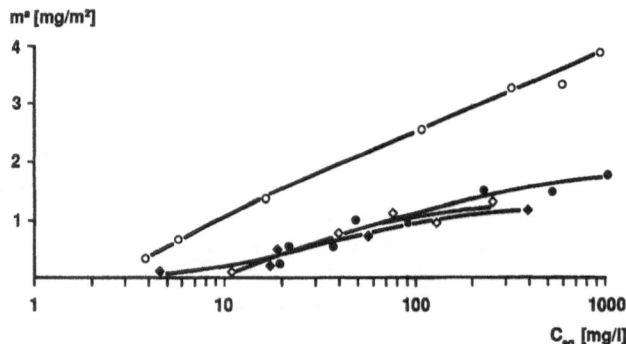

Fig. 11. Influence of water hardness on the adsorption of Na-alkylsulphosuccinate onto magnetite ○, ●, and synthetic Fe_3O_4 ◇, ◆; deionized water ●, ◆, 5.36 10^{-4} mol/l Ca^{2+} ○, ◇, pH 10

is of much higher purity ($Fe_3O_4 > 99\%$). One feasible explanation may be the fact that specific adsorption of calcium ions takes place preferentially on minerals containing calcium ions, such as hornblende or apatite. On natural Fe_3O_4, which contains silicates and phosphates, adsorption of calcium ions and, consequently, of anionic collectors, may also be possible, in contrast to the pure synthetic material. However, the different behavior of both magnetites is not yet fully understood.

Furthermore, this enhanced adsorption of alkylsulphosuccinate affects the behavior of the minerals in the microflotation tests. Thus, the recovery of apatite (Fig. 12) increases strongly at the same collector concentration if the flotation tests are performed with a calcium concentration of $3.57 \cdot 10^{-3}$ mol/l. The enhancement can be observed even more clearly in the case of hornblende (Fig. 13) and magnetite (Fig. 14). As an example, in the presence of calcium ions the recovery of hornblende at a collector concentration of 10 mg/l increases from 20% to 90%. These experiments confirm the fact that calcium ions improve the adsorption of alkylsulphosuccinate on all three minerals investigated in this study and, hence, they improve the effectiveness of this collector.

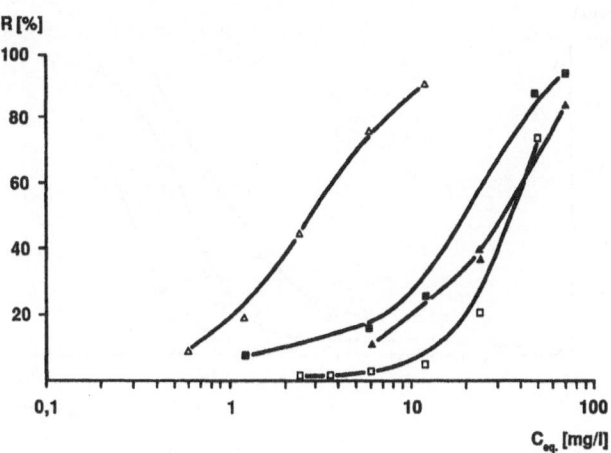

Fig. 13. Influence of water hardness on the recovery of hornblende with Na-oleate □, ■, and Na-alkylsulphosuccinate △, ▲; deionized water ■, ▲, 3.57 10^{-3} mol/l Ca^{2+} □, △, pH 10

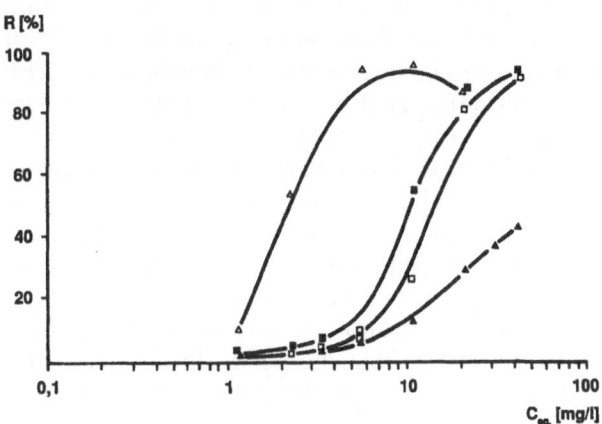

Fig. 14. Influence of water hardness on the recovery of magnetite with Na-oleate □, ■, and Na-alkylsulphosuccinate △, ▲; deionized water ■, ▲, 3.57 10^{-3} mol/l Ca^{2+} □, △, pH 10

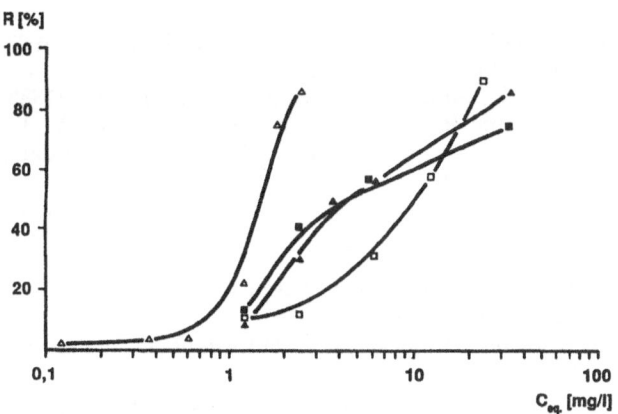

Fig. 12. Influence of water hardness on the recovery of apatite with Na-oleate □, ■, and Na-alkylsulphosuccinate △, ▲; deionized water ■, ▲, 3.57 10^{-3} mol/l Ca^{2+} □, △, pH 10

In comparison to calcium alkylsulphosuccinate, the solubility of calcium oleate is very low (solubility product, see Section 1). For this reason, the ratio of perecipitated to adsorbed amount of oleate in-

creases remarkably if calcium ions are added. If this precipitated amount is considered as a loss in reagent, i.e., not accessible to adsorption, recovery should not be improved, even in the case of a shift of the steep increase in the adsorption isotherm to lower equilibrium concentrations. Indeed, the recovery of apatite (Fig. 12), hornblende (Fig. 13), and magnetite (Fig. 14) at the same collector concentration decreases when calcium ions are added. Therefore, the collector effectiveness of oleate decreases with increasing calcium concentration.

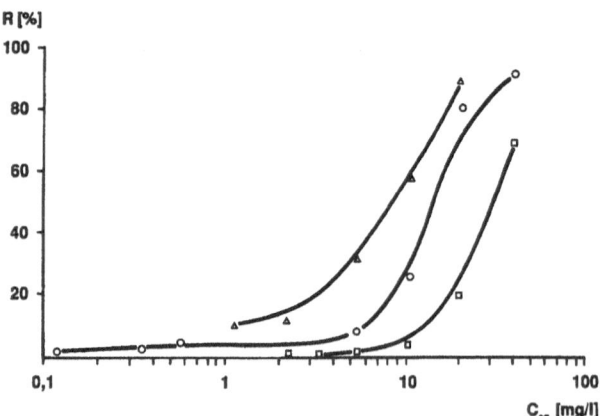

Fig. 15. Recovery of apatite △, hornblende □, and magnetite ○ with Na-alkylsulphosuccinate, 3.57 10^{-3} mol/l Ca^{2+}, pH 10

surfaces and the solubility of calcium salts. In deionized water, oleate usually adsorbs at slightly lower equilibrium concentrations than does alkylsulphosuccinate (Figs. 4—6), and the same holds for the recovery. Principally, adsorption of calcium ions on the surface of the minerals should promote adsorption of the surfactants. The actual flotation result is not only governed by this effect, but also by the amount of collector available for adsorption, i.e., by the ratio of adsorbed to precipitated amount. Thus, the alkylsulphosuccinate is the more effective collector in calcium-containing processing water. The negative effect in the case of oleate can be reduced by using mixtures of collectors or by addition of nonionic surfactants, which reduce the precipitation of the calcium salts [20—22].

The results of the flotation tests carried out with all three model minerals at a calcium concentration of 3.57 · 10^{-3} mol/l are summarized in Fig. 15 for oleate and in Fig. 16 for alkylsulphosuccinate. With both collectors, apatite can be floated at lower equilibrium concentrations than magnetite which, in turn, requires lower concentrations than hornblende. Thus, the sequence of flotation activity of the minerals (Figs. 12—14) is not changed by addition of calcium ions. Only the absolute amount of collector required for maximum recovery differs.

Comparison with the recovery results in deionized water suggests that there is a close relationship between adsorption of the collectors on mineral

References

1. Hanna HS, Somasundaran P (1976) In: Fuerstenau MC (ed) Flotation. AIME, New York, pp 197—272
2. Atademir MR, Kitchener JA, Shergold HL (1981) Int J Miner Process 8:9—16
3. Ananthapadmanabhan KP, Somasundaran P (1985) Colloids and Surfaces 13:151—167
4. Marinakis KI, Shergold HL (1985) Int J Miner Process 14:161—176
5. Hu JS, Misra M, Miller JD (1986) Int J Miner Process 18:57—72
6. Mougdil BM, Vasudevan TV, Blaakmeer J (1987) Miner Metall Process:50—54
7. Rao KH, Antii BM, Cases JM, Forssberg KSE (1988) In: Forssberg KSE (ed) Proceedings of the XVI Int Miner Process Congr, Elsevier, Amsterdam, pp 625—636
8. Kurzendörfer CP, Schlag M (1986) In: Dechema Monographien 102, VCH Verlagsgesellschaft, Weinheim, pp 561—574
9. Rao KH, Antii BM, Forssberg E (1989) Int J Miner Process 26:123—140
10. Antii BM, Forssberg E (1989) Minerals Engineering 2:217—227
11. Ananthapadmanabhan KP, Somasundaran P (1988) J Colloid Interface Sci 122:104—109
12. Rao KH, Antii BM, Forssberg E (1987) MinFo Project Nr 16, Flotation of Industrial Minerals, Progress Report No 4
13. Roberts JO, Caserio MC (1968) Chimie Organique Moderne, Ediscience, Paris
14. Leja J (1982) Surface Chemistry of froth flotation, Plenum Press, New York, p 295
15. Du Rietz C (1975) Reprint 11th int Miner Process Congr, Cagliari, pp 1—29
16. Fuerstenau MC, Miller JD (1967) Trans AIME 258:153—160

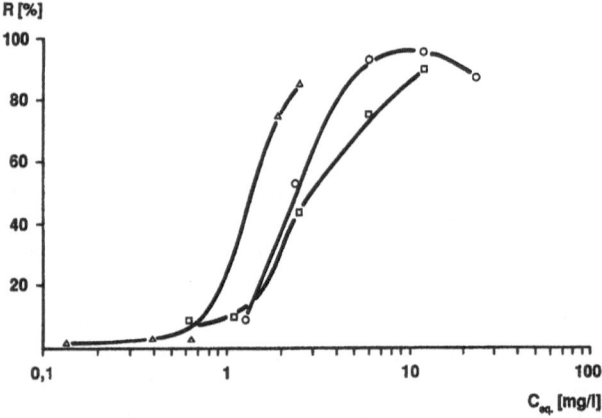

Fig. 16. Recovery of apatite △, hornblende □, and magnetite ○ with Na-oleate, 3.57 10^{-3} mol/l Ca^{2+}, pH 10

17. Miller DJ (1988) In: Proceedings of the 2nd Wolrd Surfactant Congress, Vol 1, Paris pp 399—410
18. Fan XJ, Stenius P, Kallay N, Matijevic E (1988) J Colloid Interface Sci 121:571—578
19. Pugh R, Stenius P (1985) Int J Miner Process 15:193—218
20. von Rybinski W, Schwuger MJ (1985) Aufbereitungs-Technik 26:632—639
21. von Rybinski W, Schwuger MJ (1987) Colloids and Surfaces 26:291—304

Received December 2, 1989,
accepted May 11, 1990

Authors' address:

Dr. Maria Liphard
Henkel KGaA
Postfach 1100
4000 Düsseldorf, FRG

Progress in Colloid & Polymer Science   Progr Colloid Polym Sci 83:136—145 (1990)

Specific dye adsorption at oriented monolayers

F.-J. Schmitt[a]), P. Meller[b]), H. Ringsdorf[b]), and W. Knoll[a])

[a]) Max-Planck-Institut für Polymerforschung, Mainz, FRG
[b]) Institut für Organische Chemie, Johannes-Gutenberg-Universität Mainz, Mainz, FRG

Abstract: We studied the adsorption of water soluble cyanine dyes (pseudo-iso-cyanine and "stains-all") to monomolecular layers of arachidic acid at the water-air interface in a Langmuir trough. Fluorescence microscopy was employed to on-line monitor the formation of J-aggregates upon adsorption. Large two-dimensional monodomain single crystals could be grown by this self-assembly process with optical properties reminiscent of those found and extensively studied for amphiphilic derivatives of the same chromophors [12]. In addition to the necessary Coulombic interaction between subphase-dye and target monolayer, we found that details of the crystal morphology also crucially depend on substrate parameters, e.g., the packing density during adsorption. After transfer of the monolayer to a hydrophobic solid support, either as a Langmuir-Blodgett film or in a flow cuvette, i.e., still in contact with the aqueous phase, absorption spectroscopy as well as fluorescence spectroscopy and -microscopy was employed to characterize the adsorbate stability, its desorption, re-adsorption or exchange reactions.

Key words: Adsorption; monomolecular layers; J-aggregates of cyanine dyes; self-organization; fluorescence microscopy

Introduction

The self-assembly of monolayer constituents at the water-air interface is now widely investigated for its obvious importance in fundamental as well as in applied research [1]. A major experimental break-through was achieved by the introduction of fluorescence microscopy [2] a technique that (at the water surface) allowed for the on-line characterization of such different processes as the lateral structure formation during lipid phase transition [3, 4], protein adsorption and organization [5], or polymerization of functionalized lipids [6], to mention only a few.

Another technically most relevant process [7] concerned the formation of Scheibe [8] — or J-aggregates [9] of various cyanine — dye derivatives which showed interesting features of two-dimensional crystallization [10—12].

Recently, Hada et al. [13] and Lehmann, Möbius and coworkers [14, 15] reported another mechanism to grow two-dimensional (2-D) J-aggregates: water-soluble cyanine dyes could be specifically adsorbed and organized at suitable monolayers spread and compressed at the water surface. A spectroscopic light reflection technique [16] was used to monitor the formation of these aggregates by electrostatically controlled adsorption and organization at concentrations where no aggregates in the bulk were yet found.

These studies are to be seen in connection with synchroton x-ray reflection studies by Lahav and coworkers on 2-D self-aggregation of amphililic α-amino acids and their role for crystal nucleation when prepared as Langmuir monolayers [17—19] with the general idea of "transferring structural information from 2-D domains to 3-D crystals".

In this work, we report on fluorescence microscopic investigations of the adsorption of cyanine dyes and their organization as two-dimensional crystals. We address, in particular, the question of which specific properties of the target monolayer, arachidic acid in our case, determine the growth and the optical properties of these self-assembled

J-aggregates. First results on the highly complex process of exchanging one dye (crystal) by another are reported. Finally, special emphasis is put on the question as to what extent the adsorption and organization of the dyes is changed if the target monolayer is transferred from the water-air interface to a solid support in a flow cuvette where the contact to the aqueous phase also allows for studies of desorption, re-adsorption or exchange reactions.

Experimental

Dye adsorption was investigated for pseudo-iso-cyanine iodide (PIC; 1,1'-diethyl-2,2'-cyanine iodide, structure formular given in Fig. 2) obtained from Sigma (Heidelberg, FRG) and for "stains-all" (SA; 4,5—4',5'-dibenzo-3,3'-diethyl-9-methyl-thiocarbocyanide bromide, structure formula given in Fig. 6a)) obtained from Fluka (Buchs, Switzerland), at monolayers of arachidic acid (AA; eicosanoic acid) spread from 10^{-3} M CHCl$_3$-solution. Aqueous (milli-Q quality) solutions of the dyes were buffered ($7.8 \cdot 10^{-5}$ M Na$_2$ HPO$_4$ and $1.22 \cdot 10^{-4}$ M NaH$_2$ PO$_4$) at pH 7 ($T = 20°$C). All experiments at the water-air interface were performed in a home-built Langmuir through equipped with a Wilhelmy system for the lateral pressure reading and with a fluorescence microscope for the on-line observation of the dye adsorption and organization [20]. All pictures given below show an area of 185 µm × 260 µm on the water surface or on the flow cuvette. Optical filter systems in the microscope allowed for a spectral discrimination between the PIC- and the SA-fluorescence. Desorption, re-adsorption or exchange reactions were performed by a complete exchange of the subphase with a set-up schematically shown in Fig. 1a). Typically, this procedure took about 1—2 h for complete exchange.

Transfer of the monolayers with adsorbed dye aggregates onto solid supports was done in two different ways. In one case, a microscope glass-slide was first coated with a monolayer of cadmium arachidate (CdA), and this hydrophobic substrate was then dipped nearly horizontally through the highly viscous dye-AA monolayer into the through subphase and then backed out again. These dry layers were then used to control the optical properties of the *J*-aggregates formed during the adsorption process by absorption spectroscopy in a multidiode array spectrophotometer (HP 8452 A) and by fluorescence spectroscopy (Fluorlog F 122 from SPEX). The second technique involved the use of a flow-cuvette as depicted in Fig. 1b). The upper part (again made hydrophobic by a CdA-layer) was dipped through the monolayer into the subphase, but then carefully placed to fit the bottom part of the cuvette first immersed into the through. The whole assembly was then taken out and could be further investigated by absorption spectroscopy and by fluorescence spectroscopy and -microscopy. Through the contact to the aqueous volume in the cell

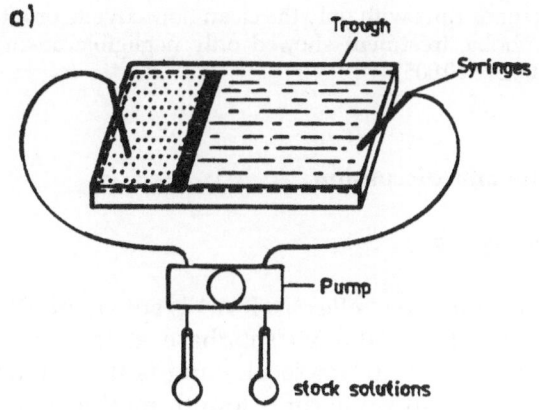

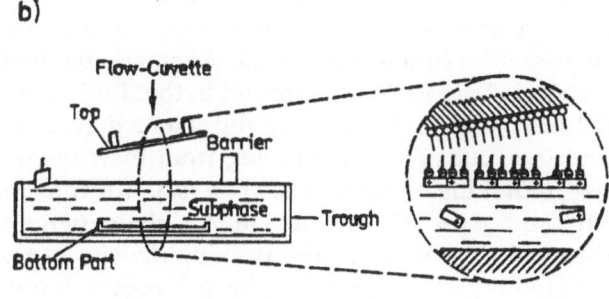

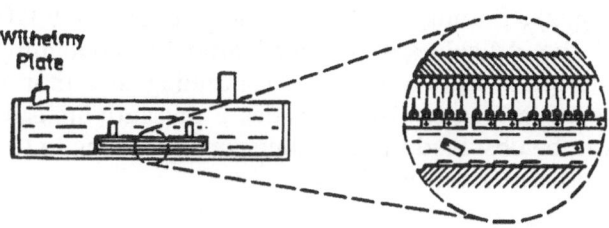

Fig. 1. a) Schematic drawing of the apparatus used to exchange the subphase in a Langmuir trough. b) Schematic drawing of the arrangement used to transfer a fatty acid/dye-aggregate monolayer onto the hydrophobic top part of a flow cuvette whose bottom part was immersed into the Langmuir through subphase

various manipulations of the organized dye-adsorbates were possible under on-line control, e.g., in the absorption spectrometer, desorption kinetics by exchanging the cell volume against dye-free buffer or exchange reactions by replacing one dye solution by another. Only strong *J*-aggregates could be followed in this way because the dye bulk solution of 1 mm optical path length in the flow cuvette dominated the spectra in the monomer and dimer spectral region (see, e.g., Fig. 4e).

Reference runs with only the clean flow cuvette or after hydrophobic treatment showed only negligible adsorption (O.D. <0.005).

Results and discussion

PIC-aggregates

Pressure-area (π-A) isotherms: If *AA* is spread on PIC-containing ($c = 10^{-4}$ M) subphase at low lateral density ($A > 2$ nm^2/molecule) and is then slowly compressed, an isotherm is found, reminiscent of the amphiphilic cyanine dye S120 which has one long hydrocarbon tail covalently attached to the chromophore: A fluid-analogue state undergoes a phase transition at $\pi \approx 5$ mN $\cdot$ m^{-1} into a condensed state (full line in Fig. 2). A remarkably high area per chain molecule is found in the fluid phase, 0.9 nm$^2 < A < 1.4$ nm^2, by the interaction of *AA* and dye, considerably differing from their respective individual dimensions: $A \approx 0.5$—0.6 nm^2 for PIC and $A \approx 0.2$ nm^2 for an *AA* chain perpendicular to the surface. The packing density of the *AA* chains in the condensed phase, however, is only slightly higher than for *AA* on pure water (see the corresponding isotherm, broken line, also given in Fig. 2) a finding which is the sharp contrast to S120. If the *AA*-solution is spread onto PIC-containing subphase in quantities corresponding to a monolayer at intermediate densities a pressure-increase can be observed up to a value corresponding to the isotherm given in Fig. 2.

Fluorescence microscopy: The adsorption of the dye is accompanied by a well-defined organization into two-dimensional *J*-aggregates at the *AA*-monolayer/water interface, as can be seen by fluorescence microscopy. In particular, if the primary adsorption step is carried out at a very low monolayer density (see Fig. 2) large single-crystalline dye-aggregates can be grown by compression with a homogeneous polarization of the fluorescence emission as in the case of pure S120 crystals. An example with polarized excitation is shown in Fig. 3a) and b), respectively.

In a few cases, other crystal morphologies are seen with radial or segmental arrangements of the chromophores. Two examples are given in Figs. 3c and d, respectively, again excited with two orthogonal polarizations. These crystals have a certain similarity with S120$_{0.33}$ stearic acid$_{0.66}$-mixed aggregates where two chains of the fatty acid, together with the dye's own chain match the areal requirements of the chromophore lattice [12]. If *AA* is first spread on pure water, compressed to the crystalline state, and then the subphase is exchanged against PIC-solution *J*-aggregates are again formed upon adsorption to the monolayer, but only as a fine polycrystalline 2-D powder. Figs. 3e and f show the development over about 2 hours. The resolution limit of the fluorescence microscope does not allow for a detailled analysis of the crystal morphologies, but one can see that they emit polarized fluorescence light.

Transferred monolayers: Further evidence that Scheibe-aggregates were, indeed, formed in all cases discussed above was obtained from absorption spectra of transferred layers which clearly showed the *J*-band peak at $\lambda = 580$ nm. An example is given in Fig. 4a. Figure 4b shows the corresponding fluorescence spectrum of the same layer, Stoke's shifted by only 1—2 nm.

More interesting was the question of whether whole crystals could be transferred to the solid wall

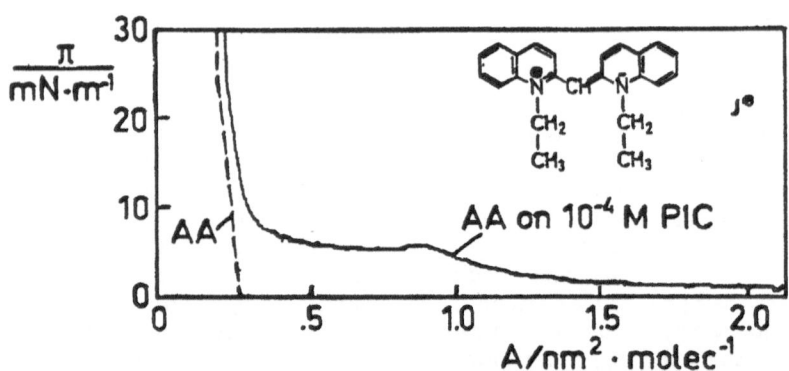

Fig. 2. Pressure-area (π-A) isotherm of arachidic acid (*AA*) on pure water (broken line) and on 10^{-4} M PIC containing subphase (full line). $T = 20$°C, pH 7. The insert shows the structure formula of PIC

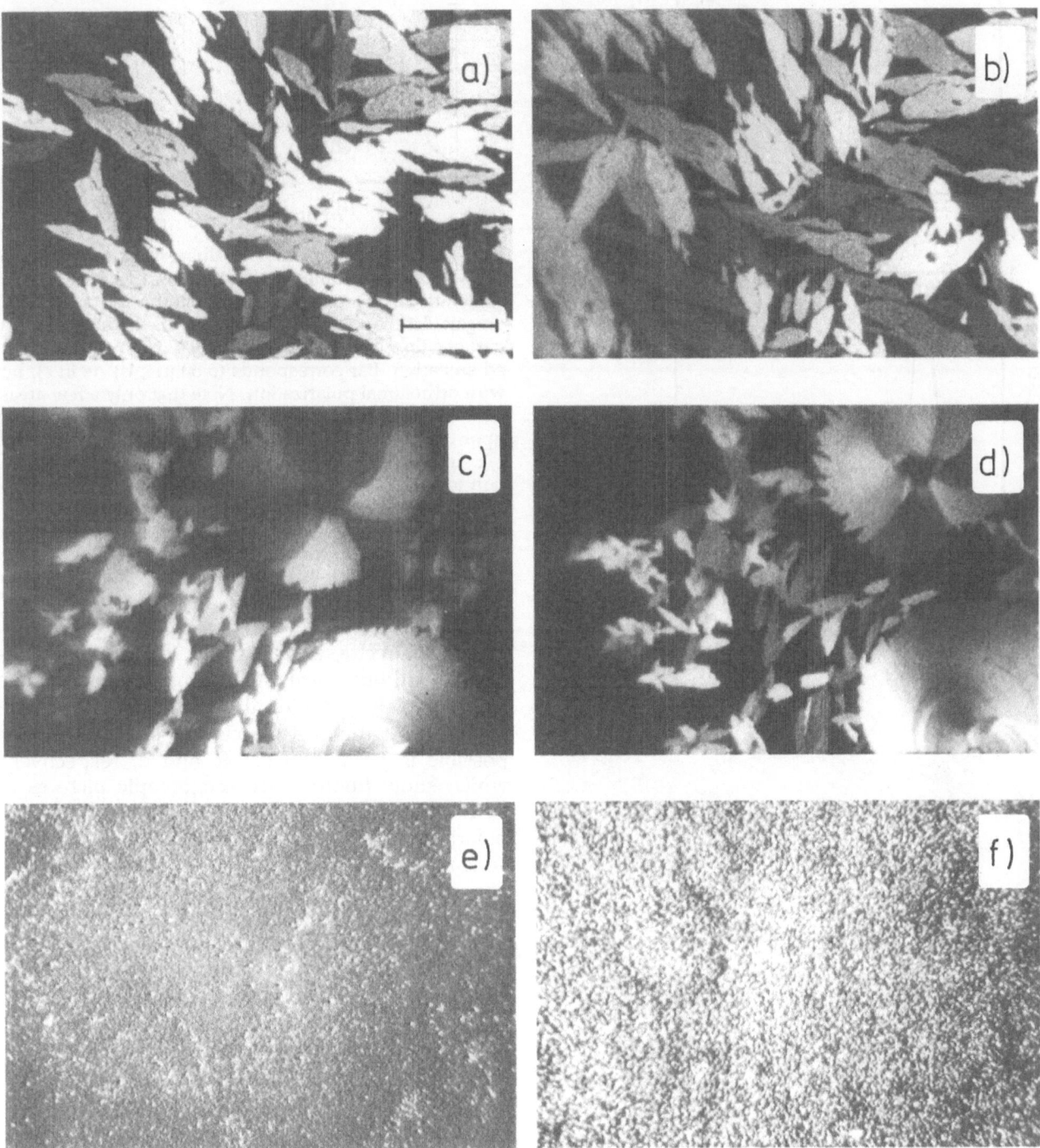

Fig. 3. a) Fluorescence micrograph of a PIC *J*-aggregate monolayer grown under conditions where large single-crystalline domains are formed by adsorption to an *AA* monolayer. The picture is taken with polarized excitations. The bar corresponds in all pictures to 50 µm on the water surface; b) As in a) but with orthogonal polarization of the excitation light; c) In fewer cases, radially arranged aggregates grown under the same conditions can be found within the same monolayer; d) as in c) but excitation polarization rotated by 90°; e) Fluorescence micrograph of a polycrystalline powder growing by adsorption of PIC to a pre-compressed monolayer of *AA* after about 1 h; f) respectively 2 h

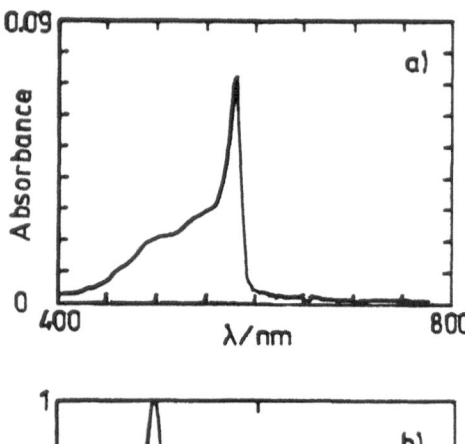

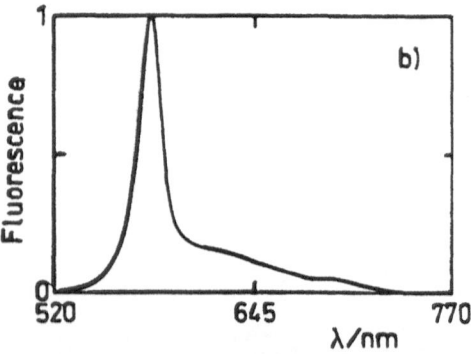

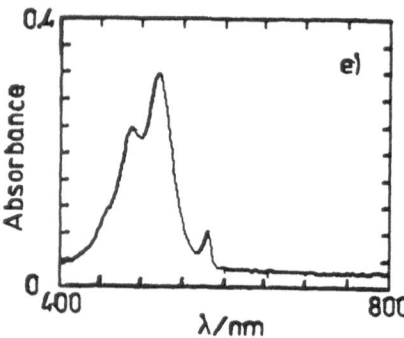

Fig. 4. a) Absorption, and b) fluorescence-spectrum of PIC aggregate monolayers after transfer to a solid substrate; c) Fluorescence micrograph of a PIC aggregate/*AA* monolayer transferred to the hydrophobic top part of a flow cuvette. The picture was taken with polarized excitation. Bar corresponds to 50 μm; d) As in c), but with orthogonal polarization. Note that only a few areas, dark in both pictures, indicate some loss of aggregates upon transfer; e) Absorption spectrum of transferred PIC-aggregate/*AA* monolayer in a flow cuvette. In addition to the strong monomer (λ = 522 nm) and dimer (λ = 488 nm) absorption mostly from the bulk solution in the cuvette a strong red-shifted absorption peak from the PIC-*J* aggregates is seen

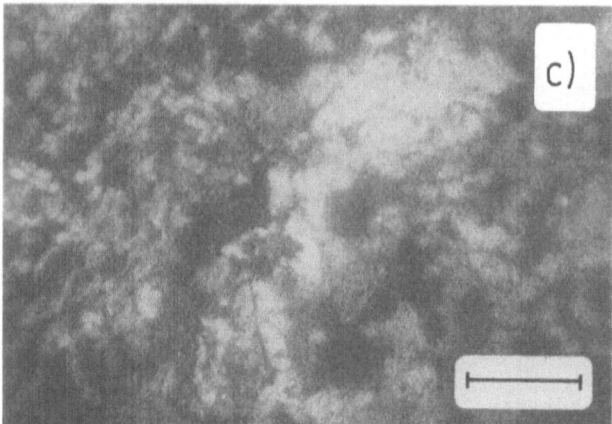

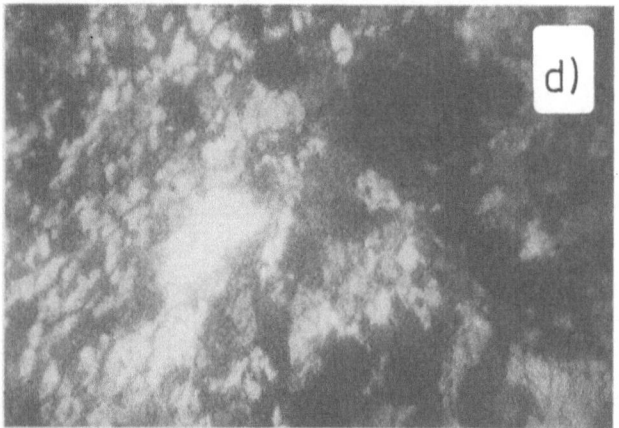

of a flow cuvette, i.e., in contact with the aqueous phase, without substantial damage of even large single-crystals and without a major change of their optical properties. Evidence that this was indeed possible is given in Figs. 4c and d, respectively, which show fluorescence microscopic pictures of such a transferred monolayer with two orthogonal polarizations. Most crystals show the typical homogeneous fluorescence intensities; only a few areas appear dark in both pictures, indicating a certain loss of aggregates during the transfer.

Further evidence for the almost intact depostion comes from absorption spectroscopy in the flow cuvette. Figure 4e shows that, in this case, the spectrum is dominated by the bulk monomer (λ = 522 nm) and dimer (λ = 488 nm) absorption, but that the aggregates, nevertheless, can be identified by their strong red-shifted absorption peak at λ = 580 nm. Its optical density of ca. O.D. = 0.05 is typical for a single monolayer.

A much weaker *J*-band absorption was found (not shown) if a condensed *AA*-monolayer was transferred from a pure buffer subphase to the flow cuvette, and then the aqueous volume exchanged against 10^{-4} M PIC-solution. O.D. $\approx$ 0.02 at λ = 580 nm was achievable, at best. This result coult not be improved by trying various packing densities of the transferred *AA*-monolayers.

Desorptoion of PIC crystals: Next, the important question of the reversibility of the dye-adsorption was to be answered for both the water-air interface, as well as for the transferred monolayer. For this aim, *J*-aggregates were grown at the *AA*-monolayer on 10^{-4} M PIC solution. Then the subphase was exchanged against pure buffer, which resulted in the time-course of the experiment actually in a 10:1 dilution of the PIC-concentration. After a certain time delay given by the slow exchange process a rapid decrease of the aggregate density was observed within a few minutes and after some more time no aggregate fluorescence could be observed. A sequence of corresponding micrographs is shown in Figs. 5a—c. The whole desorption process was accompanied by a decrease of the lateral pressure by $\Delta\pi \approx 15$ mN · m^{-1}.

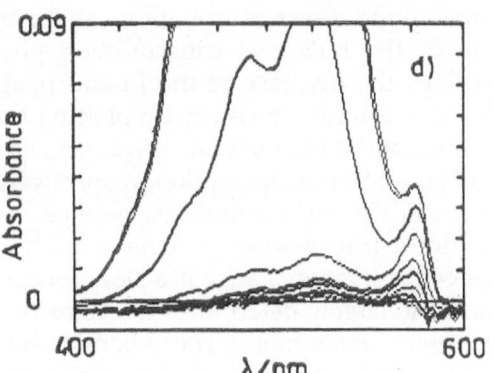

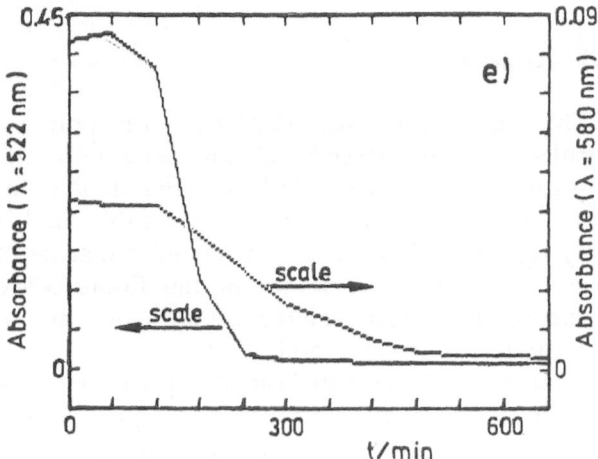

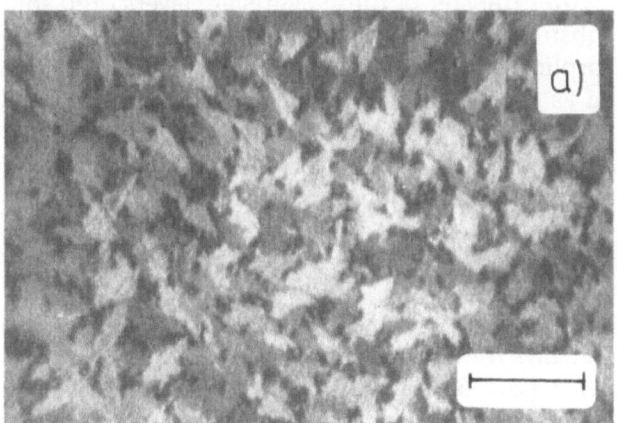

Fig. 5. a) Fluorescence micrograph of a PIC *J*-aggregate layer of an *AA* monolayer in the Langmuir trough prior to the dilution experiment; b) Same area but some time after the subphase had been diluted by ca. a factor of 10. The picture shows an intermediate stage of the desorption process. Bar corresponds to 50 μm; c) Same area after complete desorption of PIC crystals; d) Absorption spectra taken every hour after the PIC aggregate/*AA* monolayer was transferred to a flow cuvette and the (slow) exchange of the PIC-containing solution against pure buffer was started at $t = 0$; e) Desorption kinetic as seen by plotting the optical density of the *J*-band absorption at $\lambda = 580$ nm as a function of time (taken from the spectra shown in d)). The decrease of the monomer absorption at $\lambda = 522$ nm is mostly given by the bulk volume exchange, i.e., the slow PIC dilution

The corresponding experiment in the flow cuvette allowed for the spectral control of the desorption process by monitoring on-line absorption spectra as a function of time. Some examples are given in Fig. 5d. The spectra clearly show that the decrease of the monomer and dimer contribution caused by the dilution of the bulk dye concentration goes faster than does the decrease of the *J*-band peak. This can be seen even more clearly by plotting the optical densities at the monomer (λ = 522 nm) and at the *J*-band (λ = 580 nm) absorption, respectively. This is shown in Fig. 5e. It should be pointed out that this is not a true desorption kinetic analysis because the cell volume exchange is a slow process whose time scale largely determines the reduction of the monomer absorption. The *J*-band peak, however, seems to be more stable for some time before it decreases in intensity.

SA-aggregates

This paragraph summarizes the corresponding results obtained with stains-all. Figure 6a shows the π-*A*-isotherm of *AA* on 10^{-5} M buffered SA-solution. The first little plateau at $\pi_1 \approx 2$ mN $\cdot$ m^{-1} is very reproducible at areas per (chain) molecule of $A \approx .7$—.6 nm^2 and shows in the fluorescence microscope some lateral structure formation, but at a very low intensity level. Only at the second plateau at ca. 13 mN $\cdot$ m^{-1} can strongly fluorescing aggregates be observed that can be compressed to a full coverage of the water surface. An example is given in Fig. 6b, again with polarized excitation. The orthogonal polarization inverts the intensity distribution (not shown), indicating a homogeneous fluorescence from individual single-crystals, as in the case of PIC discussed above.

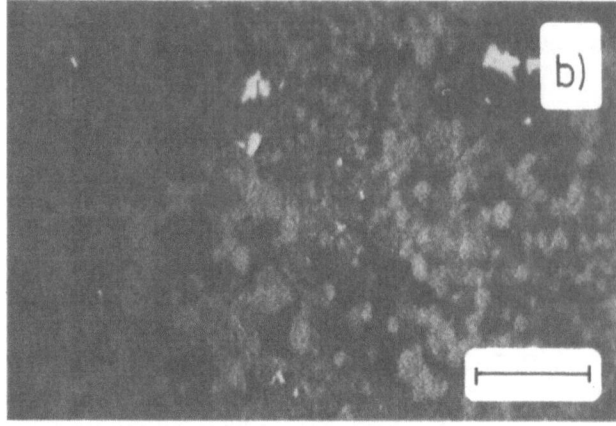

Fig. 6. a) Pressure-area (π-*A*) isotherm of *AA* on 10^{-5} M SA containing subphase. Structure formula of SA given in the insert. *T* = 20°C, pH 7; b) Fluorescence micrograph of a monolayer of large SA-aggregates grown by adsorption of SA to *AA* at large areas per molecule and subsequent crystallization by compression. The bar corresponds to 50 µm

After transfer of these aggregate layers onto solid substrates by the Langmuir-Blodgett technique absorption and fluorescence spectra could be recorded as shown in Figs. 7a and b, respectively, clearly indicating the SA *J*-band formation with a peak absorption at λ = 656 nm and a fluorescence at roughly the same spectral position.

Two-dye-system

Exchange reactions at the water-air-interface: Following the report of Lehmann [15], who found in his reflection studies a gradual exchange of the *J*-band absorption of PIC at 580 nm by that of SA at 656 nm if *AA* was spread on a subphase that contained both dyes though SA at a considerably lower concentration (1:200 relative to PIC) we, too, investigated various reaction schemes for replacing PIC by SA under fluorescence microscopic control. Generally speaking, the exchange reactions that we could observe indicated a very complex behavior with many details whose full description would be beyond the scope of the present paper.

Briefly, if *AA* was spread on a subphase containing both dyes (PIC-concentration: $c_{PIC} = 10^{-4}$ M; SA-concentration: $c_{SA} = 10^{-6}$ M) PIC crystals could be grown with different morphologies, but with virtually identical absorption and fluorescence spec-

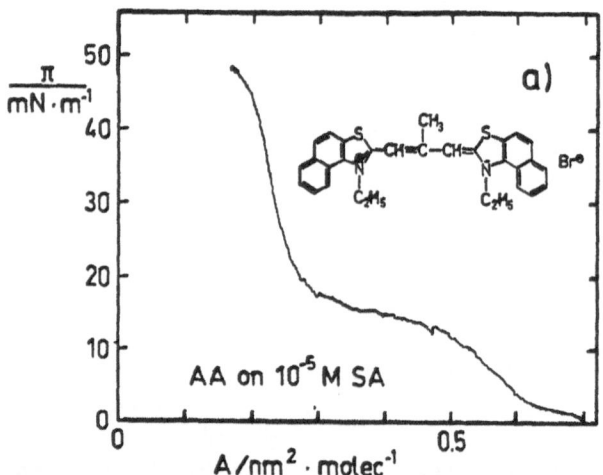

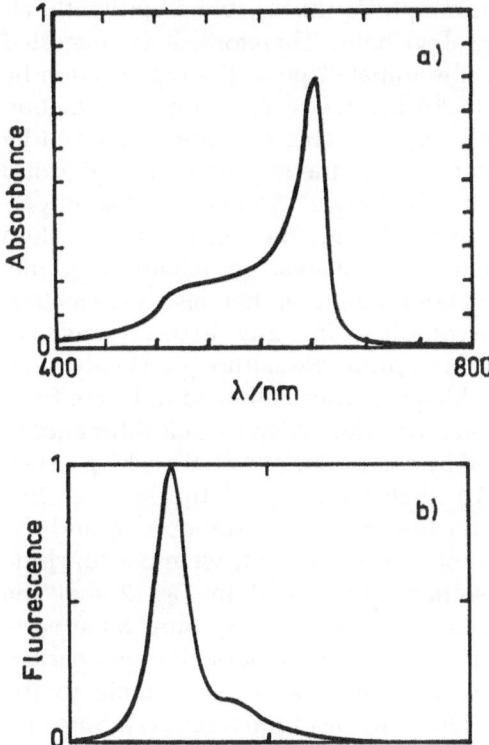

Fig. 7. Absorption a) and fluorescence b) spectrum of a SA-aggregate/*AA* monolayer taken after transfer to a solid support

tra. No exchange against SA could be seen even after 6 h. In particular, no mixed *J*-band peaks could be found.

If SA was injected at higher concentrations (c_{SA} > 1 · 10^{-6} M) a gradual replacement of PIC-crystals by SA in a complicated way could be observed: starting from the edges, both the areas between large PIC crystals and along defect lines SA-fluorescence gradually built up (see also below). At that stage both *J*-band peaks could be seen in the absorption spectra. Details of the fluorescence behavior, however, seemed to be determined also to some extent by energy transfer from the PIC- to the SA-aggregates. These observations will be reported elsewhere [21].

A more quantitative exchange process could be conducted if the pure PIC-subphase (after *J*-aggregates growth) was completely exchanged against a subphase containing both PIC and, in addition, stains-all. The respective concentrations were c_{PIC} = 5 · 10^{-5} M and c_{SA} = 10^{-6} M. An example of a

partially exchanged aggregate monolayer is shown in Fig. 8. The fluorescence micrograph shown in Fig. 8a was taken with an interference filter (λ_{max} = 650 nm) in the microscope in order to emphasize the SA fluorescence: the bright lines around the individual PIC-crystals (whose long wavelength fluorescence extends to 650 nm — see also Fig. 4b — and therefore weakly passes the interference band pass, as well) are typical for the ongoing slow exchange reaction. The same area, if observed through a longpass filter ($\lambda \geqslant$ 590 nm), mostly

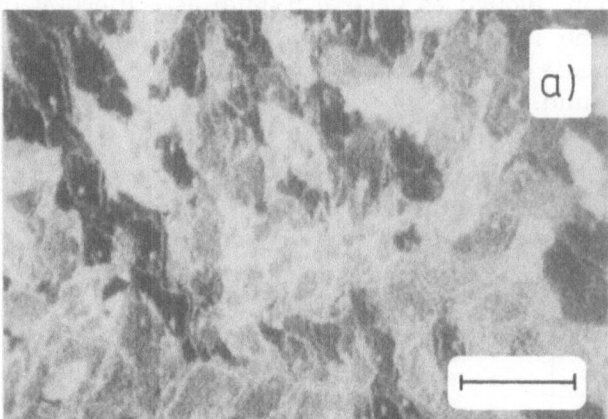

Fig. 8. Fluorescence micrographs of a PIC aggregate layer partially exchanged against SA domains. a) Picture taken with an interference filter (λ_{max} = 650 nm) in order to emphasize the SA-fluorescence: the bright lines around and between the PIC crystals (whose long wavelength fluorescence extends to λ > 650 nm and, therefore, weakly allows for their simultaneous observation) are typical for early stages of the exchange.The bar corresponds to 50 µm; polarized excitation was used; b) Same area, but observed with a longpass filter ($\lambda \geqslant$ 590 nm) so that the PIC fluorescence dominates and the weaker SA fluorescence now shows up as dark lines around the PIC crystals

shows the bright fluorescence light of the remaining PIC-crystals but, in addition, also the same PA-"decoration" lines, now darker because their fluorescence intensity is weaker (Fig. 8b) and because their fluorescence intensity was more susceptible to bleaching by the excitation light.

Exchange at the transferred layer: Finally, several attempts were made to conduct the same set of exchange reactions with transferred aggregate layers, but none of them was successful: A typical result is given in Fig. 9. The loss of PIC aggregate absorption, plotted as O.D. at 580 nm as a function of time, is obvious, but no steady and systematic increase of SA-aggregate absorption at $\lambda = 656$ nm is found.

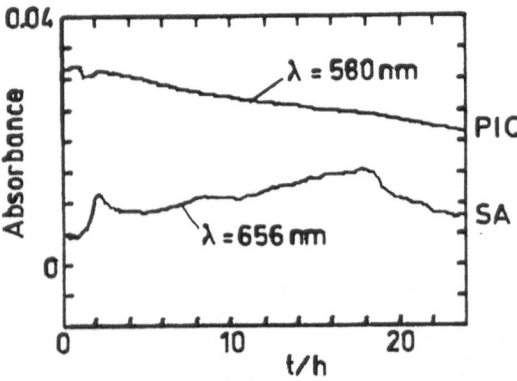

Fig. 9. Example for the unsystematic exchange of PIC against SA at a transferred monolayer in the flow cuvette. While the optical density of the PIC *J*-band absorption at $\lambda = 580$ nm steadily decreases with time, the SA peak absorption at $\lambda = 656$ nm does not increase in a corresponding way, but fluctuates only at relatively low absorbance level

4. Discussion

The most strinking result of the presented adsorption studies is the possibility to grow large, presumably single-crystalline, *J*-aggregates of PIC, provided the adsorption process is initiated by spreading *AA*-solution at a very low lateral density. Then the π-A-isotherm as well as the optical and morphological properties of the crystallites grown by compression are very similar to those found for the amphiphilic cyanine dye S 120 which has an

identical chromophore (head), but a covalently attached long alkylchain. Therefore, it is suggested that for PIC the initial stage of the organization by adsorption to *AA* involves a 1:1 complex formation between PIC and *AA*. This complex, stabilized by the ionic interaction of the negative carboxyl group and the positively charged cyanine dye has properties similar to S 120, e.g., the formation of a fluid phase at relatively large areas per (chain) molecule. The *J*-aggregate formation is then also a crystallization of these complexes into 2-D single crystals with equivalent optical properties although, clearly, their crystallographic properties still need to be confirmed by electron diffraction. A systematic difference to S 120 is found by comparing the final packing density of the *AA* chains on top of the layer of PIC crystals: given the area per chromophore and the cross-section of a fatty acid rod, we must conclude from the isotherm presented in Fig. 2 that the crystalline state is, unlike in the pure S 120 case, characterized by a more or less densely packed monolayer of alkyl chains — comparable to the crystalline S 120 stearic acid mixtures [12]. Since we see no distinct feature in the plateau of the isotherm of *AA* on PIC-solution (Fig. 2), we have to assume a continuous slip (or jump) of individual fatty acid molecules onto the PIC aggregates during compression. This is schematically shown in Fig. 10.

Fig. 10. Schematic picture illustrating the change in packing of an *AA*-cyanine dye crystal monolayer upon compression (for details see text)

In a few cases, such 3:1 *AA*:PIC complexes may have already formed before the pressure-induced crystallization process, and they might have given rise to the observed radially arranged aggregates (see Figs. 3c and d) reminiscent of the 2:1 stearic acid: S 120 mixed crystals [12].

Most interesting in view of (a very general) possible recognition reaction at organized interfaces is the finding that the self-organization of water soluble cyanine dyes to form 2-D crystals obviously requires certain properties of the target interface that a monolayer at the water-air-interface, under certain conditions, can provide, but that get lost upon its transfer to a solid support. We only can speculate about what these conditions are, but the presented results and, in particular, some re-adsorption studies with a clear hint for epitaxial growth that we reported earlier [22] seem to indicate that a general flexibility and/or compressibility is needed to match packing requirements of monolayer and adsorbate. Clearly, this problem must be solved before a more universal use of Langmuir-Blodgett layers for fundamental research, e.g., of interfacial reactions, or for device applications, e.g., in optical or electrical (bio-)sensors, can be envisioned.

References

1. See, e.g., Proceedings of the 4th International Conference on Langmuir-Blodgett Films, 1989, Tsukuba, Japan, to be published in Thin Solid Films
2. Tscharner V, McConnell HM (1981) Biophys J 36:421—427
3. Lösche M, Möhwald H (1984) Eur Biophys J 11:35—42
4. Miller A, Knoll W, Möhwald H (1986) Phys Rev Letters 56:2633—2636
5. Blankenburg R, Meller P, Ringsdorf H, Salesse C (1989) Biochemistry 28:8214—8221
6. Meller P (1989) J Microscopy 156, Pt 2:241—246
7. Steiger R, Hediger H, Junod P, Kuhn H, Möbius D (1980) Photogr Sci Eng 24:185—195
8. Scheibe G (1936) Angew Chem 49:563—564
9. Jelly EE (1936) Nature 138:1009—1011
10. Duschl C, Lösche M, Miller A, Fischer A, Möhwald H, Knoll W (1985) Thin Solid Films 133:65—72
11. Duschl C, Frey W, Knoll W (1988) Thin Solid Films 160:251—255
12. Duschl C, Kemper D, Frey W, Meller P, Ringsdorf H, Knoll W (1989) J Phys Chem 93:4587—4593
13. Hada H, Hanowa R, Haraguchi A, Yonezawa Y (1985) J Phys Chem 89:560—562
14. Orrit M, Möbius D, Lehmann U, Meyer H (1986) J Chem Phys 85:4966—4979
15. Lehmann U (1988) Thin Solid Films 160:257—269
16. Grüniger H, Möbius D, Meyer H (1983) J Chem Phys 79:3701—3710
17. Landau EM, Grayer-Wolf S, Sagiv J, Deutsch M, Kjaer K, Als-Nielsen J, Leiserowitz L, Lahav M (1989) Pure Appl Chem 61:673—684
18. Grayer-Wolf S, Deutsch M, Landau EM, Lahav M, Leiserowitz L, Kjaer K, Als-Nielsen J (1988) Science 242:1286—1290
19. Grayer-Wolf S, Leiserowitz L, Lahav M, Deutsch M, Kjaer K, Als-Nielsen J (1987) Nature 328:63—66
20. Meller P (1988) Rev Sci Intrsum 59:2225—2231
21. Schmidt F-J, Knoll W (1990) in preparation
22. Schmidt F-J, Knoll W (1990) Chem Phys Letters 165:54—58

Authors' address:

W. Knoll
Max-Planck-Institut für Polymerforschung
Ackermannweg 10
6500 Mainz, FRG

Progress in Colloid & Polymer Science   Progr Colloid Polym Sci 83:146—154 (1990)

Photodesorption of Langmuir-Blodgett multilayer assemblies

D. Johannsmann and W. Knoll

Max-Planck-Institut für Polymerforschung, Mainz, FRG

Abstract: We studied the desorption of multilayers of cadmium arachidate (CdA) deposited by the Langmuir-Blodgett dipping technique onto gold-coated, optically polished microbalance quartz plates. Thus, it was possible to monitor the UV-light (90% at λ = 254 nm)-mediated photo-desorption process by measuring simultaneously the time dependence of the mass-density of the thin film coating on the quartz oscillator as well as of its optical thickness by a compensating ellipsometer set-up. The number of CdA layers was varied between 2 (ca. 5 nm) and 54 (ca. 135 nm) layers. — In order to describe the desorption behavior, we analyzed the data according to various kinetic models among which the defect generation mechanism fits the experiments best: for thin films ($\leqslant$10 layers of CdA) we find $\dot{\Theta} \propto \Theta \cdot t$, whereas for thicker films $\dot{\Theta} \propto t$ is obtained.

Key words: Photodesorption; Langmuir-Blodgett-multilayers; ellipsometry; quartz crystal microbalance; desorption kinetics

Introduction

The growing interest in the technical application of organic multilayer assemblies [1] built-up layer by layer by the Langmuir-Blodgett dipping technique [2] from monomolecular films organized at the water-air-interface [3] has stimulated many attempts to better characterize and optimize their various physical properties, e.g., structural, mechanical, electrical or optical.

A most valuable experimental technique for the investigation of their stability is studies of the desorption behavior under the influence of UV-photons. This is generally important for a better understanding of what limits the structural integrity of these ultrathin films under the influence of, for example, mechanical stress [4], heat treatment [5] or any kind of irradiation [6]. In fact, it is the poor stability of LB films that prevents, so far, their wider use as insulating, lubricating or just protective coatings. However, the possibility of selectively photo-ablating only certain areas, thereby micro-structuring a target surface [7] is currently pursued in many laboratories as a promising technique for

photo-lithography with high-resolution in ultrathin resist materials.

In the present work, we report on photo-desorption studies with cadmium arachidate ((CH_3-$(CH_2)_{18}(COO^-)_2Cd^{++}$, CdA) multilayers ranging in thickness from a double layer ($d \approx 5$ nm) up to 54 layers ($d \approx 135$ nm). These were deposited onto gold-coated, optically polished microbalance quartz plates by the LB technique [2]. Thus, it was possible to monitor the desorption process i) by recording the frequency shift of the quartz oscillator [9], and ii) simultaneously by taking optical thickness data by a compensating ellipsometer set-up [10], both as a function of the UV-irradiation time. We will show that the data can be best described on the basis of a defect model recently proposed by Möhwald and coworkers [11] — though with certain extensions. In particular, we will demonstrate that the model needs to be modified to account for the observed thickness dependence of the derived kinetic parameters. These show a cross-over from a thickness-dependent to a thickness-independent desorption rate at about 10 layers of CdA. Special emphasis will be put on the comparison between the two techniques whose combined application can yield additional information about the samples and processes under investigation.

Experimental

The experimental set-up used in this study is schematically sketched in Fig. 1. The ellipsometer operates in the commonly used compensating mode with HeNe laser (at λ_L = 633 nm)/polarizer/quarter wave plate/sample/analyzer configuration and photodiode detection. From the measured polarizer and analyzer angles one directly obtains the ellipsometric parameters Δ and ψ [10] that describe the ratio of the complex amplitude reflectivities $\tilde{r}_p$ and $\tilde{r}_s$ for p- and s-polarized light, respectively, of our sample consisting of an Au substrate and the dielectric LB-layers:

$$\frac{\tilde{r}_p}{\tilde{r}_s} = \tan\psi \cdot e^{i\Delta} . \tag{1}$$

These reflectivities, however, can be calculated on the basis of Fresnel's theory with the known dielectric function of gold [12] and the two fit parameters of the thin film coating, i.e., the thickness Θ and the index of refraction n of the LB layer system [13]. Unfortunately, Au substrates do not allow one to analyze ψ with sufficient accuracy for very thin coatings so that no independent determination of n is possible in our case. Δ, however, is directly related to the thickness of the layers. A series expansion of Δ to first order in Θ/λ_L [14] gives a linear relation:

$$\Delta \propto \Theta/\lambda_L . \tag{2}$$

We therefore limit in the following the analysis of ellipsometric desorption data to Δ, assuming for the CdA layers a constant index of refraction n = 1.55 [13]. For the error in the thickness determination we estimate ca. 0.3 nm, mostly given by the change of the Au dielectric function with time [15] and under UV illumination.

For laterally heterogeneous samples one obtains from ellipsometry yet another parameter, namely the remaining light intensity I_0 measured at compensation behind the analyzer. This can be described by an incoherent superposition of contributions from different domains [16] which do not allow to compensate for the whole sample at one setting of polarizer and analyzer angle. Any increase in I_0, e.g., during the desorption process, therefore indicates the formation of heterogeneities on a μm-length scale, e.g., by some large scale roughnesses.

The quartz crystal microbalance (QCM) operates with 0.5″ diameter, AT-cut, optically polished quartz crystals with evaporated Au-electrodes (see front and side view schematically depicted in Fig. 1). The frequency f_{res} is measured with a network analyzer (HP 4195 A). In principle, also the equivalent resistance R can be measured.

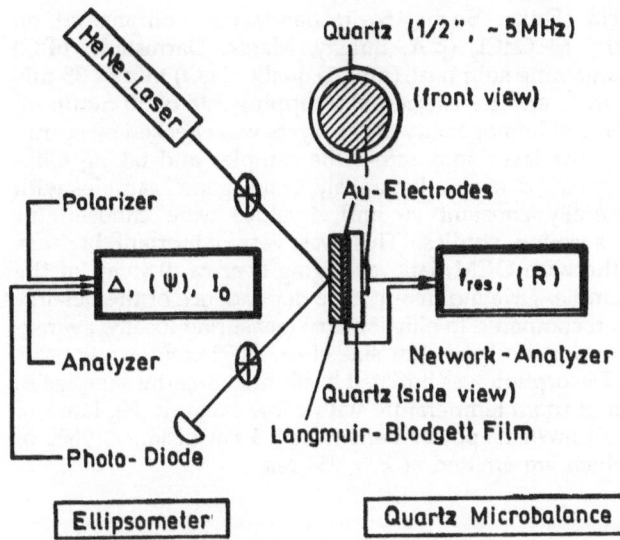

Fig. 1. Schematic of the experimental set-up consisting of a combination of compensating ellipsometer and a quartz crystal microbalance

It turned out, however, that with CdA layers deposited directly onto the quartz crystals no systematic change of the intrinsic damping of the circuit (originating from structural defects of the crystal lattice, mounting and coupling to the air) was observed, e.g., by same viscous losses in the organic films. The network analyzer is capable of measuring the whole complex impedance spectrum of the quartz crystal. In this way, also possible deviations from the ideal resonance behavior (e.g., sidebands) are detected. For the work reported here, the position and the height of the maximum of the real part of the complex admittance (conductance) were analyzed. The frequency determined in this way differs from the electrical resonant frequency as well as from the anti-resonance frequency: it is the true acoustical eigen-frequency of the mode under measurement. Thus, any possible errors originating from electrical stray capacities, are eliminated. These errors are, however, small.

The mode used for the measurement was always the fundamental mode of thickness shear oscillations. In the case of a multiplicity of modes in the vicinity of the fundamental, the mode with the lowest frequency was selected, which usually was the strongest one. (The general time-behavior of all frequencies was the same during the desorption).

From the shift of the resonance frequency Δf_{res} induced by LB layers of various thicknesses one obtains directly their mass density according to [9]

$$\sigma = \Delta f_{res} \cdot 17.7 \, \frac{ng}{cm^2 \cdot Hz} \cdot \left(\frac{5\,MHz}{f_{res}}\right)^2 . \tag{3}$$

Deposition of the CdA layers was performed in the usual way in a home-built Langmuir trough: arachidic acid (Fluka, Buchs, Switzerland) was compressed on 10^{-4} M $CdCl_2$ (p.A. quality, Merck, Darmstadt, FRG) containing subphase (milli-Q quality H_2O) to $\pi \approx 35$ mN $\cdot$ m^{-1}, and transferred at a dipping rate of 10 mm/min. Optical homogeneity of the layers was checked by scanning the laser spot across the samples and taking ellipsometric data simultaneously. Only "good" samples with laterally constant ψ- and Δ-values were choosen for desorption studies. This was very important because otherwise QCM data, averaging over ca. 0.5 cm^2 of the sample, gave a different time-dependence of the desorption compared to ellipsometry measuring locally, averaged over the laser spot size of ca. 0.002 cm^2.

Desorption was initiated by illuminating the samples in air at room temperature with a low-pressure Hg-lamp of 2.15 mW/cm^2 power density at 2.5 cm distance, 90% of which are emitted at $\lambda = 254$ nm.

Results

Figure 2 shows raw ellipsometric data as obtained during the photo-desorption of 14 layers of CdA on Au. The sigmoidal time dependence of Δ mirrors in a very direct way the thickness decrease of the LB multilayer assembly. The whole desorption process is completed after ca. 7000 s. Most remarkable is the bell-shaped curve of the residual intensity I_0 with its maximum intensity at roughly the time $t_{1/2}$ when half of the layer is desorbed. Such a behavior was consistently found for all layer systems that could be analyzed by ellipsometry (2—14 layers CdA). We will discuss this finding later in connection with the proposed desorption mechanism.

Figure 3 presents the QCM data, i.e., resonance frequency f_{res} of the coated quartz oscillator as a function of the (illumination) time, taken simultaneously with the ellipsometric data of Fig. 2. The overall form of the measured curve is virtually identical to the time dependence of Δ and

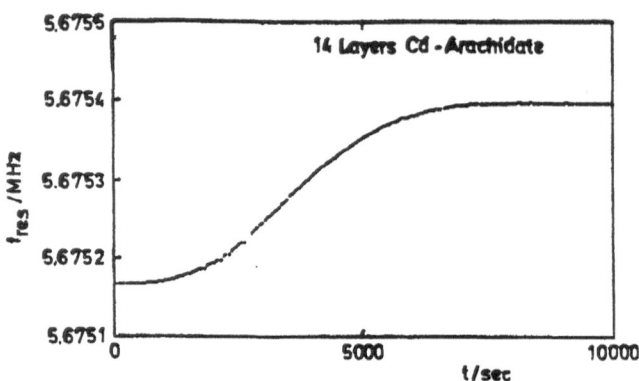

Fig. 3. Raw QCM data, resonance frequency f_{res} as a function of time, taken simultaneously with the ellipsometric data presented in Fig. 2

indicates a parallel loss in (optical) thickness and mass density of the desorbing film. As mentioned already such a behavior was only found for samples that showed no variation of the ellipsometric data if, prior to the desorption experiment, the laser spot on the LB multilayer was scanned laterally across the sample. Only such homogeneous samples were used for further measurements and analysis.

Figure 4a summarizes all thickness data as obtained by ellipsometry, assuming a constant index of refraction of $n = 1.55$ taken from the literature [13]. Shown are the desorption curves for all investigated layers as indicated. The arrows point to the $t_{1/2}$-values which, other than reported by Möhwald and coworkers [11], increase with increasing layer thickness. We will come back to this point later.

Figure 4b shows the corresponding time dependence of the mass density σ of the LB coatings as obtained from the QCM curves. In addition to these films, we also investigated the desorp-

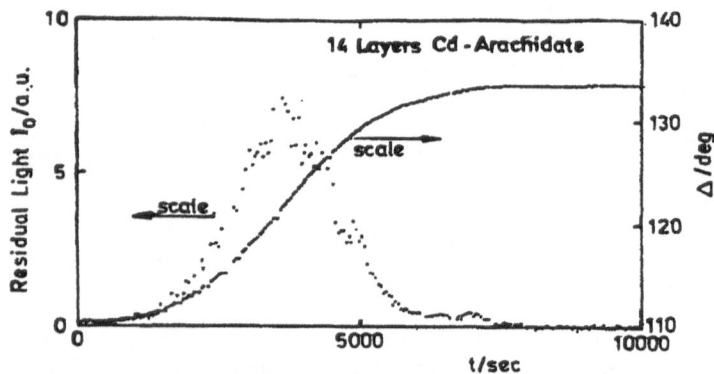

Fig. 2. Raw ellipsometric data Δ and I_0, respectively, as obtained as a function of time during the UV photo-desorption experiment with 14 layers of CdA on Au

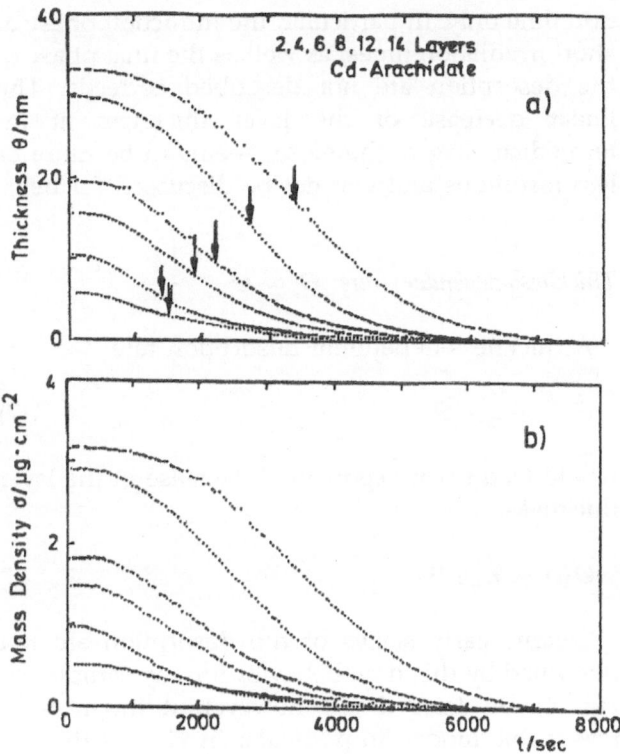

Fig. 4. a) thickness Θ, and b) mass density σ as a function of time as calculated from ellipsometric (assuming $n = 1.55$) and QCM data taken for 2, 4, 6, 8, 12, and 14 layers of CdA. The arrows point to the $t_{1/2}$-times when half of the initial layer thicknesses are desorbed

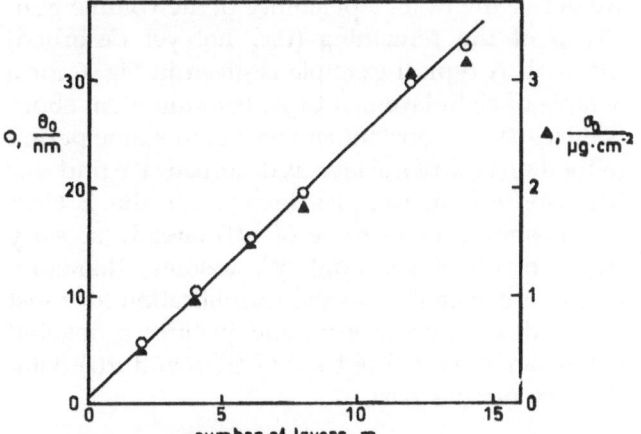

Fig. 5. Initial thickness Θ_0 (o) and mass density σ_0 (▲) as a function of the number of layers m in the samples. Data are taken from Fig. 4a and b at $t = 0$. The straight line yields $\Delta\Theta = 2.5$ nm/monolayer and $\Delta\sigma = 0.25$ µg $\cdot$ cm^{-2}/monolayer

tion behavior of 30, 50, and 54 layers of CdA with the quartz oscillator. Except for the 2 CdA layers sample the two sets of desorption curves show a very parallel behavior, indicating for these studies of the photo-desorption kinetics the equivalency of the two otherwise very different experimental techniques. It is also interesting to note that the noise level in both data sets is about equal, which also means that with our current set-ups the sensitivities of the two methods to detect changes of the film thicknesses are comparable.

Discussion

The starting conditions

First we analyzed the starting values of the layer thickness Θ_0 and their mass densities σ_0, as obtained from the curves of Fig. 4 at $t = 0$. The corresponding data are plotted in Figure 5 as a function

of the number of layers m deposited for each of these samples. Within the experimental accuracy the data points fit to a straight line, whose slope yields $\Delta\Theta = 2.5$ nm/monolayer and $\Delta\sigma = 0.25$ µg $\cdot$ cm^{-2}/monolayer. The slight deviation from the origin points to a certain contamination of the gold substrate, but is almost within the error limits of the data.

The desorption behavior

As mentioned already the geometrical thickness data of thin coatings are derived from the ellipsometric parameters Δ only if a value for the index of refraction n is put into the analysis. With $n = 1.55$, we obtain for all samples Θ_0-thickness compatible with literature data [2, 13, 15, 17]. The question arises, however, whether a constant n is a reasonable assumption also for the whole time-course of a desorption experiment. Au as a substrate for ellipsometric measurements with thin organic coatings does not allow for an independent check of this assumption (through a meaningful evaluation of ψ). We can obtain, however, an internal consistency test by combining ellipsometric and QCM data:

According to

$$\rho(t) = \sigma(t)/\Theta(t) , \qquad (4)$$

we derive the time-dependence of the volume density ρ of the remaining (i.e., not yet desorbed) material. A typical example is given in Fig. 6 for a sample of eight layers of CdA: For more than about 3500 s $\rho(t)$ is a constant and within this time period more than 80% of the layer is desorbed. We find this behavior for all samples (except for the 2 CdA which shows an increase of $\rho(t)$ already at early stages of the desorption). We assume, therefore, that n = constant is a valid simplification for most of the desorption process and justifies a detailed kinetic analysis on the basis of $\Theta(t)$- and $\sigma(t)$-data. This we discuss next.

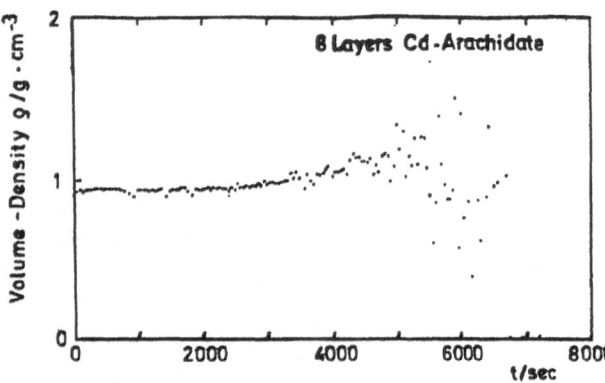

Fig. 6. Volume density $\rho(t) = \sigma(t)/\Theta(t)$ as a function of the UV-irradiation time for a sample consisting, at the beginning, of 8 layers CdA. $\sigma(t)$ and $\Theta(t)$ were taken from the desorption curves in Fig. 4

Constant desorption rate: $\dot{\Theta} = const$

The simplest assumption for a kinetic description of the desorption process is that the desorption rate is independent of time and layer thickness:

$$\dot{\Theta} = -c_1 \ . \tag{5}$$

Integration of Eq. (5) yields

$$\Theta(t) = \Theta_0 - c_1 t \ . \tag{6}$$

By inspection of the experimental data given in Fig. 4 one, indeed, can see that for all desorption curves (including the ones for 30, 50, and 54 layers not shown) there is a region where such a simple model would fit for a very limited period of desorp-

tion time only. In particular, the induction phase at short irradiation times, as well as the final phase of the desorption are not described correctly. The linear decrease of the layer thickness at intermediate stages, therefore, seems to be more or less fortuitous and will not be discussed further.

Thickness-dependent rate: $\dot{\Theta} \propto \Theta$

A thickness-dependent desorption rate

$$\dot{\Theta} = -c_2 \cdot \Theta \tag{7}$$

would lead to an exponential decrease of the layer thickness:

$$\Theta(t) = \Theta_0 e^{-c_2 t} \ . \tag{8}$$

Clearly, early stages of the desorption are not described by this time-behavior for any sample (cf. Fig. 4). At later times, however, it might be a reasonable model, in particular in view of the fact that the moment the first free (i.e., uncoated) substrate patches appear upon desorption the process *must* become dependent on the coverage. We, therefore, plotted all data in a semi-logarithmic graph, an example of which is given in Fig. 7 for the 14 layers sample. In both data sets (ellipsometric as well as by QCM) up to $t = \leqslant 6500$ s no straight line would fit the curves. Only very late stages, at best,

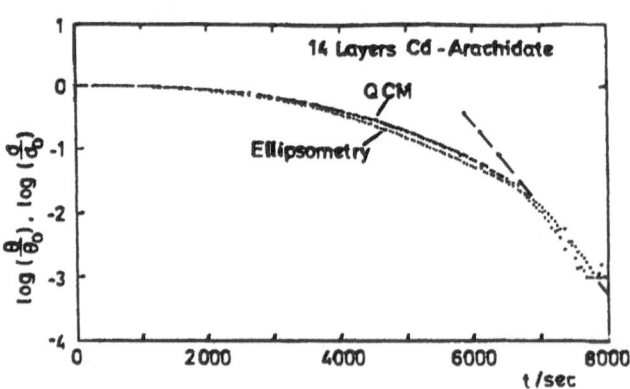

Fig. 7. Semilogarithmic plot of the thickness Θ (scaled to Θ_0) as obtained from ellipsometry and the mass density σ (scaled to σ_0) as obtained from QCM data as a function of time for the 14 layers CdA sample. Only the desorption behavior at very late stages ($t \geqslant 6500$ s) could be described by an exponential law (indicated by the dashed line)

could be described by the exponential law of Eq. (8). There, however, the layer thickness has decreased already to a few percent of its starting value (ca. 2%, corresponding to 0.8 nm, in the case of the presented example) where the noise in the data already significantly obscures the true time dependence. Moreover, in these late phases of the desorption process one can observe major deviations from a constant volume-density behavior which could result from a combination of desorption (i.e., mass losses) and index of refraction modifications. Then, however, a kinetic analysis based on Δ-values only, is meaningless. Since this behavior is found qualitatively for all samples, we disregard the exponential model as inappropriate for the observed desorption process.

The defect generation model: $\dot{\Theta} \propto \Theta \cdot t$

Möhwald and coworkers [11] recently proposed a defect generation model based on the assumption that the desorption rate depends, not only on the layer thickness, but also on the time:

$$\dot{\Theta} = -c_3 \cdot \Theta \cdot t . \qquad (9)$$

The rational for this assumption was that the action of light is twofold: it creates defects and provides energy to overcome binding. This means that desorption under the influence of a UV-photon can occur only (or predominantly) at defects which are generated by photons themselves and whose concentration therefore increases linearly in time. These authors further assumed that, in addition to what is expressed in Eq. (9), c_3 in itself depends on Θ through the proportionality to the number of photogenerated defects.

Our kinetic analysis is also based on Eq. (9), however, with two important modifications: first of all, we assume that c_3 for a given sample is a constant (only, of course, for constant photon flux as in our case) that does not depend on the layer thickness as it decreases during deposition. Furthermore, we assume that the desorption depends on the number *density* of defects which simply increases linearly in time as it is already put into the formulation of Eq. (9). This redefinition (as opposed to Möhwald and coworkers [11]) ensures that Eq. (9) can be integrated at all times (and is not limited to $t \to 0$ [11]):

$$\Theta(t) = \Theta_0 e^{-\frac{c_3}{2}t^2} . \qquad (10)$$

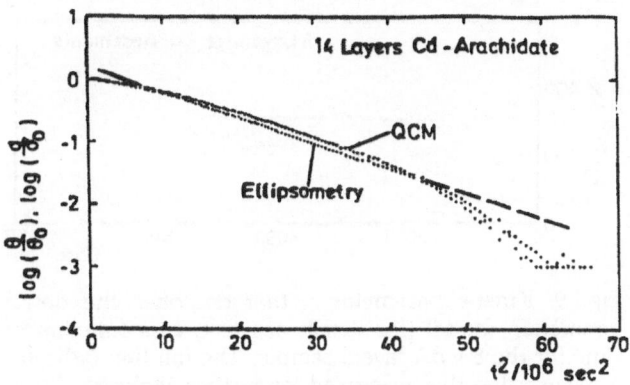

Fig. 8. Semilogarithmic plot of the thickness Θ (scaled to Θ_0) as obtained from ellipsometry, and the mass density σ (scaled to σ_0) as obtained from QCM data as a function of t^2 for the 14 CdA layers sample. The main desorption behavior ($0.80\,\Theta_0 < \Theta < 0.01\,\Theta_0$) is well described by the defect generation model which would yield a straight line (dashed line) in this representation of the data

A semilogarithmic plot of $\Theta(t)/\Theta_0$ (or, equivalently, of $\sigma(t)/\sigma_0$ vs t^2 should yield, therefore, a straight line. This is done in Fig. 8, again for the data of the 14-layer samples. One can see that by this defect generation mechanism the induction phase now can be described somewhat better, thought not perfectly, and that, moreover, the time dependence of the main desorption from ca. $\Theta = 0.8\,\Theta_0$ down to $\Theta = 0.02\,\Theta_0$ is fairly well accounted for. This is indicated by the dashed line in Fig. 8 that extrapolates the QCM data as a straight linear fit.

Obviously, however, c_3 is *not* fully independent of time, but is smaller at the beginning, and again at the end of the desorption (at least within the framework of this desorption model).

This can be better seen if one rewrites Eq. (9):

$$\frac{d\Theta}{\Theta} = -c_3 t\, dt \qquad (11)$$

to give

$$d(\ln\Theta) = -\frac{c_3}{2}\, d(t^2) . \qquad (12)$$

We therefore obtain

$$c_3(t) = -2 \cdot \frac{d(\ln\Theta)}{d(t^2)} . \qquad (13)$$

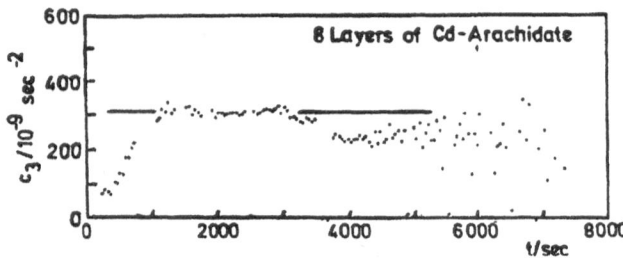

Fig. 9. Kinetic parameter c_3 that describes the defect generation model (for details see text) as a function of time for the 8 CdA layers sample. The full line indicates a mean value that was used for further analysis

This analysis of the desorption data was carried out for all samples in order to derive $c_3(t)$. An example is given in Fig. 9, this time for the 8-layer sample. As in the case of the LB film consisting of 14 layers CdA, c_3 increases during the early induction phase, then stays constant for most of the desorption and only at the end does it again decrease slightly. All c_3-values obtained for the main desorption phase are summarized in Table 1 as a function of the number of layers m that constitute the different samples. One can see that c_3 decreases rapidly with increasing initial layer thickness, varying over almost two orders of magnitude from a bilayer to 50 layers.

Further evidence that these c_3-values obtained from the analysis of (most of) the whole desorption curves are indeed "good" parameters comes from a comparison with the $t_{1/2}$-values that defines a single point in the desorption process, namely the time when half of the initial layer thickness is

desorbed. According to Eq. (10) this $t_{1/2}$ time is given by

$$ t_{1/2} = \sqrt{\frac{2 \mid \ln 0.5 \mid}{c_3}} . \tag{14} $$

In Table 1 we compare, therefore, the expression on the right side of Eq. (14) with the $t_{1/2}$-values obtained from the linear plots presented in Fig. 4 (see arrows). The agreement is remarkable. The thickness-independence of $t_{1/2}$ found by Möhwald and coworkers [11] is, at best, valid for a few layers only.

For thicker samples $t_{1/2}$ clearly increases and, hence, c_3 decreases drastically. This brings us to yet another model for the desorption kinetics: For very small c_3-values and times $t \ll t_{1/2}$, we can expand Eq. (10) to give

$$ \Theta(t) = \Theta_0 \left(1 - \frac{1}{2} c_3 t^2 \right) , \tag{15} $$

which is a simple quadratic time dependence of the desorption. The equivalent kinetic description is obtained by a pure time-dependent desorption rate.

Time-dependent desorption rate: $\dot{\Theta} \propto t$

If we assume that, for thicker layers, the desorption rate no longer depends on the initial layer thickness (which at some point should be the case, because eventually the influence of the substrate should vanish), but still increases linearly in time (e.g., through the above discussed defect generation mechanism), we obtain

Table 1. Various desorption-kinetic parameters (as defined in the text) for the CdA multilayer assemblies consisting of an increasing number m of monolayers transferred onto the substrates

Number of layer, m	2	4	6	8	12	14	30	50	54
$c_3/10^{-9}$ s^{-2}	690	690	370	320	250	180	23	14	14
$\sqrt{\dfrac{2 \mid \ln 0.5 \mid}{c_3}} \Big/ 10^3$ s	1.42	1.42	1.9	2.1	2.4	2.8	7.8	10.0	10.0
$t_{1/2}/10^3$ s	1.5	1.4	1.9	2.2	2.7	3.4	8	9.4	10
$(p + 1)$	—	—	—	2.1	2.4	2.5	—	2.2	2.3

$$\dot{\Theta} = -c_4 \cdot t \, , \tag{16}$$

and after integration

$$\Theta = \Theta_0 - \frac{c_4}{2} \, t^2 \, . \tag{17}$$

If the initial layer is sufficiently thick so that even at $t_{1/2}$ the substrate does not yet influence the desorption, we find from Eq. (17) that

$$t_{1/2}^2 = \frac{\Theta_0}{c_4} \, . \tag{18}$$

Now, Fig. 10 shows the $t_{1/2}^2$-values taken from Fig. 4 as a function of the number of layers m that constitute the various samples. For films thicker than about 10 layers, indeed, a linear increase of $t_{1/2}^2$ with Θ_0 is found as predicted by this kinetic model (see Eq. (18)). From the slope of the straight line one obtains $c_4 = 4.1 \cdot 10^{-7}$ monolayers per square second or $c_4 = 10.25 \cdot 10^{-7}$ nm/s^2.

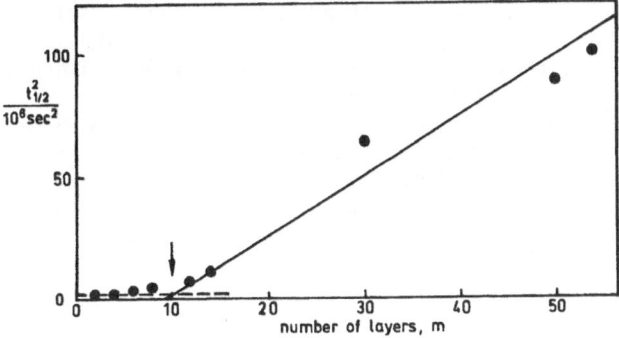

Fig. 10. $t_{1/2}^2$-values as taken from the data in Fig. 4a and QCM runs with thicker samples (30, 50, 54 layers CdA, not shown) as a function of the initial sample thickness, given as the number of layers m that constitutes the LB-film

The cross-over from a thickness-dependent to a thickness-independent desorption rate occurs at about 10 layers of CdA. From time-of-flight mass-spectrometric desorption data [6], we know that for few-layer systems molecules from lower-lying monolayers are desorbed simultaneously with those from top layers, already from the very beginn-

ing on. We can discriminate in these studies between various layers by depositing first, for example, a double layer of CdA and, subsequently, an increasing number of cadmium stearate layers. In the mass spectra we then can identify clusters of up to (7 Cd · 13 stearate)$^+$ which are easily separated from the corresponding CdA clusters. In the context of the present study we therefore can conclude that desorption preferentially occurs at deep, craterlike defects which penetrate deeply into the LB-layer structure. The increase of the residual intensity of the ellipsometric measurements (see Fig. 2) tells us that these defect structures reach dimensions on the µm-length scale. It is, therefore, well conceivable that these photon-generated defects corrode the 10 top layers, or, at least, that the defects in the first 10 layers are desorption-active. For samples with an initial thickness below this threshold the desorption rate increases with increasing layer thickness. For samples with more than ca. 10 layers the desorption rate does not change with sample thickness.

Cooperative defect model: $\dot{\Theta} \propto \Theta \cdot t^p$

Finally, we present an analysis of the desorption data which does not predict a priori a desorption rate that is strictly linearly increasing in time, but that allows for some general power p dependence of the desorption rate:

$$\dot{\Theta} = -c_5 \cdot \Theta \cdot t^p \, . \tag{19}$$

Integration of Eq. (19) yields

$$\Theta(t) = \Theta_0 e^{-\frac{c_5}{p+1} \cdot t^{p+1}} \, . \tag{20}$$

The corresponding double-logarithmic plot of the sample with 14 layers of CdA is presented in Fig. 11 for both ellipsometric and QCM data sets. The straight line obtained in this representation accounts for most of the desorption process, except for the first ca. 300 s (see dashed extrapolation in Fig. 11). From the slope of the linear regression line one obtains for this sample $p + 1 = 2.5$. The values derived from such plots for all other samples (with $m \geqslant 8$ layers) are also summarized in Table 1. The general trend to find $p + 1 > 2$ we tentatively explain by a certain cooperativity in the defect-mediated desorption process: in order to become

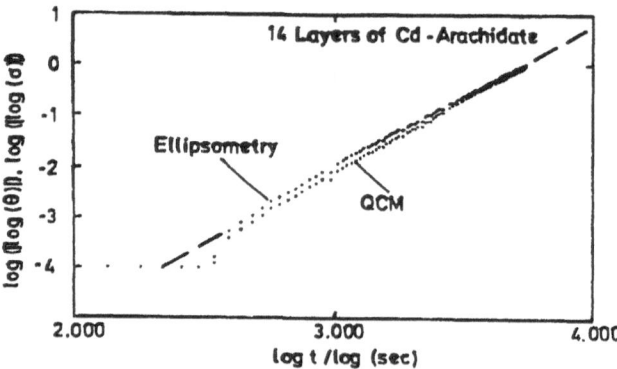

Fig. 11. Log (log(Θ)) from ellipsometry and log (log(σ)) from QCM data as a function of log t for the 14-layer sample. This plot allows the determination of the power-dependence of the desorption rate in the cooperative defect generation model (for details, see text)

an "active site" more than one defect (on average) must be photo-generated. Details of the molecular origin or mechanism of this hypothesis, of course, are completely speculative. We limited the analysis to thick samples, because for $m < 8$ layers the dominant influence of the substrate obscures any details of such a defect-cooperativity.

Final remarks

In the previous sections we presented desorption data of CdA multilayers assemblies of different thicknesses and analyzed them on the basis of various kinetic models. We found, in particular, that the proper description of the process needs to take into account that the desorption occurs at defects that are photogenerated and penetrate some 10 layers deep into the LB-film. Thus, the desorption was thickness-dependent for thin layer systems, and became independent for thicker samples.

Now, it is interesting to try to estimate the efficiency of the whole process. The employed illumination configuration was equivalent to a photon flux of ca. $3 \cdot 10^{15}$ s^{-1} cm^{-2} if referred to the 90% power at $\lambda = 254$ nm. Given the absorbance of say a 12-layer sample at that wavelength, we estimate that ca. 3% corresponding to 10^{14} photons $\cdot$ s^{-1} cm^{-2} are absorbed by the films, which means that within the 10^4 s that it takes for the whole LB-layer to desorb, some 10^{18} photons are absorbed by ca. $6 \cdot 10^{15}$ CdA molecules or, equivalently, by ca.

10^{17} CH$_2$-groups. Photodesorption, therefore, seems to be a very efficient process.

Acknowledgement

We would like to thank J. P. Rabe, H. Möhwald, R. Beck, K. G. Weil, K. Pittermann, F. Eggers, T. Funk, and H. Motschmann for helpful discussions.

References

1. See, e.g., Proceedings of the "IVth International Conference on Langmuir-Blodgett-Films" 1989, Tsukuba, Japan, published in: Thin Solid Films
2. Kuhn H, Möbius D, Bücher H (1972) Spectroscopy of Monolayer Assemblies, Chapter VII, Techniques of Chemistry, Vol. 1, Physical Methods of Chemistry Part III B, eds. A. Weissberger and B. W. Rossiter (Wiley-Interscience, New York)
3. Gaines GL Jr (1966) Insoluble Monolayers at Liquid-Gas Interfaces (Wiley-Interscience, New York)
4. Novotny V, Swalen JD, Rabe JR (1989) Langmuir 5:485—489
5. Laxhuber LA, Möhwald H (1987) Surface Science 186:1—14
6. Schmidt R, Voit H, Johannsmann D, Knoll W, in preparation
7. Barraud A (1983) Thin Solid Films 99:317—321
8. Kowel ST, Selfridge R, Eldering C, Matloff N, Stroewe P, Higgins BG, Sprinivasan MP, Coleman C (1987) Thin Solid Films 152:377—403
9. Lu C, Czanderna AW (1984) in: Wolsky SP, Czanderna AW, Methods and Phenomena 7, Elsevier, Amsterdam, New York, Tokio
10. Azzam RMA, Bashara NM (1977) Ellipsometry and Polarized Light, North Holland, Amsterdam
11. Tippmann-Krayer P, Laxhuber LA, Möhwald H (1988) Thin Solid Films 159:387—394
12. Johnson PB, Christy RW (1972) Phys Rev B6:4370—4379
13. Swalen JD (1986) J Molec Electron 2:155—181
14. Archer RJ, Gobeli GW (1965) J Phys Chem Solids 26:343—348
15. Rabe JR (1984) PhD Thesis, Technische Universität München, Munich
16. Rabe JP, Knoll W (1986) Opt Commun 57:189—192
17. Steiger R (1971) Helv Chim Acta 54:2645—2658

Authors' address:

D. Johannsmann and W. Knoll
Max-Planck-Institut für Polymerforschung
Postfach 3148
6500 Mainz, FRG

Progress in Colloid & Polymer Science   Progr Colloid Polym Sci 83:155—166 (1990)

On the mechanism of procaine penetration into stearic acid monolayers spread at the air/water interface

M. Tomoaia-Cotişel*)

Department of Physical Chemistry, University of Marburg, Marburg, FRG

Abstract: The adsorption characteristics of procaine have been studied by measuring the surface tension of aqueous procaine solutions and by recording the compression isotherms of the stearic acid monolayers in the presence of various procaine concentrations in the subphase, under controlled pH conditions. The amount of procaine penetrated into the stearic acid monolayers is derived from the molecular area increase recorded at constant surface pressure by taking into account the different possible models for the mechanism of procaine penetration into the monolayer. Comparing these results with the one obtained by using the Gibbs' adsorption equation adapted to penetration phenomena, the most probable mechanism of procaine penetration into the stearic acid monolayers is discussed. The findings show that the presence of the stearic acid monolayers entails the enhanced adsorption of procaine, especially in the liquid state of the stearic acid monolayer. In the alkaline subphase the amount of procaine penetrated is higher than its value in the acidic phase. The results are interpreted in terms of protolytic equilibria in which both surfactants, procaine and stearic acid, participate. On compression of the stearic acid monolayer the procaine adsorption increases and after attaining a maximum value it decreases and vanishes near to the collapse of the monolayer. The results are discussed in terms of penetration, conformational transitions, and subsequent expulsion of the procaine molecules during the compression of the stearic acid monolayers spread onto procaine containing subphases. At the monolayer collapse, procaine is completely squeezed out from the monolayer and it probably remains bound to the solid state of the stearic acid monolayer, entailing an important increase of its collapse pressure. This effect was interpreted in terms of dipole-ion interactions and hydrogen bondings (pH 2) or in terms of electrostatic interaction (pH 8).

Key words: Stearic acid monolayers; procaine; penetration phenomena; adsorption of procaine; surface pressure; surface tension

Introduction

The anesthetics are known to exert their action by closing the sodium channels of nerve membranes [1—3] and, hence, blocking the nerve signal propagation. The molecular mechanism of anesthetic action has been the subject of extensive studies, but it is still unclear whether this blocking is the result of a direct anesthetic-protein interaction [2] or a perturbation by the anesthetic of the lipid matrix surrounding the channels [1, 3]. Yet, it seems clear that lipids are involved either directly or indirectly via lipid-protein interaction in the mechanism of anesthesia.

Studies performed by using different experimental techniques showed some anesthetics to increase the fluidity of lipid bilayers [4], to decrease the order in hydrocarbon chains [5—9], to extend the surface area of monolayers maintained at constant

───────
*) Permanent address: Physical Chemistry Department, University of Cluj-Napoca, 3400 Cluj-Napoca, Romania.

surface pressure [10], and to increase the surface pressure of lipid film maintained at constant area [11]. Although these experiments involve different physical parameters, presumably, the molecular origin of the effects observed is the same, viz., the modification of the structural and dynamic properties of ordered acyl chains due to the binding to and penetration of anesthetics into oriented lipid systems. The quantification of the interaction of anesthetic with the membrane model, as with bilayers or monolayers, is a fundamental problem in biophysics.

Very few studies of the effects of a tertiary amine anesthetic, like procaine on membrane model have been done, and the results at the molecular level are completely different. It appears that the procaine is bound by egg phosphatidylethanolamine [7], but it is only weakly bound by the ordered acyl chains in egg phosphatidylcholine [8] aqueous lamellar dispersions. The two phospholipids used differ only in the nature of the polar head group, and have the same fatty acid composition.

In our opinion, the most conclusive approach to study the effects of procaine on highly ordered acyl chains is to use stearic acid monolayers spread at the air/aqueous interfaces. Stearic acid was chosen for several reasons. Firstly, its monolayers are stable in a large range of pH [12]. Secondly, the stearic acid monolayers are very sensitive to the subphase electrolytes [13, 14].

While in the past [10, 11], monolayer experiments were usually performed at constant monolayer area by injection of anesthetic solutions into the subphase of the already spread film (for observing the pressure increase, or at constant surface-pressure, monitoring the increase of the surface area upon incorporation of the anesthetic of interest) the present experiments were recorded in a different mode. The monolayers were directly spread on a subphase containing the anesthetic in a chosen concentration range, and after the attainment of internal equilibrium the compression isotherms were recorded in equilibrium conditions. Our monolayer experiments present the advantage that the equilibrium between the stearic acid monolayer and the procaine subphase can be followed in a very large surface pressure range, from very low surface pressures (spreading equilibrium pressures) up to the highest ones (collapse pressures) of the monolayer.

The aim of this paper is to measure the effects of procaine on stearic acid monolayers spread under controlled pH conditions and to derive the amount of procaine penetrated into the monolayers and its variation under different conditions. Attempts have been made to derive the penetration number, defined as the mole ratio of procaine and stearic acid in the mixed monolayer by using the Gibbs' adsorption equation and by means of the various proposed models for mechanism of procaine penetration into the monolayer. The penetration number is the key parameter that has to be assessed before any other characteristic of the procaine-stearic acid monolayer interaction can be fully quantified.

Materials and methods

Stearic acid (SAH) was purchased from Schuchardt. Procaine hydrochloride (PR-HCl) was obtained from Hoechst. Benzene used as spreading solvent was purchased from Merck. The subphases used for pH 2.0 ± 0.1 and 8.0 ± 0.1 were obtained from the twice distilled water by addition of the HCl solution and the mixture of KH_2PO_4 and $Na_2HPO_4 \cdot 2 H_2O$, respectively. The concentration range of procaine solutions measured was 0.001 mM—10 mM. The pH of the subphase was controlled before and after each measurement. All substances were p.a. grade and were used without further purification.

Surface tensions of procaine aqueous solutions not covered by stearic acid monolayers were measured at 20 ± 0.1 °C by using du Noüy's ring method, which has an accuracy of 0.1 mN/m.

Stearic acid monolayers were spread at the air/aqueous solution interface, both in the absence and in the presence of procaine, at pH 2 and pH 8, and left (15—30 min) to attain internal equilibrium. The monolayers obtained were then compressed. Compression speeds between 0.005 and 0.025 nm²/molecule · min ensured a good reproducibility. The equilibrium between the stearic acid monolayer and the procaine subphase establishes rapidly, and it can be evidenced by the constancy of the surface pressure value for each molecular area of stearic acid. In this case, at the air/water interface a mixed monolayer is formed that contains, besides water molecules (noted 1), also molecules of both surfactants: procaine (noted 2) and stearic acid (noted 3). Sometimes, the mixed monolayers were kept at the air/aqueous solution interface for 60 min and the results were the same, within the experimental errors. This means that we obtained in our measurements the equilibrium values, or at least the steady state conditions.

Compression of the monolayers was performed discontinuously by using the Wilhelmy method. The surface pressures were measured and the accuracy was 0.1 mN/m. In all cases, at least 10 compression isotherms were recorded under identical conditions. All measurements were performed at room temperature (22 °C).

Results and discussion

Molecular species involved

Since both the stearic acid (SAH) and procaine (PR) molecules may participate in protolytic equilibria, the fraction of the different molecular species as function of pH was calculated in our previous paper [14]. In order to study the effects of charged and non-charged procaine molecular species on the stearic acid monolayers, pH was specifically chosen as pH 2 and pH 8.

In the case of stearic acid or octadecanoic acid: $CH_3(CH_2)_{16}COOH$ at pH 2, its monolayer is an uncharged one, the molecules being completely unionized (SAH). At pH 8, stearic acid molecules are completely ionized and give a charged film containing only stearate anions (SA^-).

Procaine or novocaine: 4-aminobenzoic acid [2-diethyl aminoethyl] ester: $4-H_2NC_6H_4CO_2[CH_2-CH_2N(C_2H_5)_2]$: PR is a tertiary amine, also containing a primary amino group linked to an aromatic ring. Consequently, it may exist in three forms: uncharged molecule (PR, free base) and charged ones, i.e., monocation: $4-H_2NC_6H_4CO_2[CH_2CH_2HN^+(C_2H_5)_2]$: ($PRH^+$) and dication: $4-H_3N^+C_6H_4CO_2[CH_2CH_2HN^+(C_2H_5)_2]$: ($PRH_2^{2+}$). The calculations show, at pH 2, that the amount of PRH_2^{2+} attains about 60%, remaining at 40% for PRH^+. At pH 8, PRH^+ is the predominant species, but there are also neutral molecules, PR, in a proportion of about 24%.

Surface activity of charged and non-charged procaine in absence of stearic acid

The variation of the surface tension of procaine aqueous solutions, σ in mN/m, with the logarithm of procaine molar concentration (c_2) at the air/water interface for pH 2, pH 8 and unbuffered systems is shown in Fig. 1. It is worth mentioning that in procaine unbuffered solutions (pH 5.6—5.0) there is only the monocation form (PRH^+) of procaine [14]. For simplicity, all molecular species of procaine are noted as 2.

The procaine adsorption (Γ_2^0) can easily be obtained at different surface tension values by means of graphical derivation, according to the Gibbs' equation:

$$\Gamma_2^0 = -\frac{1}{kT}\left(\frac{\partial \sigma}{\partial \ln c_2}\right)_T, \qquad (1)$$

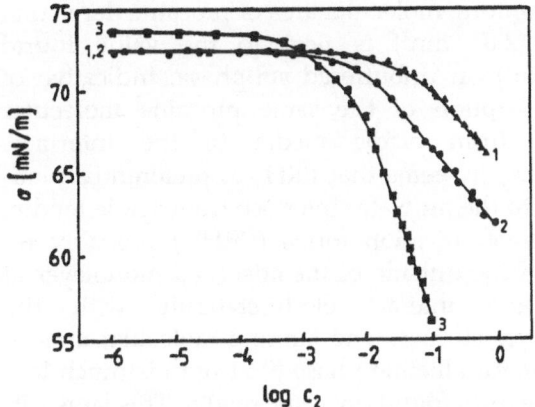

Fig. 1. Surface tension of procaine aqueous solutions as a function of the logarithm of procaine molar concentration (c_2), at various pH-values: pH 2 (curve 1), unbuffered (pH 5.6—5.0; curve 2) and pH 8 (curve 3)

where σ and c_2 have their known significance, k and T are Boltzmann's constant and absolute temperature, respectively, Γ_2^0 being expressed as the number of procaine molecules adsorbed per unit area of the air/water interface.

In Fig. 1, at higher c_2 values, the σ vs $\log c_2$ plots become linear. From the slope of these linear portions the values of maximum procaine adsorption ($\Gamma_{2,max}^0$) were calculated by means of Eq. (1) and the least-squares method and, consequently, the limiting values of procaine molecular area (A_2^0) in adsorbed monolayer at the air/water interface were obtained (Table 1).

In Table 1, with increasing pH the maximum adsorption value increases, indicating that the surface activity of the procaine molecular species increases in the order: $PRH_2^{2+} < PRH^+ < PR$. This is paralleled by the hydrophobicity of the three species since the water solubility is low for the neutral form and increases for the charged forms.

Table 1. Maximum procaine adsorption ($\Gamma_{2,max}^0$) and its limiting molecular area (A_2^0) at the air/water interface

Adsorption characteristics	pH		
	2	5.6—5.0	8
$\Gamma_{2,max}^0 \times 10^{-17}$ molecules/m²	4.831 ± 0.118	5.103 ± 0.110	11.017 ± 0.105
A_2^0, nm²/molecule	2.07 ± 0.05	1.96 ± 0.04	0.91 ± 0.01

The limiting molecular area of procaine derived at pH 2 (2.07 nm²) is near to the value found (1.96 nm²) on unbuffered subphase, indicative of the adsorption of the same procaine molecular species from acidic media to the interface. Therefore, it seems that PRH⁺ is preferentially adsorbed to the air/water interface from acidic media. In contrast, dication forms (PRH₂²⁺) probably remain on the outside of the adsorbed monolayer at pH 2 and interact electrostatically with the monolayer surface. The limiting molecular area of procaine for alkaline phase (0.91 nm²) is much less than the ones found on acidic media. This large difference might be due to the enhanced adsorption of non-charged procaine molecules from aqueous phase of pH 8 to the air/water interface. In this latter case, the electrostatic repulsion between the adsorbed molecules becomes less, allowing a closer packing.

In order to calculate the procaine area requirements, A_2, molecular models have been constructed by taking into account bond lengths, bond angles and van der Waals radii of the atoms. Area requirements have been calculated for three extreme conformations, shown in Fig. 2.

Conformation HE is a horizontal extended one presented in "vertical view, i.e., the plane of the figure coincides with the air/water interface. The horizontal linear dimensions of the molecules are

denoted by a and b, having always $a > b$. The conformations VE and VF are vertical ones, presented in Fig. 2 in "laterial view". VE-conformation may be considered as being an vertical extended conformation, with the ethyl groups lying on the air/water interface. In the case of VF-conformation the headgroup is forced to occupy the space beneath the vertically oriented aromatic ring.

Linear dimensions a and b, derived from these models, are presented in Table 2, together with the molecular areas A_4 and A_p. A_4-value is obtained as a^2 and it corresponds to a tetragonal close packing of rotating or randomly oriented molecules. A_p-value is calculated as $a \times b$ and represents the area requirement of the parallely oriented molecules, i.e., their cross-sectional area. By comparing these calculated area values for different packing degrees in procaine surface lattice (Table 2) with the molecular areas of procaine derived experimentally in pure adsorbed monolayer (Table 1), it may be concluded that procainium cations and even procaine free base molecules are adsorbed at the air/water interface in a horizontal position.

Table 2. Linear dimensions and area requirements of procaine in its extreme conformations

Confor-mation	a nm	b nm	A_4 nm²	A_p nm²
HE	1.23	0.52	1.513	0.640
VE	0.85	0.47	0.722	0.400
VF	0.52	0.47	0.270	0.244

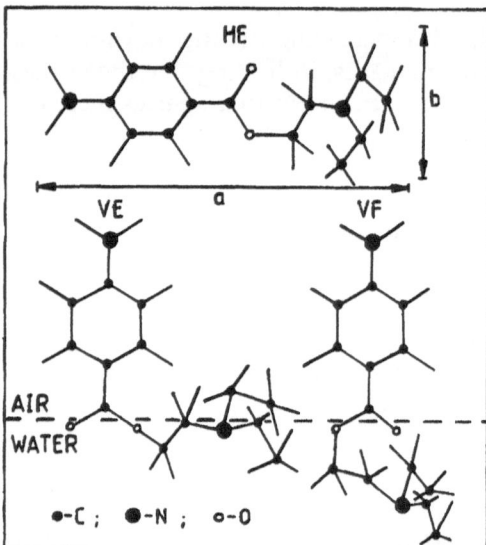

Fig. 2. Extreme conformations of procaine: HE-horizontal extended, VE-vertical extended, VF-vertical forced; dashed line indicates the air/water interface

Compression isotherms and surface characteristics of stearic acid monolayers

Compression isotherms, i.e., surface pressure (π) vs mean molecular area (A_3) of stearic acid monolayers recorded on subphases at pH 2 and pH 8 for several procaine concentrations are presented in Figs. 3 and 4, respectively. The stearic acid monolayers were spread at the same molecular area, both in the presence and in the absence of procaine in the subphase, in order to avoid the influence of spreading kinetic effects [13].

As a general feature, one may observe that all isotherms on pH 2 contain two linear portions (Fig. 3) corresponding to the liquid condensed (LC)

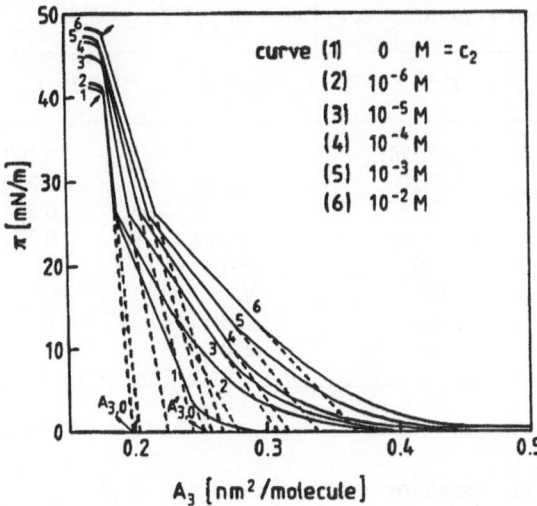

Fig. 3. Compression isotherms of stearic acid monolayers spread onto aqueous solutions of pH 2 at different procaine concentrations (c_2)

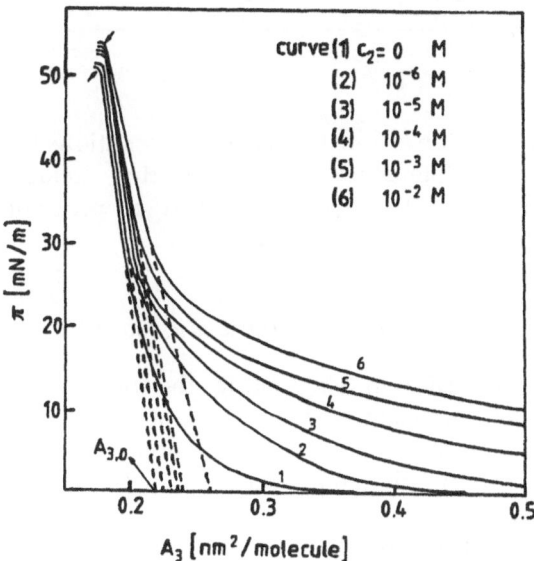

Fig. 4. Compression isotherms of stearic acid monolayers spread onto aqueous solutions of pH 8 at different procaine concentrations (c_2)

and solid (S) states, respectively. Obviously at pH 8, the isotherms present a single linear portion (Fig. 4), the monolayers remaining in a very compact liquid condensed up to the collapse. For both pH 2 and 8, with increasing procaine concentration the compression isotherms are moved to large molecular areas of stearic acid showing the expan-

sion of the monolayers. This extending effect of procaine can be characterized by the increase in surface pressure recorded at different constant A_3 values, or by the increase of molecular area corresponding to different π values.

From these compression isotherms (Figs. 3 and 4) surface characteristics of the stearic acid monolayer were derived:

$A'_{3,0}$-limiting molecular area for the liquid condensed state, obtained by extrapolating to $\pi \rightarrow 0$ of the first linear portion of the isotherm (Fig. 3) recorded at intermediate π-values;

$A_{3,0}$-limiting molecular area for the solid state (Fig. 3) or the very compact surface lattice (Fig. 4) obtained by extrapolation of the last linear portion of the isotherm at high π-values;

$A_{3,c}$, π_c-collapse area and collapse pressure, representing the A and π values, respectively, corresponding to the sudden slope change observed at high π values [13] (indicated by arrows in Figs. 3 and 4);

$A_{3,t}$, π_t-molecular area and surface pressure at the liquid condensed-solid phase transition (intersection of the two linear portions in Fig. 3);

$(C_{s0}^{-1})'$-surface compressional modulus for the liquid condensed state given by the relation:

$$(C_{s0}^{-1})' = -A'_{3,0} \left(\frac{\partial \pi}{\partial A_3} \right)_T = A'_{3,0} \frac{\pi_t}{(A'_{3,0} - A_{3,t})} .$$

C_{s0}^{-1}-surface compressional modulus for the solid state, or the very compact two-dimensional phase, given by:

$$C_{s0}^{-1} = -A_{3,0} \left(\frac{\partial \pi}{\partial A_3} \right)_T = A_{3,0} \frac{\pi_c}{(A_{3,0} - A_{3,c})} .$$

Surface characteristics of stearic acid monolayers are given in Tables 3 and 4 for pH 2 and pH 8, respectively.

Inspection of Tables 3 and 4 shows that $A_{3,c}$ is not affected by the presence of procaine in the subphase. In contrast with $A_{3,c}$ values, all the other characteristic areas, $A_{3,t}$, $A_{3,0}$, and $A'_{3,0}$ increase with increasing procaine concentration, indicating the penetration of procaine molecular species into stearic acid monolayers and their subsequent expulsion near to the monolayer collapse.

The procaine penetrated into the monolayers has an expanding and fluidizing effect upon the stearic acid monolayers, both at pH 2 and pH 8, reflected also by the enhanced compressibility of the mono-

Table 3. Surface characteristics of stearic acid monolayers on procaine containing subphases of pH 2

c_2, M	0	10^{-6}	10^{-5}	10^{-4}	10^{-3}	10^{-2}
$A_{3,c}$, nm^2	0.180	0.180	0.180	0.178	0.175	0.180
$A_{3,t}$, nm^2	0.188	0.190	0.200	0.206	0.210	0.220
$A_{3,0}$, nm^2	0.200	0.205	0.228	0.245	0.260	0.268
$A'_{3,0}$, nm^2	0.255	0.280	0.310	0.317	0.340	0.370
π_c, mN/m	40.8	41.0	44.0	46.0	47.0	47.5
π_t, mN/m	26	26	26	26	26	26
C_{s0}^{-1}, mN/m	408.0	336.2	209.0	168.2	143.8	144.7
$(C_{s0}^{-1})'$, mN/m	99.0	80.9	73.3	74.3	68.0	64.1

Table 4. Surface characteristics of stearic acid monolayers on procaine containing aqueous subphases at pH 8

c_2, M	0	10^{-6}	10^{-5}	10^{-4}	10^{-3}	10^{-2}
$A_{3,c}$, nm^2	0.176	0.178	0.180	0.180	0.180	0.180
$A_{3,0}$, nm^2	0.220	0.226	0.232	0.236	0.240	0.260
π_c, mN/m	51	51	52	52.5	53	53.5
C_{s0}^{-1}, mN/m	255	240	232	221	212	174

layers, i.e., by decreasing of the surface compressional moduli with increasing procaine concentrations (Tables 3 and 4, respectively). These results could be related to the natural membrane expansion caused by the anesthetics, which has also been proposed to be involved with the mechanism of anesthesia.

Further, it is obvious that the liquid condensed to solid phase transition occurs at the same π_t value (Table 3), irrespective of the procaine concentration in the subphase (pH 2). In contrast, the collapse pressure π_c increases systematically with increasing procaine concentration. This finding suggests that the expulsed procaine molecules remain in an adjacent layer on the outside of the stearic acid monolayer surface and π_c value increases due to a vertical interaction between the stearic acid and procaine molecular species.

The experimental results at collapse can be interpreted in terms of dipole-ion interactions between the uncharged carboxyl groups of stearic acid monolayers at pH 2 and the positively charged procaine molecules, and in terms of hydrogen bondings, especially between the carboxyl group and the primary amine group of procaine monocations (PRH$^+$). At the monolayer collapse at pH 8 the effect of procaine is less. Under these conditions the electrostatic interactions between the stearate anions (SA$^-$) and the positively charged procaine (PRH$^+$) are probably dominant.

Procaine penetration into the stearic acid monolayers

As a first step in elucidating the penetration mechanism, we estimated the amount of procaine incorporated into the stearic acid monolayers by using the thermodynamic theory of monolayer penetration [15—19].

The equilibrium between the bulk subphase and the mixed stearic acid: procaine monolayer was treated by using Gibb's adsorption equation, adapted to the penetration phenomena [15—18], which in isothermal conditions and for a constant mean molecular area of stearic acid A_3 has the form:

$$\Gamma_2 = \frac{(A_3 - \bar{A}_3)}{kTA_3}\left(\frac{\partial \pi}{\partial \ln c_2}\right)_{T,A_3}, \qquad (2)$$

where Γ_2 stands for the number of procaine molecules adsorbed on unit area of the mixed monolayer; $\bar{A}_3$ is the partial molecular area of stearic acid; π is the surface pressure defined as $\pi = \sigma - \sigma_m$, σ represents the surface tension of aqueous solutions both in the absence and in the presence of procaine, and σ_m is the surface tension of the interface with stearic acid monolayer; the other magnitudes have their known meaning.

Concerning the partial molecular area, attempts were made to approximate $\bar{A}_3$ by A_3^0, i.e., by taking the molecular area of component 3 at the same π value, but in the absence of the component 2 [15]. This approximation seems not to be perfectly satisfactory in our experiments, since the stearic acid molecule is anchored in the interface by its carboxyl group and, actually, the area requirement of the carboxyl group is less than A_3^0, especially at low surface pressures. Therefore, we propose as a more reasonable approach to take the collapse area in pure stearic acid monolayer, $A_{3,c}$, for $\bar{A}_3$ in Eq. (2).

Further, the extent of the procaine penetration is characterized by the penetration number n_p defined as the ratio of the number of procaine molecules (Γ_2) and of stearic acid molecules (Γ_3) per unit area of the mixed monolayer, i.e., by the number of adsorbed procaine molecules per one stearic acid molecule:

$$n_p = \frac{\Gamma_2}{\Gamma_3} = \Gamma_2 A_3 \ . \tag{3}$$

Combining Eqs. (2) and (3) by taking $\bar{A}_3 = A_{3,c}$, one obtains the following expression for the penetration number:

$$n_p = \frac{(A_3 - A_{3,c})}{kT} \left(\frac{\partial \pi}{\partial \ln c_2} \right)_{T, A_3} . \tag{4}$$

Therefore, the amount of penetrated procaine into the stearic acid monolayers can be estimated, by using Eq. (4), directly from the plots of surface pressure vs $\log c_2$ performed for different constant A_3 values. For this purpose a set of curves is presented in Fig. 5 (pH 2) and another set in Fig. 6 (pH 8); these curves were constructed on the basis of the compression isotherms given in Figs. 3 and 4, respectively. The derivative $(\partial \pi / \partial \log c_2)_{T, A_3}$ enclosed in Eq. (4) was obtained by means of graphical derivation from Figs. 5 and 6 for different c_2 and A_3 values. Finally, the penetration numbers of procaine into the stearic acid monolayers (n_p) are

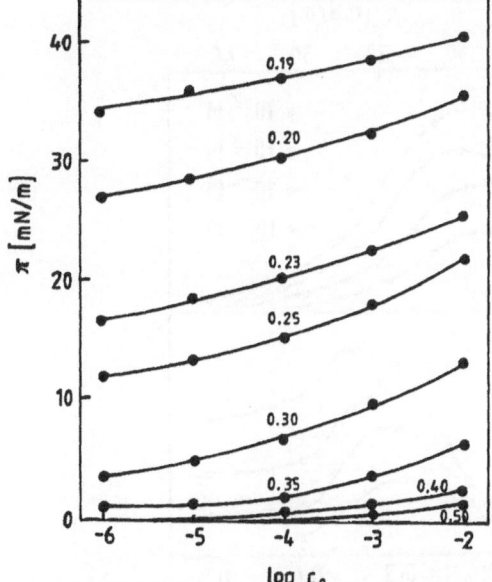

Fig. 5. Surface pressure vs logarithm of procaine concentration curves for several constant mean molecular areas (A_3) of stearic acid on pH 2. Figures near the curves indicate the A_3 values expressed in nm^2

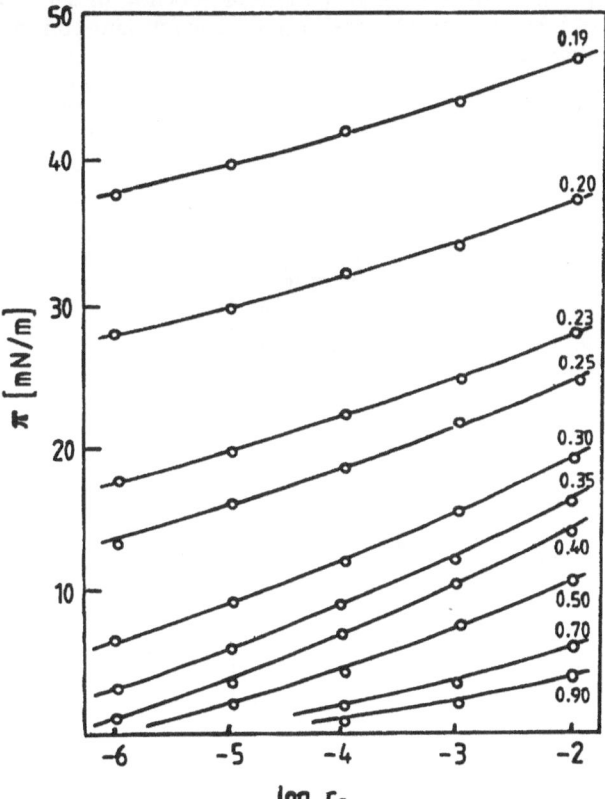

Fig. 6. Surface pressure vs $\log c_2$ curves for several constant mean molecular areas (A_3) of stearic acid at pH 8. Figures near the curves indicate the value of A_3 expressed in nm^2

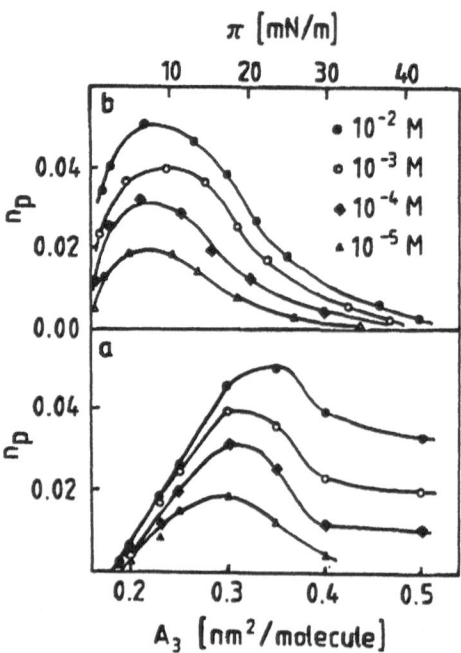

Fig. 7. Penetration numbers n_p calculated by means of Eq. (4) as function of A_3 (a) and as function of π (b) at pH 2. Figures after the symbols indicate procaine concentration (c_2)

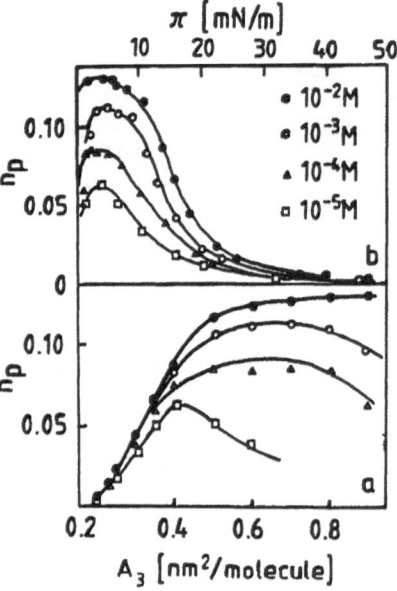

Fig. 8. Penetration numbers, n_p, calculated by means of Eq. (4) as function of A_3 (a) and as function of π (b) at pH 8. Figures after the symbols indicate procaine concentration (c_2)

visualized in Figs. 7 and 8 for both pH 2 and 8, respectively, in two different plots. In Figs. 7a and 8a, n_p values are given as function of A_3, while in Figs. 7b and 8b the same n_p values are presented as a function of the surface pressure, corresponding to different molecular areas A_3.

Inspection of Figs. 7 and 8 reveals that the penetration numbers are large at low surface pressures and very small at high surface pressures. Hence, the penetration of procaine into the stearic acid monolayers becomes increasingly difficult with increasing of packing density of the stearic acid molecules, both at pH 2 and pH 8.

The n_p values exhibit a maximum at the surface pressures of about 5 mN/m and 10 mN/m at pH 8 and pH 2, respectively. The maximum values are much higher at pH 8 (Fig. 8) than the values for subphases of pH 2 (Fig. 7), and obviously, they are dependent on the procaine concentration in subphases. The important increase of n_p with increasing pH is a cooperative effect of the protolytic equilibria in which participate both surfactants: stearic acid and procaine. On the one hand, the successive deprotonations $PRH_2^{2+} \rightarrow PRH^+ \rightarrow PR$ lead to the formation of molecular species with higher surface activity. On the other hand, deprotonation of the neutral stearic acid molecules and the formation of the stearate anions entail the appearance of important electrostatic attractions between the oppositely charged surfactant ions, which could also partially alter the interchain interactions.

On the mechanism of procaine penetration

To obtain a clearer image concerning the penetration mechanism of procaine into stearic acid monolayers, in a second step the curves of molecular area increase (ΔA) corresponding to different surface pressures (π) have been constructed for both pH 2 and pH 8 at a constant procaine concentration (Fig. 9). We have chosen 1 mM procaine concentration because the procaine effect upon the collapse of the stearic acid monolayers is more important only up to this concentration value, and over that value it becomes insignificant (see Tables 3 and 4). The ΔA value represents the difference between the A_3 values recorded in the presence (curve 5) and in the absence (curve 1) of procaine, at the same surface pressure π value (see Figs. 3 and 4). In Fig. 9, ΔA exhibits a maximum at low surface

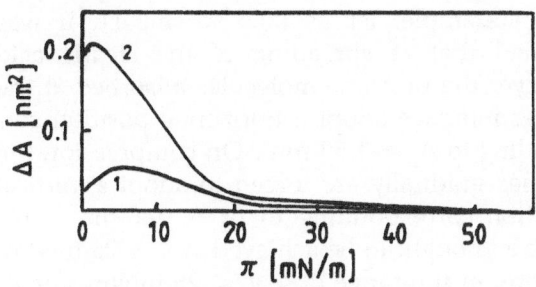

Fig. 9. The molecular area increases (ΔA) due to the procaine penetration as a function of the surface pressures, at a constant procaine concentration of 1 mM. Curve (1) = pH 2; (2) = pH 8

pressures and rapidly diminishes with increasing π.

On the basis of ΔA vs π curves (Fig. 9) the amount of penetrated procaine can be evaluated by a quantitative description of the change in surface area as follows.

The most simple approach considers the pure stearic acid monolayer to be formed of fatty acid molecules only, and their molecular area noted A_3^0 is given by:

$$A_3^0 = \frac{n_3 A}{n_3}, \tag{5}$$

where n_3 is the constant number of fatty acid molecules and A denotes the surface molecular area of stearic acid.

In the presence of procaine the mixed monolayer is treated as a binary solution and molecular area of stearic acid noted A_3 may be expressed as:

$$A_3 = \frac{n_3 A + n_2 A_2}{n_3}, \tag{6}$$

where n_2 stands for the number of procaine molecules and A_2 for their area requirements. In Eq. (6) both A and A_2 may depend on π.

By presuming that A is a unique function of π and by denoting with $\Delta A = A_3 - A_3^0$ the mean molecular area difference measured at a given π value, the penetration number n_p is given by:

$$n_p = \frac{n_2}{n_3} = \frac{\Delta A}{A_2}. \tag{7}$$

The penetration number can also be derived in a *more sophisticated manner*, by taking into account the presence of water (noted 1) molecules in the monolayer, i.e., by considering the "pure" stearic acid monolayer to be a binary solution, and the penetrated mixed monolayer a ternary one. In this approach Eqs. (5) and (6) become

$$A_3^0 = \frac{n_3 A + n_1 A_1}{n_3} \tag{8}$$

and

$$A_3 = \frac{n_3 A + n_1' A_1 + n_2' A_2}{n_3}, \tag{9}$$

respectively; where n_1 and n_1' stand for the number of water molecules from the monolayer in the absence and presence of procaine, respectively, and A_1 is the area requirement of water molecule.

The equilibrium between the bulk subphase (B) and the monolayer (M) implies the equality of the chemical potentials of water in both phases, i.e.,

$$\mu_1^{0B} + kT\ln x_1^B + kT\ln f_1^B$$
$$= \mu_1^{0M} + kT\ln x_1^M + kT\ln f_1^M + \pi A_1, \tag{10}$$

where x_1^B, x_1^M, f_1^B, f_1^M stand for the molar fraction and activity coefficient of water in the B and M phases, respectively.

In the case of pure water, one has $x_1^B = x_1^M = f_1^B = f_1^M = 1$ and $\pi = 0$. Consequently, $\mu_1^{0B} = \mu_1^{0M}$, and from Eq. (10) one obtains

$$\ln x_1^M = \ln x_1^B + \ln \frac{f_1^B}{f_1^M} - \frac{\pi A_1}{kT}. \tag{11}$$

In the righthand side of Eq. (11), the first term and the second one can be neglected before the last one, i.e., in first approximately they may be presumed to be neglectingly small. Therefore, at a given π, x_1^M may be presumed to have the same value, irrespective of the presence or absence of the procaine.

By multiplying Eqs. (8) and (9) by the molar fraction of stearic acid in the monolayer, i.e., by x_3 and x_3', respectively, for the sake of simplicity the index (M) is omitted and one obtains

$$x_3 A_3^0 = x_3 A + x_1 A_1 \tag{12}$$

$$x_3' A_3 = x_3' A + x_1' A_1 + x_2' A_2. \tag{13}$$

Taking into account that $x_1 + x_3 = 1$, Eq. (12) yields:

$$x_1 = \frac{A_3^0 - A}{A_3^0 + A_1 - A} \, . \tag{14}$$

Since for a constant π value $x_1 = x_1'$ and $x_1' + x_2' + x_3' = 1$, from Eqs. (13) and (14), one obtains:

$$n_p = \frac{n_2'}{n_3'} = \frac{A_3 - A_3^0}{A_3^0 - A + A_2} \, . \tag{15}$$

Obviously, for $A_3^0 = A$, Eq. (15) turns into Eq. (7), i.e., in the case of very condensed films both equations give practically the same result. If the monolayer is a liquid one, Eq. (15) is to be preferred.

Concerning the area requirement of stearic acid, $A = 0.18 \text{ nm}^2 = A_{3,c}$ was used, since the collapse area of pure stearic acid well approximates the actual area requirement of the carboxyl group in the monolayer, the hydrocarbon chain being raised up into the air phase.

In the case of procaine, the molecular area A_2 is not constant, but rather depends on the surface pressure applied. From inspection of molecular models (Fig. 2), the area requirement A_2 in the most extended horizontal "lying down" conformation (HE) is of about 1.50 nm^2 in the case of the "rotating" or randomly oriented molecules, and of 0.64 nm^2 for the "non-rotating", parallely oriented molecules (Table 2). By presuming a vertical orientation of the procaine molecule, i.e., by considering the tertiary amine moiety to be anchored into the aqueous subphase and the aromatic ring to penetrate into the air phase between the hydrocarbon chains of stearic acid molecules, the area is of 0.40 nm^2 in a vertical extended conformation (VE) and only 0.25 nm^2 in a vertical forced conformation (VF).

The experimental ΔA vs π curves presented in Fig. 9 were processed according to Eqs. (7) and (15) by using different hypotheses concerning the dependence of A_2 upon π. The variants tested are summarized in Table 5. In Table 5 the A_2 values are indicated for several π values characteristic for stearic acid monolayers viz. at spreading of the monolayer ($\pi \approx 0$), at the phase transition (26 mN/m), and at the collapse of the monolayer (π_c). Between two constant A_2 values indicated in Table 5, A_2 was presumed to be a linear function of π.

As an example, let us take variant 11. It was presumed that at spreading of the stearic acid monolayer the procaine molecules adsorbed at the air/water interface adopt a horizontal position corresponding to $A_2 = 1.50 \text{ nm}^2$. On compression the molecules gradually are forced to adopt a vertical orientation corresponding to $A_2 = 0.40 \text{ nm}^2$. This process is thought to be achieved at $\pi = 26$ mN/m. Therefore, in the range $0 \leqslant \pi \leqslant 26$ mN/m, for A_2 the following expression is used:

$$A_2 = 1.50 - 0.0423 \, \pi \, . \tag{16}$$

Beginning from 26 mN/m, the compression of the monolayer is presumed to further reduce A_2, the procaine molecular area reaching 0.25 nm^2 at monolayer collapse, obeing the linear law

$$A_2 = 0.40 - \frac{(0.40 - 0.25)}{(\pi_c - 26)} (\pi - 26) \, . \tag{17}$$

As an example, several n_p vs π curves calculated for subphases of pH 2 by means of Eq. (15) and by using the variants indicated in Table 5 are presented in Fig. 10.

In order to choose the best variant for procaine penetration, the n_p values calculated from ΔA are compared with the n_p values obtained by applying Gibbs' equation (Fig. 7) and given in Fig. 10 as circles. The best agreement with adsorption measurements is obtained by means of variant 11. This

Table 5. Molecular area A_2 (in nm^2) of procaine penetrated into the stearic acid monolayer as function of π, as presumed for calculating n_p

Variant	π, mN/m		
	0	26	π_c
1	0.25	—	0.25
2	0.40	—	0.25
3	1.00	—	0.25
4	1.00	0.64	0.25
5	1.00	0.40	0.25
6	1.30	—	0.25
7	1.30	—	0.40
8	1.30	0.40	0.25
9	1.50	—	0.25
10	1.50	—	0.64
11	1.50	0.40	0.25

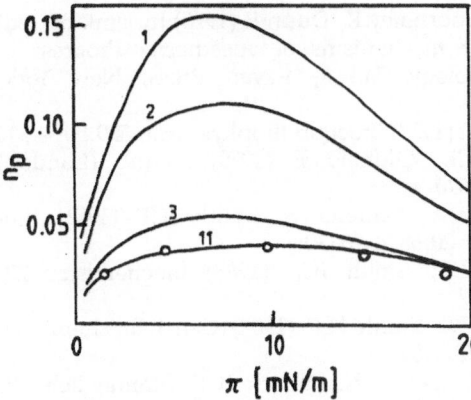

Fig. 10. Deriving of the penetration number, n_p, values at pH 2 by means of Eq. (15). Figures near the curves indicate the variant used for $A_2 = A_2(\pi)$ (see Table 5). Circles are n_p values obtained by using Gibbs' equation (see Fig. 7)

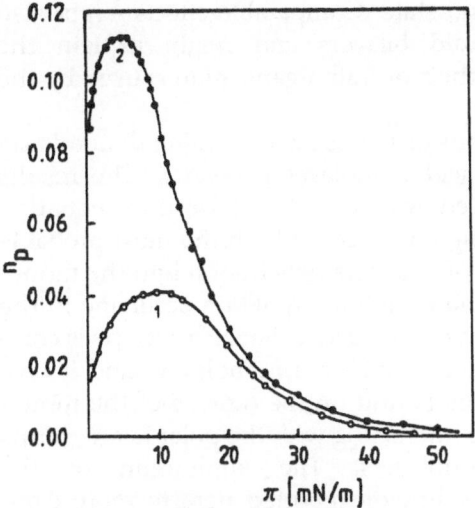

Fig. 11. The n_p values derived by means of Eq. (15), according to variant 11 in Table 5, from the ΔA vs π curves given in Fig. 9. Curve (1) = pH 2; (2) = pH 8; ○ and ● represent n_p values obtained by using Gibbs' equation for pH 2 and pH 8, respectively (see Figs. 7 and 8)

finding suggests the idea that the most probable mechanism of the procaine penetration into the stearic acid monolayers follows the variant 11, with the function $A_2 = A_2(\pi)$ illustrated by Eq. (16). Thus, the procaine molecular species are adsorbed at the spreading of the stearic acid monolayer in a horizontally lying down position according to the data of procaine adsorptions (Table 1). On compression of the monolayer the procaine molecules are gradually forced to adopt a vertical orientation (Fig. 2) in which the hydrophobic interactions with the stearic acid hydrocarbon chains are increased, and this process is completely achieved at about 26 mN/m.

The same mechanism of procaine penetration into the stearic acid monolayers is also found at pH 8, according to the data in Fig. 11 where the n_p vs π curve obtained from ΔA, by using the penetration modell 11 from Table 5, is well described by the n_p values determined by means of Gibbs' equation (see Fig. 8) and represented as full circles.

It is worth mentioning that in Fig. 11 the penetration numbers vs π at pH 2 are also represented for the whole interval of surface pressures, until the collapse of the stearic acid monolayer. Obviously, in the range 26 mN/m $\leqslant \pi \leqslant \pi_c$ for the function $A_2(\pi)$ Eq. (17) was taken for both pH 2 and pH 8. The data in Fig. 11 show that it is a good agreement between the n_p-values obtained by both methods reported here, even at high surface pressures. The use of Eq. (7) instead of Eq. (15) does not modify the general picture.

Inspection of Fig. 11 further reveals that upon compression of the stearic acid monolayers, spread on subphases of pH 2 and pH 8, the penetration number increases at the beginning and it attains a maximum value at surface pressure of about 10 mN/m and 5 mN/m, respectively. Further, it begins to decrease, indicating a gradual squeezing out of procaine molecules from stearic acid monolayers and it vanishes near to the collapse of the monolayers.

As the amount of intercalated procaine depends critically on the monolayer pressure (Fig. 11) it seems that procaine molecular species penetrates preferentially into the liquid state of the stearic acid monolayers. At the monolayer collapse, procaine is completely squeezed out from the stearic acid monolayer and it probably remains bound to the solid or to the very compact state of stearic acid monolayer, entailing an enhanced collapse pressure.

Conclusions

The results presented show that procaine molecular species both expand the packing of ordered acyl chains of stearic acid monolayers and increase the fluidity of the monolayers, especially in their liquid state. Comparable effects might also occur in lipid bilayers and might explain the changes in their overall organization caused by the anesthetics.

The amount of penetrated procaine depends on the stearic acid monolayer pressures. The results are interpreted by using Gibbs' adsorption equation and by taking into account both the most probable mechanism of procaine penetration into the monolayers and the protolytic equilibria occurring in the system. At the monolayer collapse, procaine is completely squeezed out from monolayers and it probably remains bound on the outside of the monolayer surface interacting with the polar head groups of the monolayers. The application of the monolayer technique reported here to more complex systems, including phospholipids and even proteins in interaction with procaine and other anesthetics, should add significant further understanding of their possible role in the mechanism of anesthesia.

Acknowledgement

I would like to acknowledge the financial support of the Alexander von Humboldt Foundation in the completion of this work. I would also like to express my thanks to Professor W. A. P. Luck, Dr. D. Möbius, and Dr. P. J. Quinn for their useful suggestions.

References

1. Trudell JR (1980) In: Fink BR (ed) Molecular mechanism of anesthesia: Progress in anesthesiology. Vol 2, Raven Press, New York, pp 261—270
2. Boggs JM, Roth SM, Yoong T, Wong E, Hsia JC (1976) Mol Pharmacol 12:136—143
3. Hille B, Courteney K, Dunn R (1975) In: Fink BR (ed) Molecular mechanisms of anesthesia: Progress in anesthesiology. Vol 1, Raven Press, New York, pp 13—20
4. Trudell JR (1977) Biochim Biophys Acta, 470:509—510
5. Turner GL, Oldfield E (1979) Nature (London) 277:669—670
6. Boulanger Y, Schreier S, Smith ICP (1981) Biochemistry 20:6824—6830
7. Kelusky EC, Smith ICP (1983) Biochemistry 22:6011—6017
8. Kelusky EC, Smith ICP (1984) Can J Biochem Cell Biol 62:178—184
9. Auger M, Jarrell HC, Smith ICP, Siminovitch DJ, Mantsch HH, Wong PTT (1988) Biochemistry 27:6086—6093
10. Seelig A (1987) Biochim Biophys Acta 899:196—204
11. Vilallonga FA, Phillips EW (1979) J Pharm Sci 68:314—316
12. Tomoaia-Cotişel M, Zsako J, Mocanu A, Lupea M, Chifu E (1987) J Colloid Interface Sci 117:464—476
13. Gaines Jr GL (1966) In: Insoluble monolayers at liquid-gas interfaces. Wiley-Interscience, New York
14. Tomoaia-Cotişel M, Chifu E, Mocanu A, Zsako J, Sálájan M, Frangopol PT (1988) Rev Roum Biochimie 25:227—237
15. Pethica BA (1955) Trans Faraday Soc 51:1402—1411
16. Nakagaki M, Okamura E (1982) Bull Chem Soc Japan 55:1352—1356
17. Nakagaki M, Okamura E (1982) Bull Chem Soc Japan 55:3381—3385
18. Nakagaki M, Okamura E (1983) Bull Chem Soc Japan 56:1607—1611
19. Davies JT, Rideal EK (1963) In: Interfacial phenomena. Seconded. Academic Press, New York

Author's address:

M. Tomoaia-Cotişel
Dept. of Chemistry
State University of New York
Buffalo, New York 14214, USA

Progress in Colloid & Polymer Science   Progr Colloid Polym Sci 83:167—175 (1990)

Iridescent phases in aminoxide surfactant solutions

G. Platz, C. Thunig and H. Hoffmann*)

*) Lehrstuhl für Physikalische Chemie I der Universität Bayreuth, Bayreuth, FRG

Abstract: Surfactant systems that show iridescent colors are reported. The systems consist of dimethylaminoxide surfactants ($RN(CH_3)O$), n-hexanol and water. Lamellar phases are formed at room temperature which can be diluted with water saturated with n-hexanol to iridescent phases with brilliant colors ranging from blue to red. The mole ratio of hexanol molecules to surfactant molecules in the bilayers is 1.40, 1.38, 1.62, and 1.67 for R = $C_{12}H_{25}$, $C_{14}H_{29}$, $C_{16}H_{33}$ and Oleyl ($C_{16}H_{31}$) respectively. The areas of the lamellae per surfactant molecule and hexanol are 67 Å^2, 62 Å^2 and 83 Å^2. Analogous lamellar phases are found with the cosurfactants n-pentanol, n-hexanol, n-heptanol, and n-nonanol. n-alcoholes with n < 5 or n > 10 are not able to form such iridescent phases at room temperature. The double chain surfactant dodecyl-octyl-methylaminoxide forms iridescent phases in the binary system. The colors of these phases are intensified with small amounts of $C_{14}H_{25}(CH_3)_2PO$. In the single phase regions the colors of all systems do not depend on the temperature. The observed phase inversion with small amounts of electrolyte is shown to be determined by the ionic strength. The electrostatic repulsion is necessary, but is too small to explain the stability of the systems. The bilayers are strongly fluctuating and are stabilized by steric repulsion.

Key words: Iridescent phases; liquid crystalline phases; surfactants; L$_3$-phases; aminoxides

Introduction

Iridescent surfactnat phases have been extensivly studied since 1984. Suzuki and Tsutsumi [1] found that 1—2% aqueous solutions of certain diglycerol-ethers show coloration that changes according to surfactant concentration. The origin of this phenomenon was found to be the interference of the incident light with the interfaces of lamellar liquid crystalls with huge interlamellar spacings. The distances have been estimated by scanning electron microscopy, reflexion and transmission spectroscopy. In 1986, Porte et al. reported that the system p-octyl-phenyl-sulfonate-pentanol-water is able to form lamellar phases whose interlamellar distances can be increased by addition of decane with 8% decanol until two well-defined Bragg peaks appear in light-scattering measurements [2]. The large spacings are stabilized by entropic fluctuations of the extremely flexible bilayers [3].

Papers concerning iridescent phases have been published since then by Satoh and Tsuji [4], Porte et al. [5], Hoffmann et al. [6], and Strey [7]. Satoh et al. used dilute solutions of alkenylsuccinic acid and measured reflection spectra, x-ray- and UV-diffraction of the phases. They give a complete review of the literature on iridescent phenomena.

We found another type of surfactants that form such phases. N-C14-dimethylaminoxide forms a lamellar phase with n-hexanol and water. This phase can solubilize a large amount of hydrocarbon-like n-decane. The four-component system can be diluted with water so that all colors from blue to red can be easily observed at room temperature [7]. We assumed that the stabilization of the large distances in our systems is influenced essentially by electrostatic forces due to the partial autoprotonation of the aminoxide group.

In this paper, we show a series of new iridescent phases with brilliant colors at room temperature.

These are three- or two-component systems containing aminoxide, water, and hexanol. We find evidence that weak electrostatic forces are present that could act as stabilizers at very low ionic strengths. It is shown that fluctuation forces dominate at higher salt concentrations.

Experimental

The surfactants were a gift of Hoechst AG, Gendorf. They have been twice crystallized from acetone. Critical micelle concentration-values have been measured with the surface tension method. The melting points and CMC-values are:

n-$C_{12}H_{25}(CH_3)_2NO$	119—120°C	1.7 mMol/l
n-$C_{14}H_{29}(CH_3)_2NO$	130.2—130.5°C	.14 mMol/l
n-$C_{16}H_{33}(CH_3)_2NO$	130—130.5°C	$2.5 \cdot 10^{-5}$ Mol/l
Oleyldimethyl-aminoxide		$5.25 \cdot 10^{-6}$ and $5.25 \cdot 10^{-5}$ Mol/l
n-dodecyl-n-octyl-methylaminoxide	99—100.3 C	.019 mMol/l.

Oleyldimethylaminoxide and n-dodecyl-n-octyl-methylaminoxide are strongly hygroscopic.

The n-alcanoles (C5—C10) were Fluka p.a. chemicals and were used without further purification.

Bidistilled water (1.5 µS/cm at 25°C) was used for preparing the solutions.

Results and discussion

Phase diagrams

The phase boundaries of the three-component systems were estimated by titration of the surfactant solutions with hexanol. The region of the lamellar phase is identified by polarization microscopy. The iridescent phases with the dimethylaminoxides form quickly. In contrast to these systems, it is necessary to wait 1 or 2 days until the phases with $C_{12}C_8$-aminoxide are formed. The process can be easily observed in standing glass tubes.

A lamellar phase can generally be diluted with a solution whose composition is that of the bulk phase surrounding the lamellae. In a binary system this simply means a dilution with water. In a ternary or higher system, we must take into account the solubility of the different components in water. In our case this is the n-hexanol, because the solubility of the surfactant monomers can be neglected.

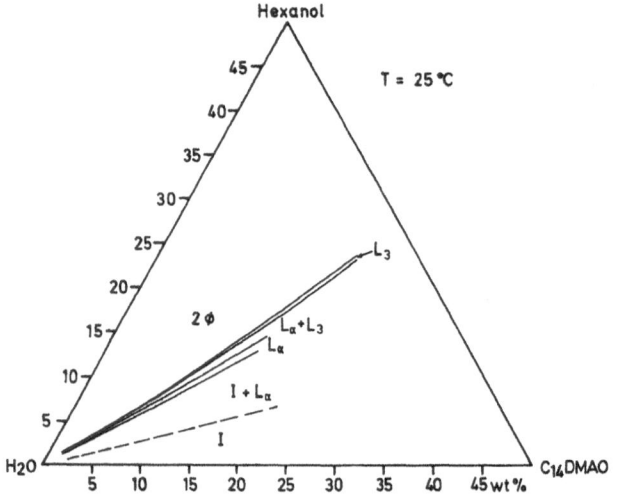

Fig. 1. Lamellar phase region of C_{14}DMAO water and hexanol (25°C)

Figure 1 shows the phase diagram of the ternary system $C_{14}H_{29}(CH_3)2NO$ water and hexanol. The iridescent monophasic region looks like a straight line that points to the lamellar phase. The properties of the phase-diagrams that are of interest can be better presented in a rectangular plot, as in Fig. 2. Figure 3 shows that the system with oleyldimethyl-

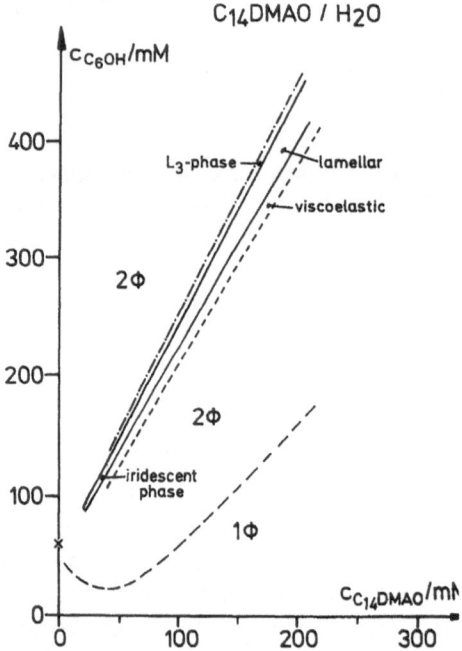

Fig. 2. Lamellar and iridescent phase region of C_{14}DMAO water and hexanol (25°C)

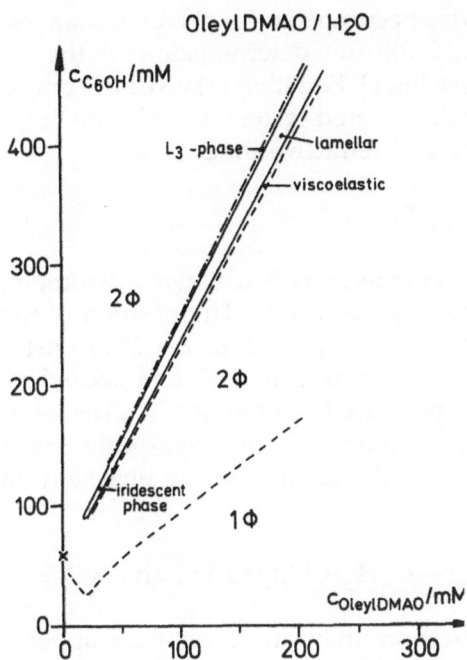

Fig. 3. Lamellar and iridescent phase region of oleyl-dimethylaminoxide water and hexanol; 25 °C

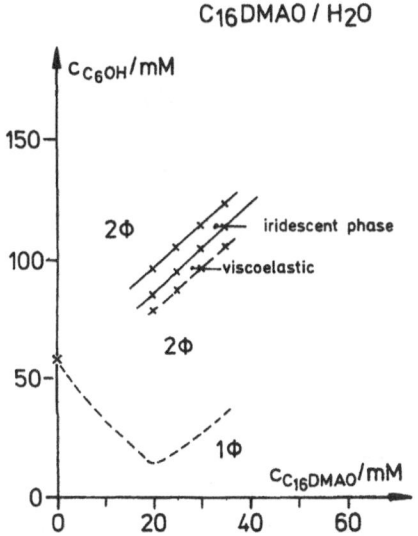

Fig. 4. Iridescent region of hexadecyldimethylaminoxide water and hexanol; 25 °C

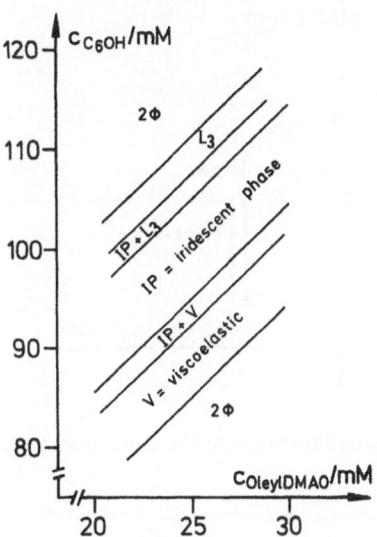

Fig. 5. Phase behavior of oleyldimethylaminoxide water and hexanol around the iridescent region as a function of the hexanol concentration; 25 °C

aminoxide contains a double bond in the side chain and behaves analogously. Figure 4 gives the iridescent region of the system hexadecyldimethylaminoxide, hexanol, and water. Figure 5 shows in enlarged scale the typical phase behavior that is found with increasing concentrations of hexanol around the iridescent phase; the oleyl system is taken as an example.

We recognize that the lamellar phases can be diluted with water that is saturated with n-hexanol. The saturation concentration of n-hexanol at 25 °C is 58.3 mMol/l. For each surfactant concentration it is necessary to have a clearly defined concentration of hexanol. The linear relationship shows that the dilution does not change the lamellae.

In most cases, we find a small region with a clear phase at the lower and at the higher boundaries of the iridescent phase. The clear phase at the lower boundary is viscoelastic. The upper clear phase is isotropic and has a very low viscosity. This is a typical L3-phase [8]. The addition of the hexanol changes the natural curvature of the surfactant-water interface. We observe a transition from (probably) rodlike aggregates that form the viscoelastic system over a flat interface, giving the lamellar phase a curvature in the opposite direction, and leading to large isotropic aggregations in the L3-phase that have a very low viscosity.

The double-chain surfactant $n\text{-}C_{12}H_{25}\text{-}nC_8H_{17}\text{-}N(CH_3)O$ forms an iridescent phase even in its binary solution. Figure 6 shows the important parts of the phase diagram. A lamellar phase is found around 80 weight% of surfactant in water. The dilution with water leads to phase separation. It is

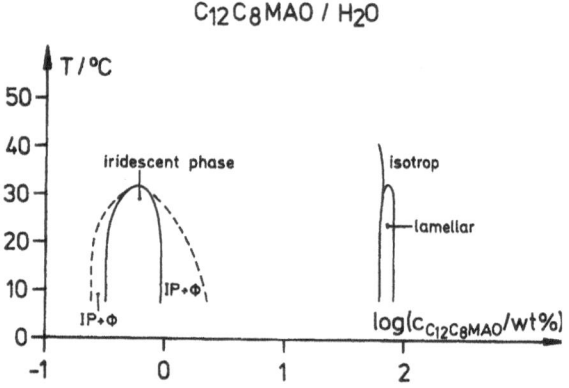

Fig. 6. Iridescent and lamellar phase in the binary system

simple UV-Vis-spectrometer is, therefore, an excellent method for the determination of the interlamellar spacing D. Equation (1) gives the Bragg-condition for the ordered system; m and n are Bragg order number and refractive index:

$$m \cdot D = \lambda_{max}/2n \ . \tag{1}$$

We used a lambda-17 Perkin Elmer instrument. Figure 7 shows typical results. The position of the peaks is obtained from plots E or $E \cdot \lambda^4$ against λ (Figs. 7a, b). The primary ($m = 1$) and secondary ($m = 2$) Bragg-peak are found in nearly all cases. It should be pointed out that three Bragg-peaks &$m = 1, 2, 3$) can be found with the turbity detection on the system:

$$C_{12}H_{25}C_8H_{15} + C_{14}H_{25}(CH_3)_2PO \ 9:1 \ \text{and water.}$$

The fact that more than one Bragg-peak appears shows that the bilayers in the lamellar liquid crystalline state are highly ordered. The wavelength scale in the figures shows how it is possible to extend the measurements beyond the region of the visible light.

surprising that large regions with neither lamellar nor iridescent phases are passed. Mainly, two isotropic phases are present. Finally, at very low concentrations of surfactant, we find the iridescent phase. It is interesting that the highest temperature where the iridescent phase is stable is equal to the temperature where the lamellar phase disappears. If the bilayers in the concentrated state and in the extremely diluted solution have the same composition and the same intralamellar energies, the temperature of the decomposition should also be the same.

The brightness of the colors of the $C_{12}C_8$-system and the temperature region of the iridescent phase is increased if mixtures with n-$C_{14}H_{25}(CH_3)_2PO$ or n-$C_4H_{25}(CH_3)_2NO$ are used with 9 weight parts of the double chain surfactant and 1 part of the phosphinoxide or aminoxide.

It should be noted that the colors never change as a function of the height in the glass tubes, as long as the systems are in the single-phase region of the lamellar phase. However, strong sedimentation effects are observed in all regions in which the iridescent phase is in coexistence with an isotropic phase. The upper regions are shifted to blue in this cases, the lower regions to red.

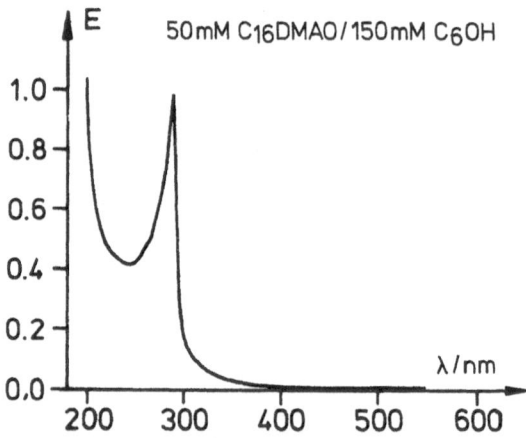

Fig. 7a. Bragg-peak (Bragg order $m = 1$) 50 mMol/l C_{16}DMAO, 150 mMol hexanol 25 °C

Interlamellar spacing

We found that the lamellae of our iridescent phases are nearly perfectly ordered parallel to the surface of flat glass cells. In this case the Bragg-peaks can be only observed at a scattering angle of 180°. The turbidity detection with the aid of a

Figures 8a—f present the interlamellar spacing as a function of the surfactant concentration. $1/D$ is proportional to the surfactant concentration. This result shows that the model of dilution of a lamellar phase is valid. Only the system $C_{12}H_{25}(CH_3)_2NO$ seems to give small deviations, but these are near

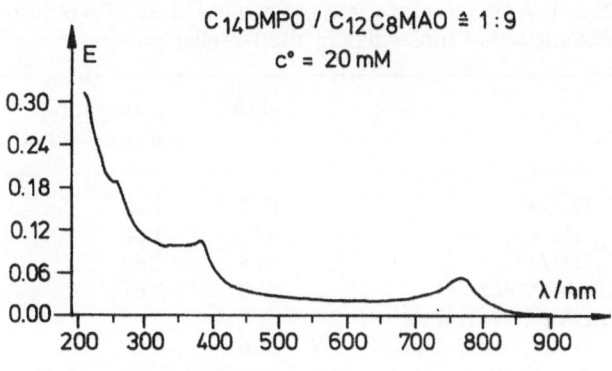

$C_{14}DMPO \, / \, C_{12}C_8MAO \triangleq 1:9$

$c^\circ = 20 \, mM$

Fig. 7b. Three Bragg-orders ($m = 1, 2, 3$)

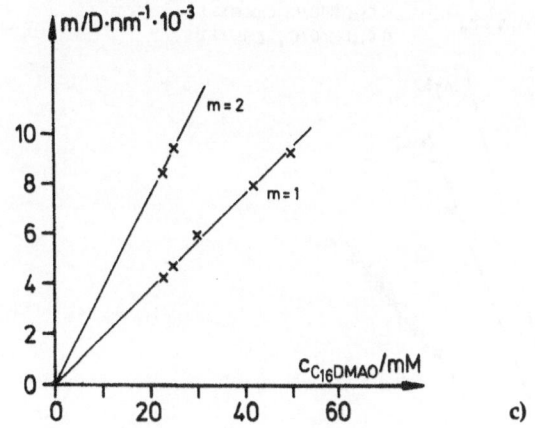

c)

the experimental errors. It should be noted that it is useful to wait some hours before starting the optical measurement; the quality of the measurement is increased in this way.

It is interesting that the colors of the systems change very little within the composition region of

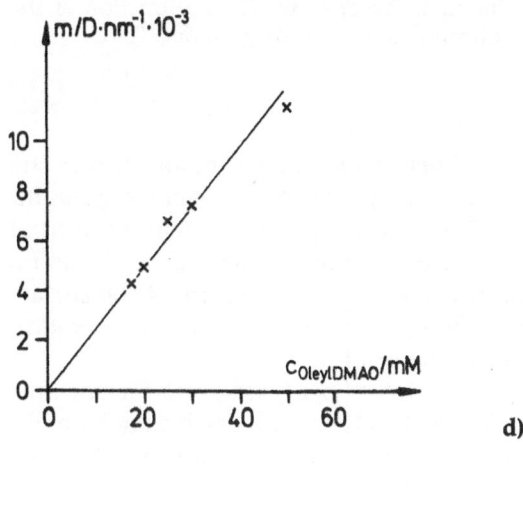

d)

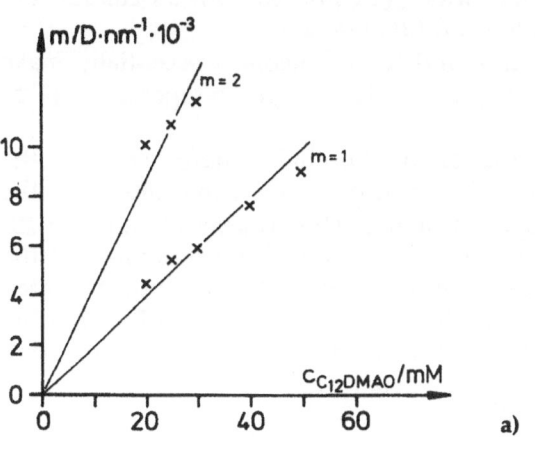

a)

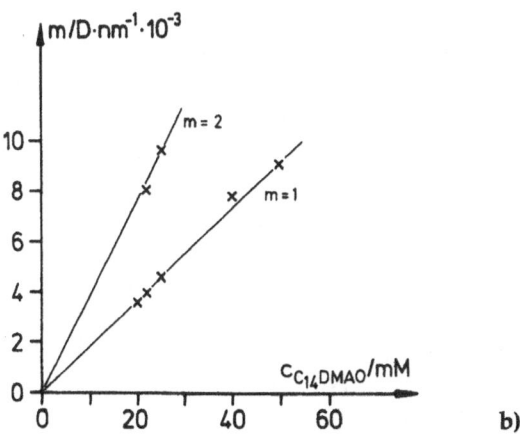

b)

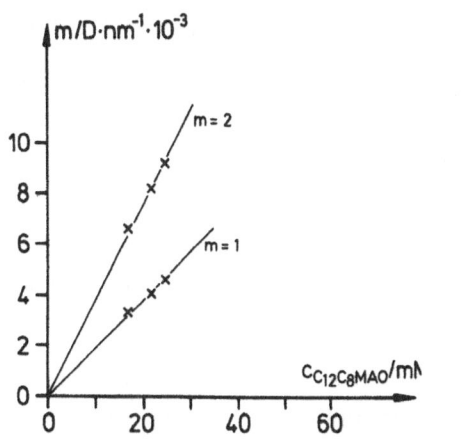

e)

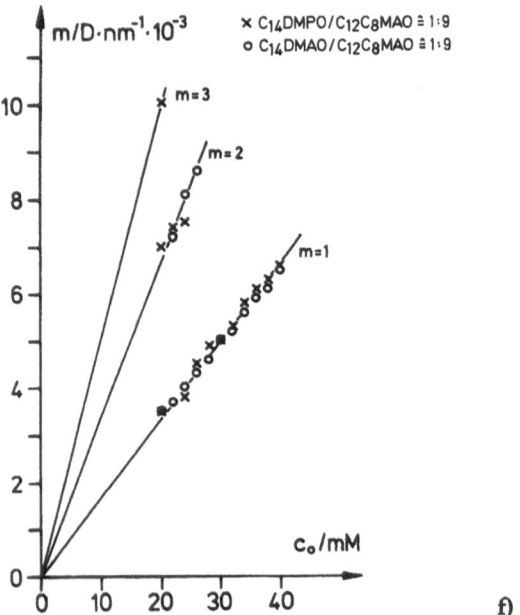

Fig. 8a—f. Interlamellar spacings D as a function of the surfactant concentration. m = Bragg-order; 25°C

Table 1. Area per surfactant molecule A_s and ratio of hexanol/surfactant molecules in the lamellar phases

	$A_s/\text{Å}^2$	n (hexanol)/ n (Surfactant)
C_{12}DMAO	67.3	1.40
C_{14}DMAO	62.0	1.38
C_{16}DMAO	64.8	1.62
Oleyl-DMAO	83.0	1.67
C_{12}DMAO + n-decane 1:1	59.0	
C_{14}DMPO + $C_{12}C_8$MAO 1:9	55.6	0

the hexanol. This result can be understood if the small composition region of the hexanol is taken into account. The lamellae need a certain amount of the alcohol; if there is more or less, there exists no flat interface. The area per headgroup of the surfactant is not influenced by these small concentration changes of the hexanol.

We obtain the area A_s per headgroup for one surfactant molecule together with its hexanol neighbors in each side of the bilayer from the D-values and the surfactant concentration c_c:

$$A_s = 2/(c_s \cdot N_L \cdot D) . \qquad (2)$$

The number of hexanol molecules per surfactant molecules is directly obtained from the linear slope of the iridescent phase region in Figs. 2—4. The small variation (ca. 10%) of the hexanol concentration at fixed surfactant concentration, which is given by the thickness of the iridescent region, has been neglected in this paper. The results are given in Table 1.

The geometric model of the iridescent phases predicts that no color change with temperature should be observed in the single-phase region, if the lamellae are not influenced by the temperature. We have measured the Bragg-peaks from 15° to 25°C and found no change of their position.

At room temperature, iridescent phases are formed with n-pentanol, n-hexanol, n-heptanol, and n-nonanol. Alcohols with n smaller than five do not make such phases. It is interesting that n-octanol gives no iridescent phase within the observed time, although its neighbors with n = 7 and n = 9 do so. However, if a small amount of hydrocarbon — like n-decane is added to the C_{12} solutions with n-octanol, clear iridescent phases with all colors from red to blue can be obtained.

N-decanol and higher alcohols essentially make iridescent phases only at higher temperatures ($T > 50°C$).

The variation of the chain-lengths changes the amount of alcohol necessary for the stability of the flat lamellar structure. This concentration is the difference between the actual concentration and the saturation concentration of the alcohol in water. Table 2 gives the D-values, numbers of Bragg order, and concentration data. The spacings have been measured in the usual way. The values change only within 15% for the different alcohols.

Salt effects

The electrostatic repulsion decreases strongly with the salt concentration. Therefore, we have investigated the influence of very small amounts of electrolyte in order to recognize the importance of such forces. The iridescent solutions were prepared in the usualy way with water containing no electrolyte. Small amounts of 1—10 mM NaCl, Na_2SO_4 or $MgCl_2$ solutions were added using a 20-microliter syringe. The systems were allowed to equilibrate for 3 days. No further changes could be found after this time.

Table 2. 30 mMol/l $C_{14}DMAO$ + $C_nH_{2n+1}OH$; 25°C. c_S = Saturation concentration of $C_nH_{2n+1}OH$ in water; c_{ROH} = total concentration of ROH

n	D/nm	m	c_S/mMol/l	c_{ROH}/mMol/l
4	—	—	1040	
5	526	—	250	242
6	417	1	58.3	45
7	414	1	15	42
8		—	4.5	
9	540	1.2		12
10	—.	—		
11	—	—		

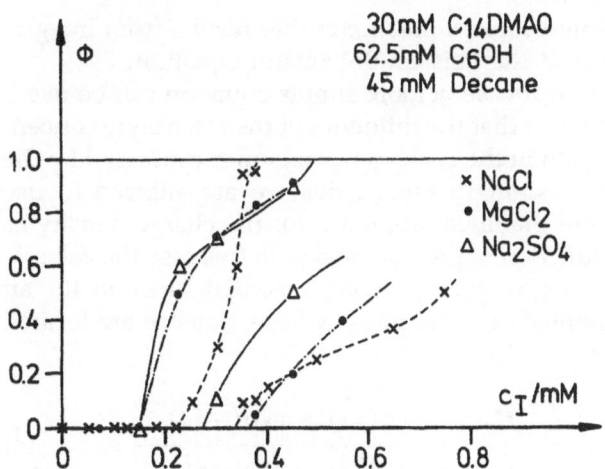

Fig. 10. Phase volumina as a function of the ionic strength I; 25°C

We used different systems for the investigation and found the following typical behavior (Fig. 9). Extremely small amounts of sodium chloride lead to phase separation. The iridescent phase is the upper phase, the lower phase is a strongly scattering L3 phase. The iridescent phase disappears completely at some higher electrolyte concentrations. Phase inversion is observed. A third low scattering phase fills the bottom of the glass tubes.

This behavior leads to the assumption that a coalescence of the lamellae is observed that can be described by laws similar to the Schulze-Hardy rule. The experiments with Na_2SO_4 and $MgCl_2$ clearly show that there is no preference for highly charged counterions. The coalescence of the iridescent phase is an effect that depends on the ionic strength of the system; Fig. 10 illustrates this.

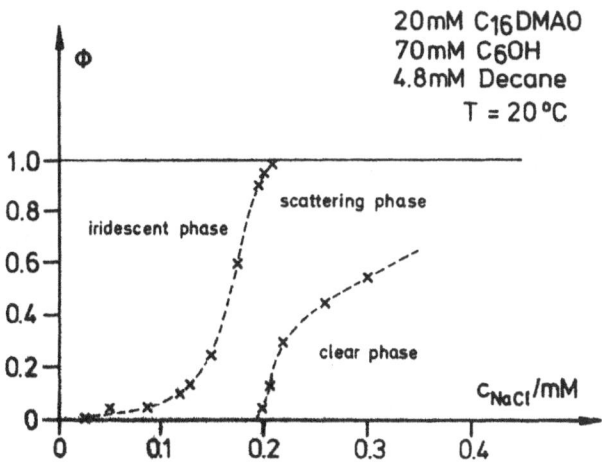

Fig. 9. Phase volumina as a function of the NaCl concentration; 25°C

Starting from systems without salt, the position of the first Bragg peak and, therefore, the color of the system are unchanged with increasing salt concentration until the phase separation occurs. However, the ordered state of the lamellae seems to rapidly go down with the salt concentration, although neither a phase separation nor a change of the interlamellar distance D is observed. Light-scattering measurements are in preparation for this phenomenon.

The influence of fluctuation pressure and electrostatic repulsion on the structure factor of lamellar phases has been studied by C. R. Safinya [10].

The electrostatic repulsion pressure $p_{I=0}$ in charged lamellar phases has been calculated for the simplified case that the ionic strength is zero [11]. This result can be used approximately for our systems in the case of no added salt, when we assume that all hydrogen ions are located on the surface of the bilayers and that the hydroxylions are the only ions present in the water phase:

$$p_{I=0} = k_B \cdot T \cdot \frac{2a}{A_s \cdot D} \frac{q}{tg(q)} \,, \tag{3}$$

with

$$q \cdot tg(q) = \frac{e_0^2 \cdot a}{4\sigma\sigma_0 \cdot k_B \cdot T} \frac{D}{A_s} \,.$$

a is the dissociation degree of the surfactant in the bilayers; A_s is the area defined in Eq. (2); q is a dimensionless parameter that results from integration of the Poisson-Boltzmann equation.

In our case, a more simple equation can be used. The fact that the influence of the electrolyte concentration in the coalescence region is expressed by the ionic strength means that we are allowed to use there the linear approach for the charge density as a function of the potential φ. In this case the calculations give the following potential function for an oriented lamellar phase whose lamellae are located a $x = -D/2$ and $x = +D/2$:

$$\varphi = \frac{D_0}{\varepsilon\varepsilon_0\kappa} \frac{\exp(\kappa x) + \exp(-\kappa x)}{\exp(\kappa D/2) - \exp(-\kappa D/2)} . \tag{4}$$

$$\kappa^2 = 2e_0^2 \cdot N_L \cdot I/(\varepsilon\varepsilon_0 \cdot k_B T) . $$

D_0 and I are electric field replacement and ionic strength; $1/\kappa$ is the Debye length; D_0 is given by the density of the electric charges at one side of the bilayer:

$$D_0 = e_0/A_s \cdot a . \tag{5}$$

Equation (3) is an excellent representation for the potential if the layer potential $\varphi_{D/2}$ (Eq. (6)) is lower than 10 mV, and it is also a good approximation for potentials lower than 30 mV:

$$\varphi_{D/2} = \frac{D_0}{\varepsilon\varepsilon_0\kappa} \cdot \tanh^{-1}(\kappa D/2) . \tag{6}$$

It must be noted that Eqs. (4) and (6) are valid for the potential function in the inner region of a lamellar phase. A two-plate system or a system with only two lamellae has a different potential due to the different boundary conditions:

$$\varphi D/2 = \frac{D_0}{\varepsilon\varepsilon_0\kappa} \cdot (1 + \exp(-\kappa D)) . \tag{7}$$

We can approximate the layer potential. The pH-value of a 30 mMol/l $C_{14}H_{29}(CH_3)_2NO$ solution with 1 mMol/l NaCl is 8.6. The hydroxyl-ion concentration is, therefore, 10^{-6} m. The salt was added to avoid problems with diffusion potentials during the pH measurements. At a surfactant concentration of 30 mMol/l one of 7500 surfactant molecules must be protonated. We can assume that the same ratio of protonated to unprotonated surfactant molecules exists also for the molecules in the bilayer. With the

typical values $F_s = 60$ Å^2 and ionic strength $I = 0.1$ mMol/l, we obtain $\varphi_{D/2} = 31$ mV for $D = 3$ nm. This shows that the assumption of very weak electrostatic forces is valid for all interlamellar distances. The approximation also shows the principal importance of electrostatic forces in nonionic lamellar systems, and makes clear that the repulsion energy can be calculated starting from Eq. (4).

The repulsion energy per area $E_p(D)/A$ is obtained by calculating the electric energy that would be necessary to charge the layers starting from $D = 0$ and taking into account that $E_p(\infty) = 0$:

$$E_p/A = \frac{2D_0^2}{\varepsilon\varepsilon_0\kappa} \frac{1}{\varepsilon\chi\pi(\kappa D) - 1} . \tag{8}$$

The electrostatic repulsion pressure p_I is, therefore,

$$p_I = \frac{D_0^2}{2\varepsilon\varepsilon_0} \cdot \text{Sinh}^{-2}(\kappa D/2) . \tag{9}$$

The total repulsion pressure is the sum of the electrostatic and the fluctuation pressure. In the single-phase region of the iridecent systems no sedimentation effects are observed. This means that the interlamellar distances do not change with the height in the glass tubes. This means that the hydrostatic pressure must be much smaller than the repulsion pressure.

The hydrostatic pressure p_h is given by

$$dp_h = \Delta\rho \cdot \frac{d}{D(h)} \cdot g \cdot dh . \tag{10}$$

d is the thickness of the lamellae. The $\Delta\rho$-values are obtained from density measurements. With $\Delta\rho = -150$ kg/m^3 and $h = 0.1$ m, $d/D = 0.01$ the following hydrostatic pressure is obtained: $p_h = 15$ Pa. Sedimentation effects have been neglected for this approximation.

For $D = 200$ nm, $I = 0.001$ mol/l, and pH = 8.6, the electrostatic repulsion pressure is 0.006 Pa. If the ionic strength is reduced by a factor of 10 or 100, the pressure will rise to 0.65 or 8.6 Pa. The upper limit, 8.3 Pa, of the electrostatic repulsion pressure is obtained with $I = 0$ from Eq. (3). This result shows that by far, the electrostatic repulsion is not the only stabilization force in our iridescent phases that exist up to ionic strength values of 0.1 mMol/l.

The fluctuation pressure p_{fluct} is

$$P_{fluct} = 0.84 \cdot \frac{(k_B \cdot T)^2}{k_c} \cdot D^{-3} . \qquad (11)$$

k_c is the elastic bending modulus for a single bilayer.

It is a realistic assumption that the repulsion pressure must be greater than 25 Pa in order to avoid sedimentation. Thus, $k_c/k_B \cdot T$ must be smaller than 0.015. It is a problem to understand such small k_c values. There is a rough approximation which allows to qualitatively calculate the fluctuation behavior. If parts of the bilayers could fluctuate like free disklike micelles, the aggregation number of such micelles must be smaller than 3000 in order to obtain a repulsion pressure of 25 Pa. The diameter of such micelles would be only 17 nm. If the disks are part of bilayers, the repulsion pressure will be decreased. It is, therefore, not clear that the entropic fluctuation pressure of extremely flexible bilayers will be the main stabilizer of these iridescent phases.

This problem could be solved if the bilayers would have a mean curvature of zero and a mean square curvature greater than zero in their unperturbed state. In this case, a strong repulsion pressure could be obtained without the necessity of entropic fluctuations. In contrast to the entropic fluctuations the repulsion pressure would increase with the bending modulus.

If we assume that fluctuations or periodic changes of the interlamellar distance D occur, the electrostatic repulsion could dominate at positions where the bilayers come close together. The electrostatic repulsion forces seem to be necessary to prevent the bilayers from collapsing at the moment when they would touch another. This assumption could explain the phase behavior as a function of the ionic strength.

In the two-phase region the repulsion forces are so small that we observe sedimentation effects. In this case the repulsion pressure is too low for direct measurements by osmometric methods, methods that are successful with typical lamellar phases [9, 11]. It is interesting to note that it is easily possible to obtain the total repulsion pressure p_{ww} as a function of the interlamellar distance D from measurements of the interlamellar distance as a function of the height in the glass tubes:

$$\frac{dp_{ww}}{dD} = \frac{\Delta\rho \cdot d \cdot g}{D(dD/dh)} . \qquad (12)$$

The fact that the sedimentation effect appears shows that the dominating repulsion pressure disappears near the phase boundary of the iridescent phases.

References

1. Suzuki Y, Tsutsumi H, Yukagaku (1984) Journal of Japan Oil Chemists Society 33:48—54
2. Larche FC, Appell J, Porte G, Bassereau P, Marignan J (1986) Phys Rev Letters 56:1700—1703
3. Helfrich W (1978) Z Naturforsch 33a:305—315
4. Satoh N, Tsujii K (1987) J Phys Chem 91:6629—6632
5. Appell J, Bassereau B, Marignan J, Porte G (1989) Colloid Polym Sci 267:600—606
6. Thunig C, Hoffmann H, Platz G (1989) J of Colloid Polym Sci 79:297-307
7. Strey R, presented at Bunsentagung 1989, Siegen
8. Ghosh O, Miller CA (1987) J of Phys Chem 91:4528. Miller CA, Gradzielski M, Hoffmann H, Krämer U, Thunig C, J of Phys Chem in press
9. Cowley AC, Fuller NL, Rand RP, Parsegian VA (1978) Biochemistry 17:3163—3168
10. Safinya CR (1989) Mat Rec Sec Symp Proc Vol. 143:9—18
11. Parsegian VA, Fuller N, Rand RP (1979) Proc Natl Acad Sci USA 76:2750—2754

Authors' address:

G. Platz
Institut für Physikalische Chemie
Universität Bayreuth
8580 Bayreuth, FRG

Progress in Colloid & Polymer Science

Progr Colloid Polym Sci 83:176—180 (1990)

Reversible light-induced phase transition in the system cetyltrimethylammonium bromide — water containing a crown-ether-bearing azobenzene

T. Wolff, B. Klaußner and G. von Bünau

Physikalische Chemie, Universität Siegen, Siegen, FRG

Abstract: Phase transition temperatures pT (nematic ⇌ isotropic) in the system cetyltrimethylammonium bromide — water (CTAB-H_2O) were measured in the presence of small amounts of trans AB15C5, an azobenzene derivative that bears two crown-ether moieties. pT decreased as a function of the AB15C5 concentration. This pT decrease can be ascribed to interactions of the crown-ether moieties with CTAB. The effect was reverted upon irradiation, i.e., upon photochemical formation of the water soluble cis form of AB15C5. Thereby in situ changes of pT are possible, which are reversible due to thermal reisomerization. The cis-trans reisomerization proceeds with a rate constant of 10^{-4} s^{-1} at 33°C.

Key words: Cetyltrimethylammonium bromide; crown ether; lyotropic liquid crystals; light-induced phase transition

Introduction

In previous studies, we found that phase transition temperatures (pT) in lyotropic liquid crystalline systems are sensitive to the presence of small amounts of certain aromatic compounds. The observed pT shifts differ from common impurity effects on colligative processes (such as melting point depression) and they are specific with respect to the solubilized compound. This specificity allows in situ switching of phases at a given temperature via photoreactions of solubilized compounds, provided reaction products have influences on pT differing from those of the educts. The photochemical induction of phase transitions in the system cetyltrimethylammonium bromide — water (CTAB-H_2O, cf. Fig. 1) has been shown for two different photoreactive systems: anthracene carboxylic acid present at a few permille by weight in CTAB-H_2O gives rise to an increase of pT (nematic ⇌ isotropic, cf. Fig. 1) by several degrees. Irradiation, i.e., photochemical formation of dimers, decreased pT to values very close to those of pure CTAB-H_2O

[1, 2]. Similarly, the solubilization of N-methyl-N,N-diphenylamine increased pT, whereas its photochemical conversion to N-methylcarbazole reverted the effect [3].

In this paper we present a new system that allows reversible switching of phases via photochemical trans-cis photoisomerization and thermal cis-trans reisomerization of a crown-ether-bearing azobenzene AB15C5:

trans-AB15C5   cis-AB15C5

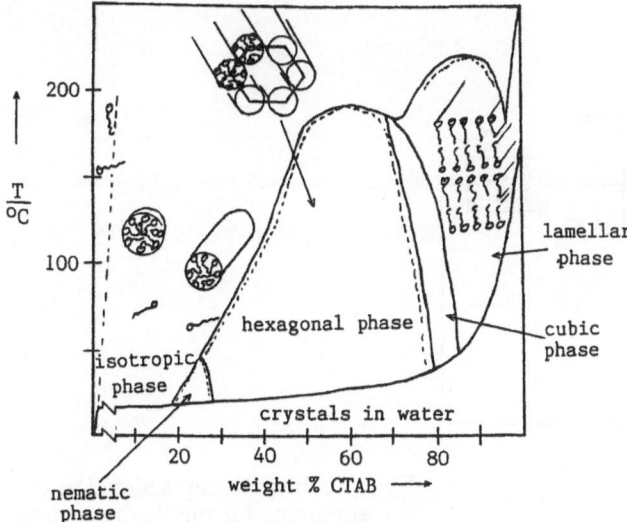

Fig. 1. Phase diagram of the cetyltrimethylammonium-bromide-water system (redrawn after [7] and [8])

Experimental

Materials

Cetyltrimethylammonium bromide (Merck, p.a.) was recrystallized from acetone-methanol (9:1 mixture). Benzo15C5 and azobenzene were purchased from Fluka (purum grade) and Merck (99%) and used as supplied. AB15C5 (bis(2,3,5,6,8,9,11,12-octahydro-1,4,7,10,13-benzo-pentaoxacyclopentadecin-15-yl)-diacene) was prepared from 4'-nitrobenzo15C5 that was synthesized according to the literature [4] and agreed with the spectroscopic properties reported there; mp 96—96.5°C (lit. 84—85°C). To a suspension of 1.2 g (0.03 mol) of LiAlH$_4$ in 200 ml of dry THF (that had been stirred for 1 h under argon before in a dry glass apparatus) 6.26 g (0.02 mol) of 4'-nitrobenzo15C5 in 15 ml of dry THF were added dropwise within 3 h. This mixture was stirred overnight at room temperature. After hydrolyzing the excess LiAlH$_4$ by slow addition of 2 ml of water in 20 ml of THF the mixture was filtered using a G3 glass filter plate covered by a 1-cm thick layer of SiO$_2$. The filter cake was washed with THF until the filtrate became colorless. The combined filtrates were concentrated in vacuo to a volume of ca. 50 ml and, after cooling to −28°C, precipitation of the product occurred. After recrystallization from ethanol, we obtained 2.0 g (3.6 mmol; 36%) of AB15C5 (A different literature procedure yielded only 9% [5].); mp 168—188°C (lit. 187—188°C; the wide melting range is caused by a high content of cis-AB15C5 in the product converting to trans-AB15C5 while melting); NMR (CDCl$_3$) 3.7—4.3 ppm (m, 32H), 6.8—8.0 ppm (m, 6H).

Determination of phase transition temperatures

pT was measured in an apparatus drawn schematically in Fig. 2. The method makes use of the fact that optically anisotropic phases (such as the nematic or the hexagonal one) are capable of rotating the plane of linearly polarized light. Therefore, in the presence of an anisotropic phase, light from a helium-neon laser can pass through a thermostatted sample (between 0.2- and 10-mm thickness) placed between two perpendicularly crossed filters for linearly polarized light. The transmitted light is registrated by the detector as photodiode voltage U_D. The temperature of the sample and the thermostat were measured using a miniature thermocouple and a Pt-100 thermometer, respectively, connected to a 68000 CPU microcomputer (kws, model SAM 68K) that controls the thermostat via a relay, allowing for temperature program runs. The sample temperature can be kept constant within 0.05°C and measured with an accuracy of 0.01°C. A typical diagram for a pT-determination is reproduced in Fig. 3 (left side). The upper part of the figure displays the sample temperature during the pT determination program. The lower part shows the corresponding photodiode signal U_D. The onset of U_D is taken as the phase transition temperature pT.

Irradiations

Samples were irradiated in the apparatus for pT determination employing a 150 W Xenon lamp (Osram XBO), as indicated in Fig. 2.

Thermal cis-trans isomerization rate constant

A sample of 24% CTAB-H$_2$O containing trans-AB15C5 was irradiated in the apparatus shown in Fig. 2 and then placed inside a UV-VIS spectrophotometer (Beckman Acta M VII), together with the thermostatted copper block. The absorption at 377 nm, i.e., at maximum absorption of the trans form, increased as a function of time, which allows the determination of the rate constant. As the spectrometer light source is sufficient to cause photoisomerization [5], the absorption was measured discontinuously, i.e., every 10 min for a few seconds.

Results

Between 0.2 and 0.6 weight-% (corresponding to 3 through 11 × 10^{-6} mol/g) of trans-AB15C5 were added to CTAB-H$_2$O mixtures and phase transition temperatures pT (nematic $\rightleftharpoons$ isotropic) were measured. Compared to pure CTAB-H$_2$O, decreases of pT between 2 and 6 degrees were observed as comprized in Fig. 4. The pT decrease appeared as a linear function of the AB15C5 concentration, as long as solubility limits were not exceeded (e.g., 0.8% of AB15C5 in 24% CTAB-H$_2$O at 27.4°C).

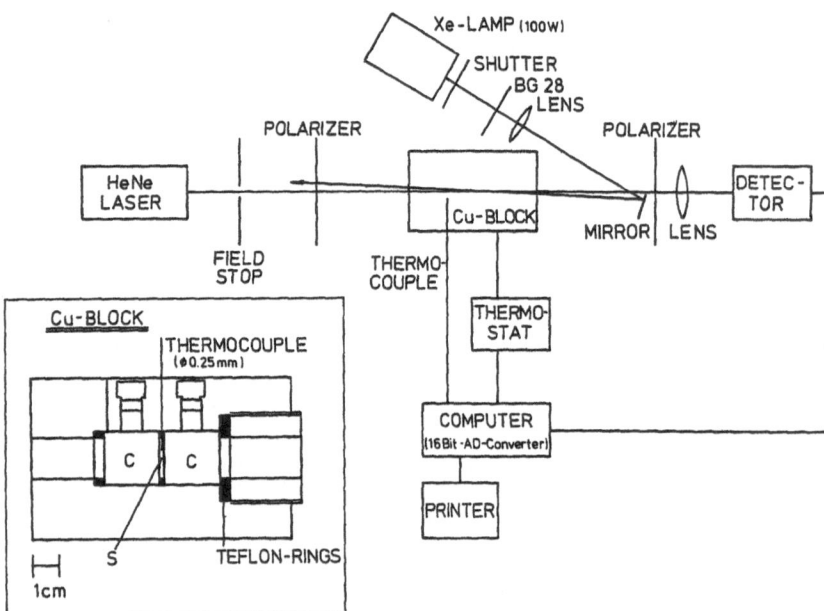

Fig. 2. Schematic representation of the apparatus for the determination of phase transition temperatures. S = sample; C = cuvettes filled with water

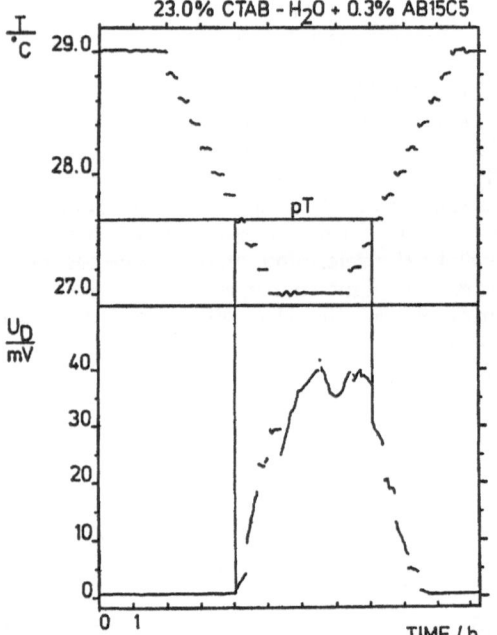

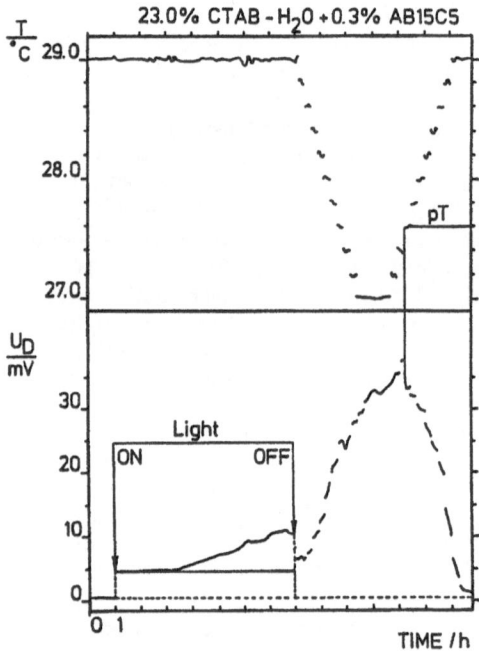

Fig. 3. Left side: Determination of the phase transition temperature pT of a 23% cetyltrimethylammonium-water mixture containing 0.3 weight-% of trans-AB15C5; upper part: sample temperature as a function of time during the pT determination program; lower part: photodiode signal U_D (cf. Fig. 2) as a function of time during the pT-determination program, the onset of U_D increase while decreasing the temperature and of U_D decrease upon increasing temperature can be taken as pT. Right side: Light-induced phase transition in the system described above; change of U_D as a function of irradiation time (light on) and subsequent pT-determination program

In situ phase transitions were induced by irradiating samples in the isotropic phase near pT, e.g., a sample containing 0.3% (5.33 × 10⁻⁶ mol/g) trans-AB15C5 at 29°C in 23% CTAB-H₂O. During the irradiation (i.e., when photoconverting trans-AB15C5 to cis-AB15C5) the phase transition to the anisotropic nematic phase occurred as illustrated in Fig. 3 (right side): upon switching on the irradiation source a small offset due to scattered irradiation light was measured by the photodiode. After ca. 2 h, pT had raised to 29°C and the formation of the nematic phase was revealed by the increasing photovoltage U_D. The curved shape of the U_D increase gives an idea of the time constant of the phase transition. When, after switching off the light the usual pT determination program was run, the thermal cis-trans reisomerization took place so that a phase transition temperature agreeing with that of the trans-AB15C5-containing sample was obtained (27.6°C, Fig. 3 (right side)), indicating the reversibility of the photochemically induced pT shift. A reisomerization rate constant $k_{ct} = 10^{-4}$ s⁻¹ was determined at 33°C (cf. $k_{ct} = 1.2 × 10^{-3}$ s⁻¹ in dry orthodichlorobenzene and $k_{ct} = 1.5 × 10^{-4}$ s⁻¹ in water-saturated orthodichlorobenzene at 30°C [5]).

For comparison, pT of 24% CTAB-H₂O was determined in the presence of 9 × 10⁻⁶ mol/g trans azobenzene and of 18 × 10⁻⁶ mol/g benzo15C5, respectively. While trans azobenzene had a pT-increasing effect, benzo15C5 induced a pT decrease to a value very close to that in the presence of an equimolar amount of AB15C5 (with respect to the crown-ether moiety of AB15C5), as shown in Fig. 4. Irradiation of the azobenzene-containing sample, i.e., formation of cis-azobenzene, did not change pT.

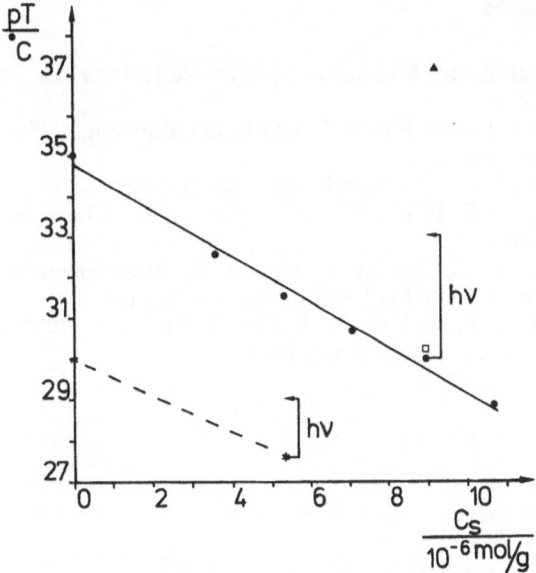

Fig. 4. Phase transition temperature (pT) decrease upon addition of increasing amounts of AB15C5 in 23% (∗) and 24% (●) cetyltrimethylammonium bromide-water mixtures. hv = pT increase as a consequence of irradiation, i.e., trans-cis isomerization of AB15C5. ▲ = pT in the presence of azobenzene; □ = pT in the presence of benzo15C5 (at doubled concentration corresponding to the two crown-ether moieties in AB15C5)

crown-ether moieties of AB15C5. Solubilization of trans-AB15C5 in the micellar phase can be expected since trans-AB15C5 is not soluble in water [5] and it complexes ammonium ions (even alkyltrimethylammonium ions, though not so strongly as alkylammonium ions, as revealed by NMR studies [6] and photostationary cis/trans ratios [5]). Cis-AB15C5, however, is water soluble [5]. We therefore tentatively ascribe the pT increase after irradiation to the formation of cis-AB15C5 diffusing off the micellar aggregates into the bulk aqueous phase, thus minimizing micelle — solubilizate interactions and raising pT towards the value of the pure CTAB-H₂O system.

azobenzene   benzo15C5

Discussion

The control experiments using benzo15C5 and azobenzene clearly show that the observed pT decrease induced by trans AB15C5 can be ascribed to interactions of surfactant aggregates with the

Acknowledgment

Financial support by the Deutsche Forschungsgemeinschaft and by the Fonds der Chemischen Industrie is gratefully acknowledged.

References

1. Wolff T, von Bünau G (1984) J Colloid Interface Sci 99:299
2. Wolff T, von Bünau G (1984) Ber Bunsenges Phys Chem 88:1098
3. Wolff T (1989) Colloid Polym Sci 267:345
4. Ungaro R, El Haj B, Smid J (1976) J Am Chem Soc 98:5193
5. Shinkai S, Nakaji T, Ogawa T, Shigematsu K, Manabe O (1981) J Am Chem Soc 103:111
6. Shinkai S, Yoshida T, Manabe O, Fuchita Y (1988) J Chem Soc Perkin Trans 1:1431
7. Götz KG (1961) Dissertation Universität Göttingen
8. Hertl G, Hoffmann H (1988) Progr Colloid Polym Sci 76:123

Authors' address:

Prof. Dr. Thomas Wolff
Physikalische Chemie
Universität Siegen
Postfach 101240
5900 Siegen, FRG

Solubilization and conformation of protamines in reverse micelles

G. Ebert, U. Zölzer and N. Nishi*)

Fachbereich Physikalische Chemie der Philipps-Universität Marburg, FRG
*) Department of Polymer Science, Faculty of Science, Hokkaido University, Sapporo, Japan

Abstract: Protamines are strong basic nucleoproteins containing approximately 70% L-arginine. In aqueous solution they usually exist as an extended coil as a consequence of the electrostatic repulsion of the positively charged guanidinium side-chain of arginine. Their conformation in the native complex with DNA is, as of yet, unknown. However, it is suspected to at least partially be an α-helix, due to the electrostatic interactions with the phosphoric acid anions of the DNA and a low degree of hydration. For studying these influences, studies on the conformation of protamines in reverse micelles, e.g., of the system AOT/i-octane/H_2O depending on the H_2O/AOT-ratio (w_0-value) by circular dichroism (CD)-spectroscopy should be useful. Indeed, at low w_0-values, α-helix spectra have been obtained, as well as a remarkable dependence of the conformation of the protamines on w_0. The component YI of clupein shows a significant difference in conformation against YII and Z.

Key words: Protamines; conformation; reverse micelles; circular dichroism (CD) spectra; AOT; w_0-value; poly-L-arginine

Introduction

Protamines are strong basic nucleoproteins with an arginine content of $\approx 70\%$; they occur in the nuclei of fish spermatozoa [1] (where they were found at the beginning of protein research) and were recently detected in that of mammals. The molecular weight of protamines is usually rather low and amounts to 4000—8500 m.w. in general.

The biological role of protamines is not yet completely clear. Isolating them from the native occurring DNA-complex leads to a denaturated state, probably due mainly to cancelling the electrostatic and other intermolecular interactions between the basic arginine side groups and the phosphoric acid anions of the DNA, accompanied by an increase in the degree of hydration. As a consequence, in aqueous solution the protamines attain the conformation of an extended random coil. In the native state they are assumed, by several authors, to exist in an at least partially ordered conformation that forms α-helices. But because of the rather complicated spectra obtained from the native pro-

tamine-DNA complex, it is very difficult to get information in this regard by spectroscopic methods, e.g., by circular dichroism (CD)-spectra.

Considering the primary structure of the components of two well-studied fish protamines — clupeine from herring, and salmine from salmon — (shown below) it becomes obvious that these molecules should be able to attain (at least partially) the α-helix structure.

Clupeine YI ARRRRS SSRPIRRRR PRRRTTRRRR AGRRRR

       YII PRRR TRRASRPVRRRR PRR VSRRRR A RRRR

       Z ARRRRSRRASRPVRRRR PRR VSRRRR A RRRR

Salmine AI PRRRRS SSRPVRRRRRPR VSRRRRRRGGRRRR

      AII PRRRRRRSSSRPIRRRR PRR ASRRRRR GGRRRR

(A: alanine, G: glycine, I: isoleucine, P: proline, R: arginine, S: serine, T: threonine, V: valine)

For studying the conformation of protamine molecules underlying strong electrostatic interactions at a low level of hydration by CD measurements, model systems like reverse micelles, e.g., of sodium (bis-ethylhexyl)sulfosuccinate (AOT)/H$_2$O/isooctane should be very useful, because in these systems only the protamine is an optically active component with a CD spectrum and with no possible overlapping with the spectra of any other component.

Reverse micelles were investigated as early as a decade ago, and also extensively in regard to the solubilization of polypeptides and other biopolymers [2—5].

Studying the CD spectra of polypeptides solubilized in reverse micelles is a rather convenient method to obtain information on the influence of electrostatic interactions, on the degree of hydration, etc., and on the conformation of these biopolymers. Seno et al. [6, 7] have shown that poly-L-arginine — which is usually used as a model polypeptide for protamines — and other basic polypeptides attain ordered periodical conformations in reverse micelles of these systems. It should be emphasised that $(Arg)_n$ is an α-helix forming polypeptide in aqueous solutions, only if the strong electrostatic repulsive forces of the positively charged guanidinum side-groups (pK: 12.2) are cancelled by interaction with some specific interacting anions such as perchlorate or thiocyanate, as was shown by Sugai and coworkers [8, 9].

Also in the case of clupein and salmine such an effect of perchlorate was observed by Toniolo et al. [10, 11].

Ichimura et al. [12] showed that very low concentrations of sulfate induce α-helix-formation of poly-L-arginine in contrast to poly-L-lysine, which forms α-helices only at very high (e.g., methylsulfate) concentrations [13]. However, sulfonates like methanesulfonate promote α-helix-formation of poly-L-arginine, only at concentrations as high as 2—6 mole/l [14].

From these results one can conclude that the interactions between sulfonate anions and the guanidinium groups of arginine are much less pronounced than in the case of sulfate, and also that they depend strongly on hydration.

Materials and methods

Clupeine and salmine sulfate were obtained from Sigma. Clupein-components YI, YII, Z were isolated by chromatography on a Sephadex CM 25 column (150 × 1.2 cm) in 1.5 m NaCl, 0.05 m Na-acetate, pH 6.0, flow rate 20 ml/h.

Sodium bis-(ethylhexyl-)sulfosuccinate (AOT), research grade from Serva was used without further purification.

For preparing the reverse micellar solution the amount of water (for trace analysis, Fluka) necessary for obtaining the w_0-value desired was added to a 0.05 m solution of AOT in isooctane (Fluka, distilled over Na-Pb) by an Eppendorf pipette under nitrogen for avoiding contact with atmospheric CO$_2$. After shaking, a definite amount of protamine was added to this solution of "empty reverse micelles" as a powder and stirred overnight for solubilizing the protamine.

The CD spectra were run on a J-20 spectrophotometer, JASCO-Corp., Tokyo, using thermostated quartz-cells with an optical path length of 0.02 cm.

Results

It could be shown that solubilization of clupeine and salmine occurs in reverse micelles (RM) with a w_0 (water: AOT-ratio)-value ≥ 3.

The CD spectra obtained are shown in Fig. 1 (A and B). As one can see, up to w_0 8 they show the characteristic features of the α-helix with a double-minimum at 222 and 208 nm. The specific ellipticities are in the order of $-(10 \pm 3) \times 10^3$ deg · cm^2/dmole.

With increasing w_0 the CD spectra change, not only quantitatively, but also qualitatively — obviously, as a consequence of the overlapping of different conformation types. at $w_0 \approx 20$ they resemble the CD spectra of a random coil structure, such as in aqueous solution.

If the specific ellipticities at 208 and 222 nm are plotted against w_0, sigmoid-type curves are obtained (Fig. 2) up to $w_0 \approx 20$; above that they reach a plateau.

By plotting $[\Theta]_{208}$ against $[\Theta]_{222}$ of clupeine and salmine sulfate a linear relationship is found (Fig. 3) that is characteristic for a helix-coil transition, according to Maeda et al. [15].

The temperature dependence between 0° and 70°C is shown for clupein sulfate and w_0-values 4.6, 10.1, and 17.1 in Fig. 4.

As can be seen from Fig. 5, the specific ellipticity of clupeine sulfate depends on the AOT-concentration at a constant protamine concentration. Correspondingly, there is also a dependence on protamine concentration at a constant AOT concentration (Fig. 6).

Because of some characteristic differences between primary structure of the components, the

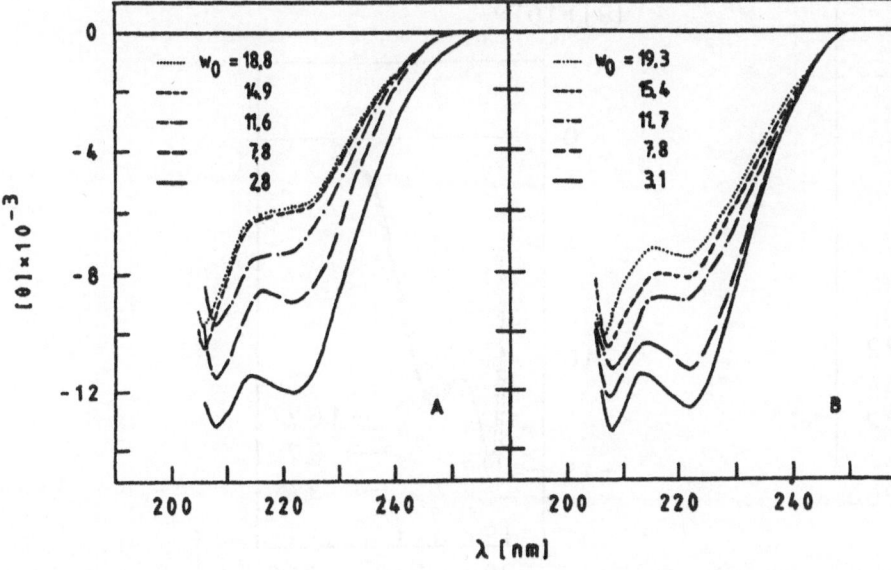

Fig. 1. ((CD) spectra of clupeine sulfate (A) and salmine sulfate (B) in reverse micellar (RM) systems of AOT/isooctane/water at different w_0-values. C_{AOT}: 0.05 m; $C_{Protamines}$: 0.006 m; T: 25 ± 3°C. Molar ellipticities in deg × cm²/dmole

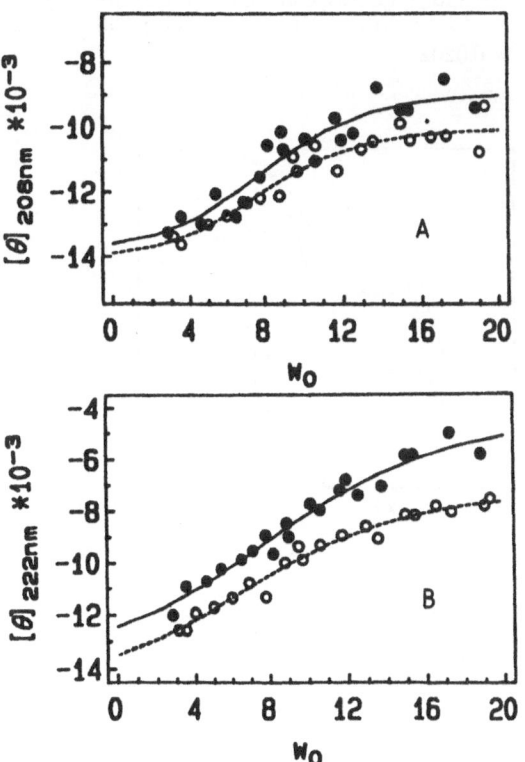

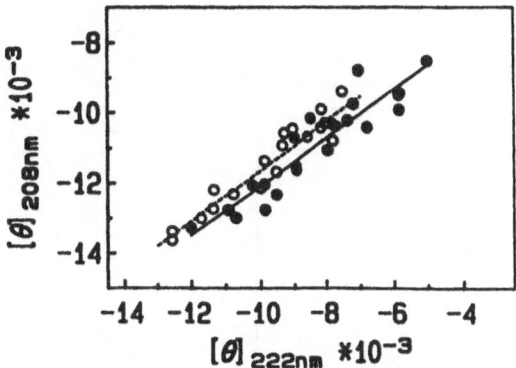

Fig. 3. Molar ellipticities of clupeine sulfate (—●—) and salmine sulfate (—○—) in RM at 208 nm as a function of the molar ellipiticities at 222 nm

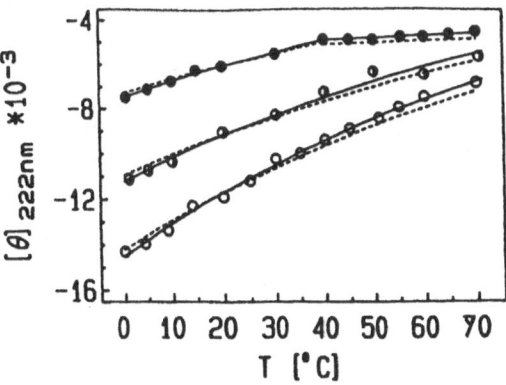

Fig. 2. Molar ellipticities of clupeine sulfate (—●—) and salmine sulfate (—○—) at 208 nm (A) and 222 nm (B) in RM as a function of w_0. Molar ellipticities in deg × cm²/dmole. T: 25°C

Fig. 4. Temperature dependence of $[\Theta]_{222}$ of clupeine sulfate solubilized in RM at w_0 4.6 (○), 10.1 (◐), 17.1 (●). C_{Clup}: 6.31—6.78 · 10⁻³ m, C_{AOT}: 0.05, path-length 0.1 cm

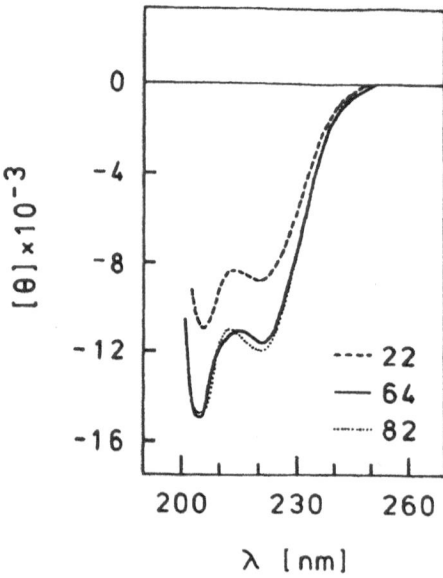

Fig. 5a. CD spectra of a clupeine sulfate solubilized in RM at different AOT-concentrations (22, 64, 82 $\times$ 10^{-3} mole/l) and $w_0 = 4$

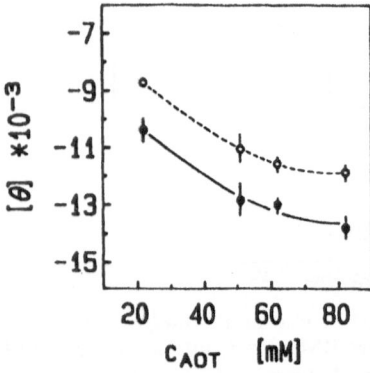

Fig. 5b. The specific ellipticities of clupeine sulfate in RM at 208 ($\bullet$) and 222 ($\circ$) nm, depending on the AOT-concentration, $w_0 = 4.0$. Molar ellipticities in deg $\times$ cm²/dmol

CD spectra of the YI, YII, and Z-component solubilized in RM at w_0 were measured. The results are shown in Fig. 7.

For comparing the CD spectra of the two protamines in RM with those in aqueous solutions of AOT and ethanesulfonate, the results obtained are shown in Fig. 8.

Discussion

Considering Fig. 1, it becomes obvious that with increasing w_0-value, not only a quantitative,

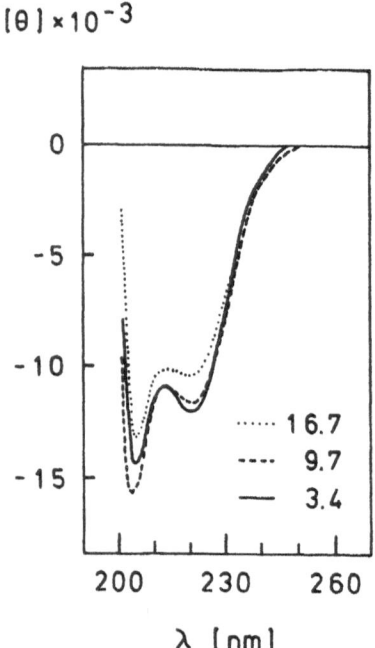

Fig. 6. CD spectra of clupeine sulfate in RM, at $w_0 = 3.9$ and different clupeine sulphate concentrations (—) 3.4; –– 9.7; ... 16.7 $\times$ 10^{-3} mole, c_{AOT}: 5.56 $\times$ 10^{-2} m, T: 22.5°C, l = 0.0202

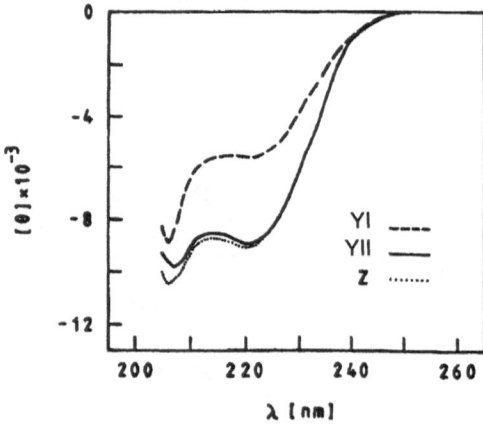

Fig. 7. CD spectra of the YI, YII, and Z-components of clupeine (all as hydrochlorides) in AOT-reverse micelles at $w_0 = 4.0$. C_{Clup}: 0.006 m; C_{AOT}: 0.05 m; T: 25 $\pm$ 0.3°C; molar ellipticities in deg $\times$ cm⁻²/dmole

change of the CD-spectra occurs. The quantitative change, i.e. the decreasing specific ellipticities are assumed to be due to a loss in molecular order with increasing hydration of the solubilized protamin molecules. This agrees with the results obtained in

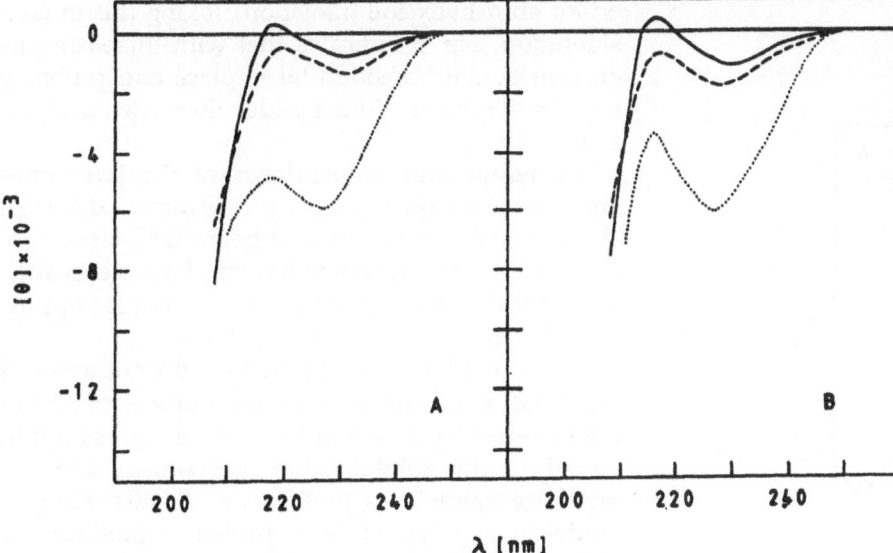

Fig. 8. CD spectra of clupeine sulfate (A) and salmine sulfate (B) in water (——) and in AOT solutions of different concentrations (———); 3.17×10^{-5} m, ...: 1.6×10^{-3} m); pH 5, 7 $\pm$ 0.3; T: 25 $\pm$ 0.2°C. Protamine concentration:« 2.4 $\times$ 10^{-4} m. Molar ellipticities in deg $\times$ cm²/dmol

aqueous solutions of sulfonates (Fig. 8) [14]. Concomitantly, other conformations occur with different CD spectra overlapping with the α-helix spectrum, and causing a qualitative change, i.e., a frequency-shift of the CD peaks.

The sigmoid shape of the $[\Theta]$-w_0-curves in Fig. 2 indicates, not only that the degree of α-helicity decreases with w_0, but also that above $w_0 \approx 12$ (together with reaching a plateau) only negligible conformation changes take place at higher w_0-values. But, as one can see from the frequency shift of the minimum at 208 nm at $w_0 \approx 12$ (especially in Fig. 1A), a conformation change from α-helix to an extended coil or a mixture of various conformations is observed.

This may be due to the fact that (according to several authors [16—21]) up to $w_0 \approx 12$—13, as a consequence of the strong interactions between the water molecules and the sulfonate anions of the AOT, their counterions, and the protamine molecules solubilized, the water molecules in the core of the RM have to be considered as hydrate water with a structural order and a mobility very much different from water in bulk. At higher w_0 values water populations with physicochemical properties approaching those of water in bulk occur, which is probably responsible for the conformation changes observed.

The specific ellipticities at 208 and 222 nm are remarkably lower than the values reported for a complete α-helix. These values $(-10 \pm 3) \times 10^3$

deg · cm²/dmol, which are approximately one-third of that of a long-chain α-helical polypeptide (according to Greenfield and Fasman [22]) indicate that the length of the α-helical regions is a rather short one. Yaron et al. [23] showed that the ORD of (L-lysine)$_n$ oligomers up to a 22mer and polymers depends remarkably on the chain length. According to our results of CD studies on poly-L-arginine that determined chain length, $[\Theta]_{208}$ and $[\Theta]_{222}$ of (Arg)$_{18}$ in RM are of the same order as those of the protamines in RM. For (Arg)$_{62-2320}$, $[\Theta]$-values between —35000 and —40000 deg $\times$ cm²/dmol have been found, which are almost independent of chain length of the polypeptide (Fig. 9) agreeing with [23]. Therefore, one can conclude that the low $[\Theta]$-values observed are due to the short chain-length of the protamines. Moreover, it is questionable if the whole protamine molecule will be able to attain the α-helix conformation or only some parts of it, because as one can see from the primary structure, the α-helix forming arginine (R) sequences are interrupted by non-α-helix-compatible amino acid residues, such as proline, isoleucine, serine, threonine, and valine. Besides arginine, only alanine in the N- and C-terminal sequences fits into an α-helix. As one can see, the longest uninterrupted α-helix forming sequence is the C-terminal nonamer R$_4$AR$_4$ of YII and Z, which is able to form at least two α-helix turns, each containing 3.6 amino acid residues.

As was reported by Maeda et al. [15], a linear relationship between $[\Theta]_{208}$ and $[\Theta]_{220}$ should be obtained for an α-helix coil transition. Taking this in consideration, Fig. 3 indicates that with increasing w_0 an α-helix-coil transition takes place and probably no other conformations besides these two are present.

The temperature-dependence of the [Θ]-values and, therefore, the conformation is remarkable (Fig. 4) and one can conclude that below 0°C further increase of α-helix degree will occur, because in such small entities the freezing point of water is supposed to be much below 0°C [18, 19].

The dependence of [Θ] on the concentration of AOT and, concomitantly, on the number of RM in the system (Fig. 5 (a and b) is rather interesting in regard to the solubilization mechanism and the model of solubilized protamines. A very popular model of a polypeptide or protein solubilized by RM is that of a macromolecule swimming in the water pool of one RM, or at least surrounded more or less completely by one RM. However, as we could already show for $(Lys)_n$ [24, 25], this model does not hold for large polypeptide molecules at low and medium w_0-values, because in this case, one single RM is too small to enclose in tube fashion one of these rather long, rodlike polypeptide molecules with molecular weights of approximately 100000. It revealed that not only several empty RM are involved in forming one big "filled" RM, but also that the number of the AOT- and water-molecules available in these "filled" RM does not fulfill the conditions for a polypeptide molecule swimming in the waterpool of such an RM.

Therefore, it becomes obvious that in the case of extended molecules several RM are arranged like pearls in a necklace along the chain. This assumption seems to be supported by the AOT-concentration dependence of [Θ], because it indicates a partial solubilization at, for example, 22 m mole ATO/l, with a lower degree of molecular order than after complete solubilization at above 50 mmol AOT/l. Probably for the same reason, a clupeine concentration dependence of the CD spectra has been observed. Assuming that a protamine consists of 30 amino acids, the length of a complete α-helical molecule would be 45 Å. Because the radius of a RM with $w_0 = 4$ has a radius of only ≈ 6 Å, it becomes clear that also such a rather small rodlike protein molecule does not fit in one RM without severe perturbation of the dimensions of the original "empty" RM.

The influence of the primary structure on the conformation of the protamines solubilized in RM is

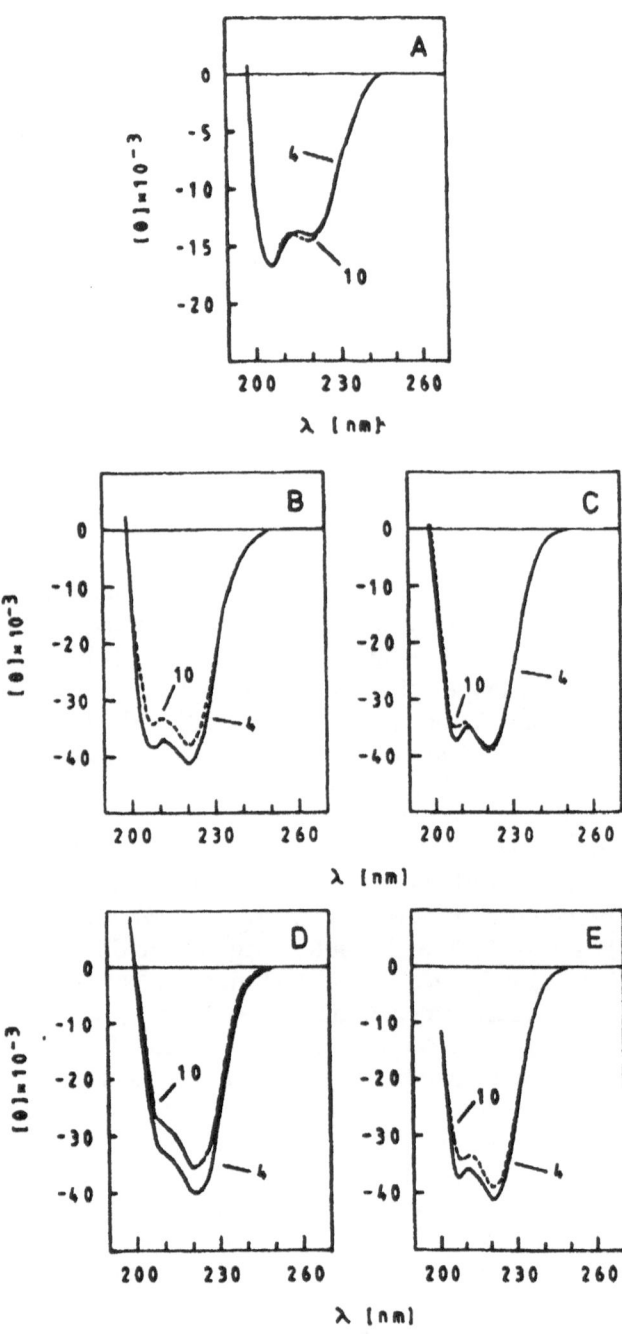

Fig. 9. CD spectra of poly-L-arginine HCl with different degrees of polymerization solubilized in RM at $w_0 = 4.0$ and 10. A: DP = 18; B: DP = 62; C: DP = 12.5; D: DP = 597; E = 2320, $C_{Polyarg}$: 6.1 ± 0.2 · 10^{-3} m, C_{AOT}: 5.0 ± 0.1 · 10^{-2} m, T: 22.5 ± 0.5°C, : 0.02 cm optical path lengt, molar ellipticities in deg · cm²/dmol

shown by the CD spectra of the three components of clupeine (Fig. 7). The main differences between the YI component vs the YII and Z occur

1) in the N-terminal region with AR_4S_3R, instead of PR_3TR_2AS and AR_4SR_2ASR; and
2) in the C-terminal region one G is inserted between A and the terminal RRRR-sequence.

These differences in primary structure may be responsible for the differences in the CD spectra. However, detailed investigations on the influence of the primary structure on the conformation of protamines in RM studies on sequence model polypeptides are necessary for elucidating these effects, as was done by Nishi et al. [26] on sequence polypeptides in aqueous solautions. Corresponding experiments in reverse micelles are being carried out at present.

References

1. Ando T, Yamasaki K, Suzuki K (1973) Protamines, Springer-Verlag, Berlin
2. Eicke HF (1980) Top Curr Chem 87:85
3. Luisi PL (1985) Angew Chem 97:449
4. Luisi PL, Magid LJ (1986) CRC Crit Rev Biochem 20:409
5. Waks M (1986) Proteins: Structure, Function and Genetics 1:4
6. Seno M, Noritomi H, Kuroyanagi Y, Iwamoto Y, Ebert G (1984) Colloid & Polymer Sci 202:897
7. Noritomi H (1986) Thesis, University of Tokyo
8. Murai N, Miyazaki M, Sugai S (1976) Nippon Kagaku Kaishi 1976:569
9. Miyazaki M, Yoneyama M, Sugai S (1978) Polymer 19:995
10. Toniolo C, Bonora GM, Marchiori F, Borin G, Filippi B (1979) Biochim Biophys Acta 576:429
11. Bonora GM, Bertanzon F, Moretto F, Toniolo C (1981) Z Naturforsch 36c:305
12. Ichimura S, Mita K, Zama M (1978) Biopolymers 17:2769
13. Ebert G, Kim YH (1983) Progr in Colloid & Polymer Sci 68:113
14. Ebert G, Lukasch H (1986) Int J Biol Macromol 8:142
15. Maeda H, Kato H, Ikeda S (1984) Biopolymers 23:1333
16. Wong M, Thomas JK, Nowak T (1977) J Am Chem Soc 99:4730
17. Lim YL, Fendler JH (1978) J Am Chem Soc 100:7490
18. Sunamoto J, Hamada T, Seto T, Yamamoto S (1980) Bull Chem Soc Jpn 53:589
19. Hauser H, Haering G, Pande A, Luisi PL (1989) J Phys Chem 93:7869
20. Jain TK, Varshney M, Maitra A (1989) J Phys Chem 93:7409
21. Douzou P, Keh E, Balny C (1979) Proc Natl Acad Sci USA 76:681
22. Greenfield N, Fasman GD (1969) Biochemistry 8:4108
23. Yaron A, Katchalski E, Berger A (1971) Biopolymers 10:1107
24. Plachky M (1987) Thesis, University of Marburg (Lahn)
25. Ebert G, Plachky M, Seno M, Noritomi H (1988) Progr in Colloid & Polymer Sci 77:67
26. Nishi N, Tsunemi M, Hagiwara K, Tokura S, Tsutsumi A (1986) Peptide Chemistry (Kiso Y (Ed)), 246

Authors' address:

G. Ebert
Fachbereich Physikalische Chemie
Philipps-Universität Marburg
Hans-Meerwein-Str., Gebäude H
3550 Marburg, FRG

Progress in Colloid & Polymer Science   Progr Colloid Polym Sci 83:188—195 (1990)

Phase behavior and interfacial phenomena in microemulsions used for enhanced oil recovery

D. F. Anghel

Institute of Chemical Physics, Department of Colloids, Bucharest, Romania

Abstract: Surfactant flooding (micellar or microemulsion) is one of the enhanced oil recovery (EOR) methods developed to recover residual oil left after water-flooding. Both low-concentration and high-concentration surfactant solutions have evolved in practice and both imply reduction of interfacial tension to levels of about 10^{-3} mN/m. The purpose of this paper is to study the phase equilibria existing in the oil-water-surfactant plus cosurfactant systems and to clarify the role played by interfacial forces appearing at microemulsion/oil and microemulsion/water interface in high-concentration surfactant systems used for EOR. The influence of surfactant, cosurfactant, salinity, oil composition, and temperature was considered. Optimal salinities were obtained by means of interfacial tension or solubilization parameter measurements. The results show that optimal salinity decreased with increasing angionic surfactant hydrophobicity and concentration, increasing temperature, increasing the radius of the electrolyte cation, and when changing from *n*-alkane to *iso*-alkane to *cyclo*-alkane to aromatic oils. At the same time, optimal salinity increased with the radius of electrolyte anion, the molecular mass of the *n*-alkane oil phase, the hydrophilicity of the alcohol cosurfactant and the hydrophile-liphophile balance of nonionic surfactant. Optimal salinity from phase behavior was found to agree well with the one obtained from interfacial tensions. This research allowed us to understand immiscible aspects of microemulsion flooding and to develop screening procedures for optimal floods.

Key words: Microemulsion; phase behavior; interfacial tension; surfactant; cosurfactant; enhanced oil recovery

Introduction

In a microemulsion flood, the injected fluid banks interact with one another and with the reservoir rock, brine, and crude oil. Initially, the microemulsion fluid may be miscible with crude oil and reservoir brine. However, because of dilution and surfactant adsorption the flood can degenerate to an immiscible displacement [1]. If low interfacial tension (IFT), or more specifically, high capillary number (velocity × viscosity/IFT) is maintained between all the phases, the displacement efficiency is good. For an ordinary water flood the capillary number is of the order of 10^{-6}. Laboratory studies indicated that the displacement efficiency of a microemsulion flood approaches 100% of the residual oil when the capillary number is increased to 10^{-2}—10^{-1}. This can be achieved by decreasing the interfacial tension between oil and water by a factor of 10000, in order to obtain ultra-low IFT values (10^{-3}—10^{-4} mN/m) [2].

This paper suggests some considerations that must be included in the design of an optimal microemulsion fluid. Phase behavior and interfacial tension data were used to determine the optimal salinity. The dependence of optimal salinity on oil, brine, surfactant, cosurfactant, and temperature was examined.

Materials and methods

Sodium didodecylbenzene sulphonate (NaDiDBS) was supplied by Dero Chemicals (Ploieşti). It had been purified of inorganic salts and unsulphonated oil. The sample had an equivalent molecular weight of 480, which denotes the presence of some lower molecular weight monoalkylbenzene sulphonates and of polysulphonates. Ethoxylated nonylphenols, NP_n ($n = 3$, 8, 12, and 16, respectively) and sodium dodecylthreeoxyethylene sulphate were supplied by Timişoara Detergent Factory and used as received.

The cosurfactants were reagent grade *iso*-propanol, *iso*-butanol and *iso*-pentanol.

Reagent grade hydrocarbons of *n*-alkane (hexane, heptane, octane, decane, dodecane, tetradecane and hexadecane), *iso*-alkane (isooctane), *cyclo*-alkane (cyclohexane), and aromatic (benzene, toluene, ethylbenzene) type made up the oil phase. A kerosene sample dried on anhydrous sodium sulphate was also used.

The aqueous phase consisted of solutions of sodium fluoride, sodium chloride, and a mixture of sodium chloride-calcium chloride in the molar ratio of 10:1. All aqueous soloutions were prepared on the basis of weight.

The surfactant-cosurfactant mixture was made at a weight ratio of 2:1. Throughout the experiments the oil/water ratio was kept constant (1:1, v/v). After putting the immiscible phases together, the surfactant-cosurfactant mixture was added. The systems were vigorously shaken and left to attain equilibrium on standing. Optimal salinity values were obtained using the approach described by Healy and Reed [3]. The effect of nonionic surfactants on the surfactant formulations was studied for $NaDiDBS-NP_n-iso$-pentanol (1:1:1, wt.) mixtures. The effect of ether sulphate on the properties of mixed surfactant formulations was studied by replacing NaDiDBS with sodium dodecylthreeoxyethylene sulphate.

Viscosity was determined by means of an Ubbelhode viscometer (K = 0.01001). The density of the different phases after equilibration was measured using a 1-ml density bottle. Electrical conductivity was measured with the aid of a radiometer conductivity meter fitted with a CDC 314 bright platinum electrode cell; the nominal constant of the cell was of 0.316 cm. Interfacial tensions were measured using the spinning drop tensiometer developed by Cayias et al. [4]. Except for the data obtained at 52 °C all phase volume and interfacial tension data were obtained at room temperature (i.e. 25 °C ± 1 °C).

Results and discussion

Water and oil are practically immiscible liquids. However, their mutual solubility can be improved by means of a surfactant or surfactant-cosurfactant mixture. At constant temperature, such mixtures may separate into three liquid phases: an aqueous lower phase, a surfactant-rich middle phase, and an oil-rich upper phase. In the middle phase, frequently referred to as microemulsion, one finds (for thermodynamic reasons) a maximum of the mutual solubility between water and oil and a minimum of the interfacial tension between them. In the case of a particular reservoir, the temperature, the chemical nature of the oil, and the salt concentration of water are known. The challenge is to find the surfactant which is the most efficient in solubilizing these reservoir fluids at the given temperature.

Consider first a mixture of equal volumes of oil and water in the presence of a constant amount of surfactant-cosurfactant mixture. If the salt content of the water is increased, the system will undergo a series of phase changes, as in Fig. 1. At low electrolyte levels the mixture will separate into two phases, an aqueous surfactant-rich phase in equilibrium with an excess oil phase. This is the so-called Winsor-I (O/W + O) type [5] or Knickerbocker's $\underline{2}$ [6], because the system has two phases and the surfactant is mainly dissolved in the lower phase. By increasing the electrolyte concentration, the mixture will separate into three phases, a lower aqueous phase containing almost all the salt and molecularly dispersed surfactant, a middle surfactant-rich phase, and an upper oil-rich phase. The middle phase contains almost all the surfactant and cosurfactant, and large amounts of water, including electrolyte and oil. In the above notations they are denoted by Winsor-III type (W + microemulsion + O), and Knickerbocker-3 type, respectively. The salt concentration at which the middle phase appears is the lower salinity critical end point (ε_l) [7]. With further increase of the salt concentration, the three phases will coexist within a salinity range that stretches up to the upper salinity critical end point (ε_u), when the middle phase merges with the upper phase. From now on, one again finds two phases, but the surfactant is in the upper one. This is Winsor's type II (W/O + W) or Knickerbocker's $\bar{2}$.

Another way to present such results is by means of the modification of the volume fraction of the phases with respect to salinity, as depicted in Fig. 2. In this case, the lower and the upper salinity end points are more evident.

Figure 3 shows solubilization parameters, V_0/V_s and V_w/V_s, as a function of salinity, corresponding to the same surfactant-cosurfactant mixture, but here the oil phase was kerosene. V_0/V_s is plotted only for lower- or middle-phase microemulsions. V_w/V_s for lower-phase microemulsion would be of less interest since all water in the system is always in the microemulsion phase and, therefore, V_w/V_s is plotted only for upper- or middle-phase microemulsions. Simultaneous occurrence of

NaDIDBS–i–C$_5$OH (2:1, w); n-Heptane

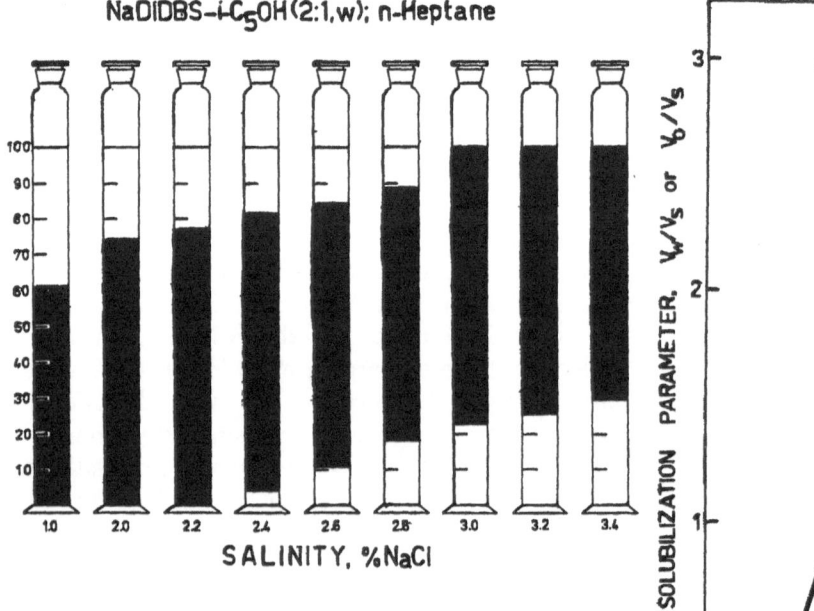

Fig. 1. Phase behavior as function of salinity

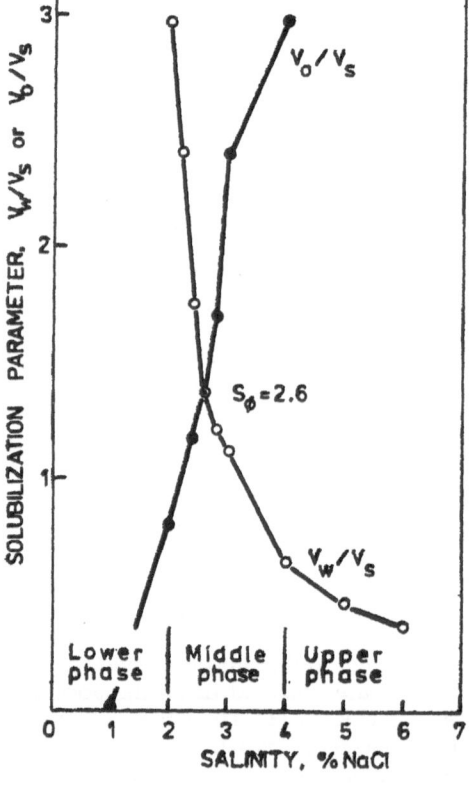

Fig. 3. Solubilization parameters vs salinity. The system is similar to that in Fig. 1, except for the oil, which is kerosene

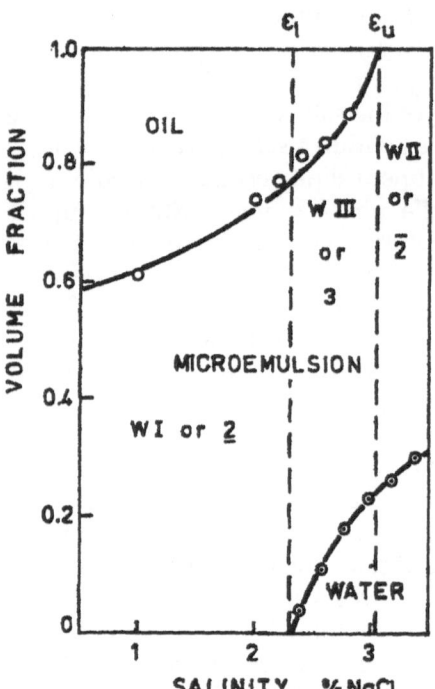

Fig. 2. Volume fraction of phases of the system presented in Fig. 1

V_0/V_s and V_w/V_s for a given salinity implies a middle-phase microemulsion, and the two graphs are thereby mutually dependent. For spherical water-external micelles, the effect of increasing salinity is to decrease the water-hydrophile interaction, permitting the micelle to attain a larger radius and accomodate more oil in its interior. V_0/V_s and V_w/V_s intersect at $S_\emptyset$, the optimal salinity for phase behavior. For this system, $S_\emptyset = 2.6\%$ NaCl.

Since both microemulsion-oil and microemulsion-water interfaces occur, tensions at both interfaces γ_{mo} and γ_{mw}, respectively, are of interest. These tensions are plotted as a function of salinity in Fig. 4. Whenever both γ_{mo} and γ_{mw} are plotted at a given salinity, a middle-phase exists. When lower-phase microemulsions are mixed with water, they always take up all the water and often reject oil, hence, $\gamma_{mw} = 0$. Similarly, for upper-phase microemulsions, $\gamma_{mo} = 0$. Definite trends are evident: γ_{mo} decreases as salinity increases, whereas

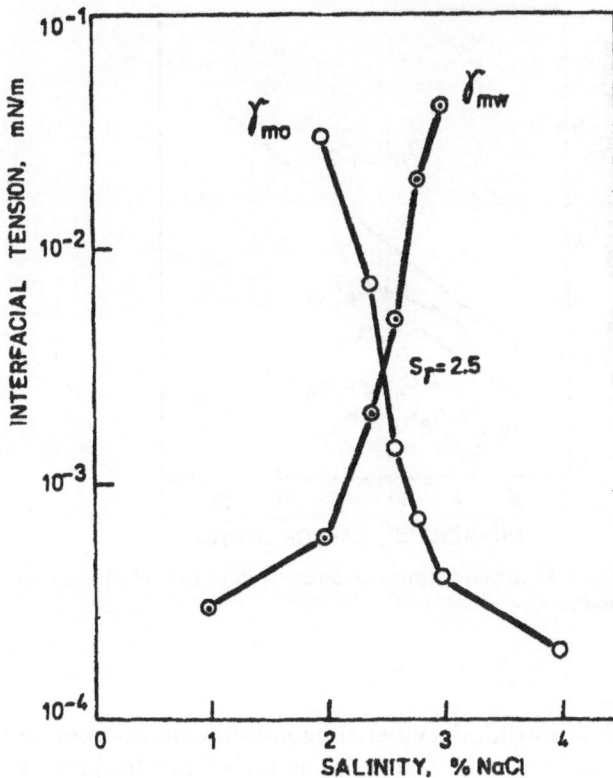

Fig. 4. Interfacial tension vs salinity for the system presented in Fig. 3

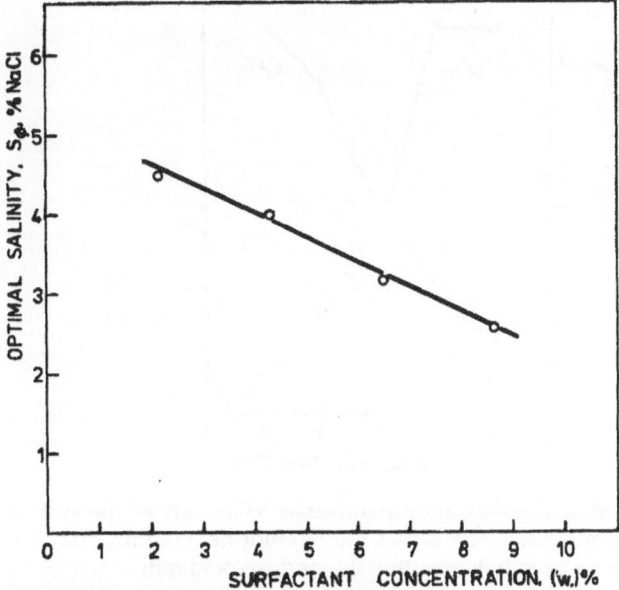

Fig. 5. Optimal salinity as a function of overall surfactant concentration

γ_{mw} increases with increasing salinity. The salinity at which the two functions intersect is termed optimal and is written S_γ; the tension there is called optimal interfacial tension. For this system $S_\gamma = 2.5\%$ NaCl and $S_\varnothing = 2.6\%$ NaCl are nearly the same, and it may be useful to explore sensitivity of optimal salinity to various parameters of the system.

Figure 5 shows $S_\varnothing$ as a function of overall surfactant concentration at constant 1:1 water-oil ratio (WOR) for the system NaDiDBS-*iso*-pentanol (2:1 wt.), kerosene, brine. As overall surfactant concentration increases, $S_\varnothing$ decreases, a trend that is in a good agreement with previously reported data [8].

Surfactant formulations based on anionic sulphonates exhibit low salt tolerance. Usually, the reported optimal salinity values are in the range 1—3% NaCl [8, 9]. This did turn out to be the case with our sodium didodecylbenzene sulphonate. To improve the salt tolerance of the surfactant formulations, nonionic surfactants [10, 11] and ethoxylated sulphonates [12, 13] were proposed.

In our surfactant formulations, we have substituted half of the sulphonate quantity for a nonionic surfactant one. Table 1 shows the $S_\varnothing$ values as a function of ethoxylation degree and hydrophile-lipophile balance (HLB) of the ethoxylated nonylphenols. Optimal salinity increases with the hydrophilic character of the nonionic surfactant. The increasing ratio is very small in the case of hydrophobic surfactants (ethoxylation degree of 3 and 8 respectively). It is considerably higher for the hydrophilic surfactants ($n = 12$ and 16) when an HLB increase of about one unity almost doubles the $S_\varnothing$ value.

In the system consisting of kerosene, water plus sodium chloride, and surfactant plus *iso*-pentanol,

Table 1. The effect of nonionic surfactants on optimal salinity of the NaDiDBS-NP$_n$-i-C$_5$OH (1:1:1 w.)-kerosene-water-NaCl

Nonionic surfactant	HLB	$S_\varnothing$
NP$_3$	7.5	0.5
NP$_8$	12.3	0.6
NP$_{12}$	14.1	7.7
NP$_{16}$	15.2	13.5

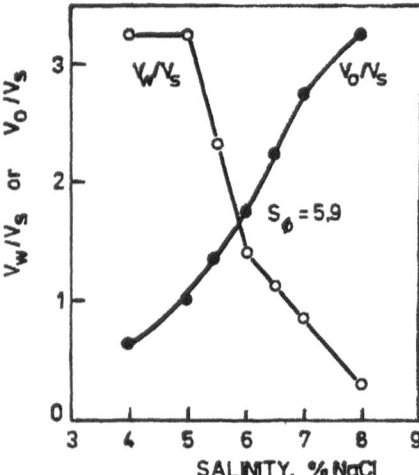

Fig. 6. Solubilization parameters vs salinity of the system kerosene, water plus sodium chloride, sodium dodecyl-threeoxyethylene sulphate, and *iso*-pentanol

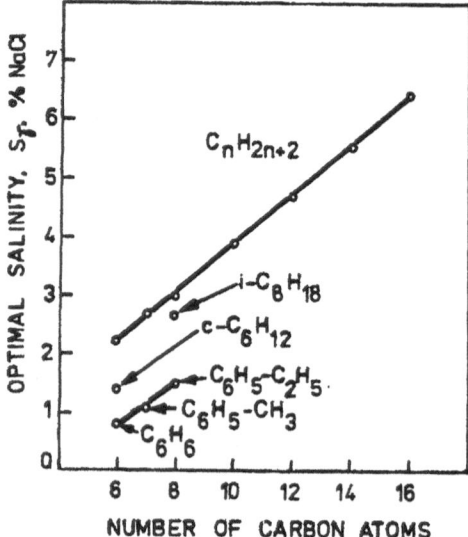

Fig. 7. Optimal salinity as a function of the oil phase composition

the sulphonate surfactant was completely substituted for the ether sulphate one. Consequently, an increase of the optimal salinity from 2.6 to 5.9% NaCl was obtained (see Figs. 2 and 6).

The influence of the oil phase on optimal salinity was investigated for the system containing a surfactant formulation based on sodium didodecylbenzene sulphonate-*iso*-pentanol. The obtained results are presented in Fig. 7. In the *n*-alkane series, S_y increased with increasing the hydrocarbon molecular weight in a manner that obeyed Traube's rule. Of the hydrocarbon studied, *n*-alkanes had the highest optimal salinity values. Branching of the hydrocarbon chain lowered the optimal salinity. When comparing aromatic, cycloaliphatic, and aliphatic hydrocarbons having the same number of atoms of carbon in the molecule, the lowest S_y value was recorded for aromatics. Grafting methyl and ethyl chains on the benzene ring entailed an increase of optimal salinity. Thus, the knowledge of the influence of the oil phase on optimal salinity is of utmost importance when referring to crude oil that is always a complex hydrocarbon mixture.

Another parameter affecting the phase behavior is the composition of the water phase. The influence of some inorganic electrolytes (i.e., NaF, and NaCl/CaCl$_2$) was considered. It was found that when decreasing the electrolyte anion radius from chloride to fluoride a lowering of $S_\emptyset$ from 2.6% to 1.8% was noted for the system based on

sodium didodecylbenzene sulphonate-*iso*-pentanol and kerosene. The result is not of practical importance since fluoride is not usually present in the connate water. However, calcium and magnesium ions are often present at various levels in connate water. They have an adverse effect on microemulsion stability since they precipitate the sulphonate surfactants.

The influence of calcium ions on optimal salinity was investigated by adding calcium chloride into the brine. Of the investigated systems, the one based on kerosene, water, and sodium didodecylbenzene sulphonate-*iso*-pentanol had an optimal salinity of 2.6% NaCl and of 1.5% NaCl/CaCl$_2$, respectively. The presence of calcium ions in the reservoir water considerably decreases the optimal salinity.

It was pointed out above that substitution of sulphonate surfactant for an ether sulphate one considerably increased the optimal salinity. It was also shown that the ether sulphate can tolerate higher concentrations of calcium and magnesium ions [12, 13]. This finding was checked by us and the obtained results are shown in Fig. 8. For the investigated system, optimal salinity lowers from 5.9% NaCl to 5.0% (NaCl/CaCl$_2$), the adverse effect of calcium ions being less pronounced than in the case of didodecylbenzene sulphonates.

The influence of the alcohol cosolvent on optimal salinity of the system kerosene, water plus sodium

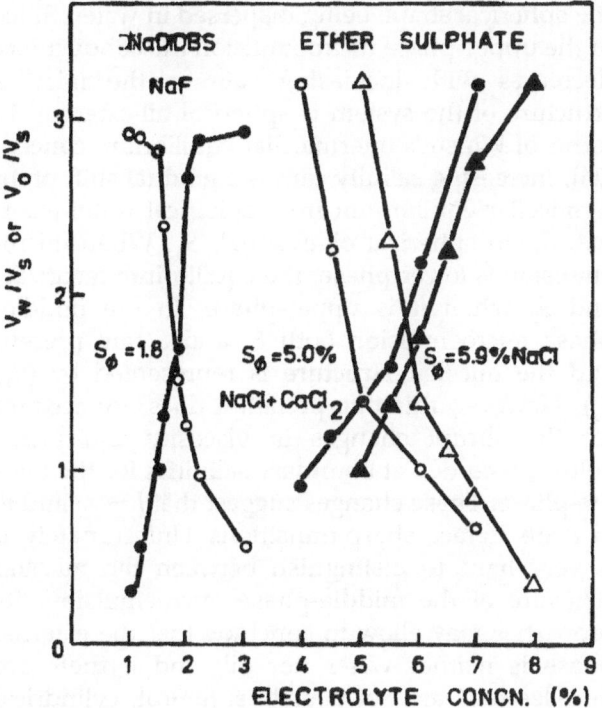

Fig. 8. The influence of inorganic salts on solubilization parameters and optimal salinity

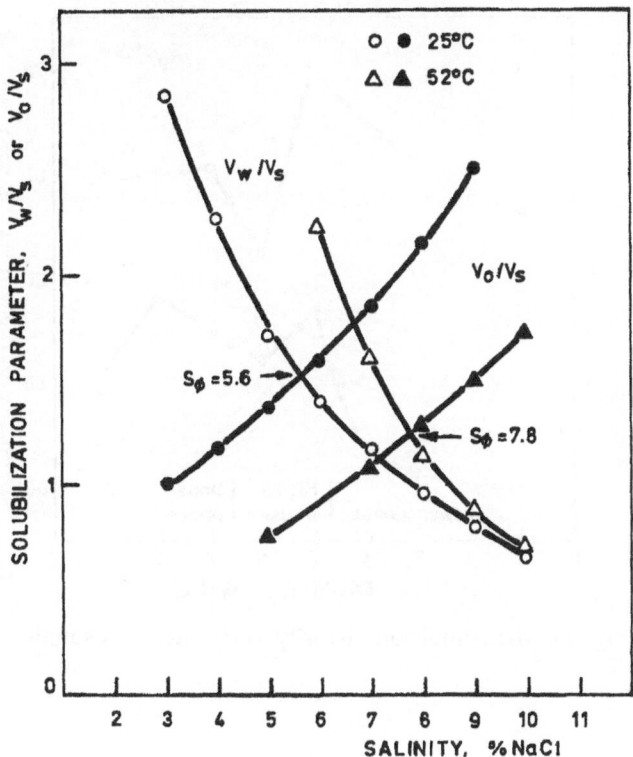

Fig. 9. Effect of temperature on solubilization parameters

chloride, sodium didodecylbenzene sulphonate-alcohol was studied. Table 2 shows optimal salinity values for three different cosurfactants: *iso*-propanol, *iso*-butanol and *iso*-pentanol. An increase in the alcohol molecular weight entails a decrease of optimal salinity.

Figure 9 shows the dependence of V_0/V_s and V_w/V_s on temperature and salinity for NaDiDBS-*iso*-butanol-kerosene-brine mixtures. Increasing the temperature causes phase behavior to change in the direction $u \rightarrow m \rightarrow l$ (upper $\rightarrow$ middle $\rightarrow$ lower phases). As temperature increases at constant salinity, V_0/V_s decreases and V_w/V_s increases. This

Table 2. The influence of cosurfactant on optimal salinity. Surfactant: sodium didodecylbenzene sulphonate, Oil: kerosene

Cosurfactant	$S_\varnothing$ (% NaCl)
Isopropanol	15.0
Isobutanol	5.7
Isopentanol	2.6

suggests that $S_\varnothing$ increases with increasing temperature because lipophile-oil interaction energy decreases and hydrophile-water interaction energy increases.

In order to draw conclusions about the influence of salinity on microemulsion type, determinations of viscosity and electrical conductivity were undertaken. The systems under consideration were the ones containing water with various amounts of NaCl or NaCl/CaCl$_2$, kerosene and ether sulphate-*iso*-pentanol. Figure 10 shows the dependence of viscosity and microemulsion phase volume on salinity. As salinity increases, the microemulsion phase undergoes the transition $l \rightarrow m \rightarrow u$. At these transitions, both viscosity and microemulsion phase volume change abruptly. The minimum point recorded in both properties in the microemulsion at 6.0% NaCl agrees well with optimal salinity ($S_\varnothing$ = 5.9% NaCl).

When sodium chloride is substituted for sodium chloride/calcium chloride mixture the viscosity and electrical conductivity have the behavior presented in Fig. 11. In this case neither the minimum of viscosity nor the maximum of electrical conduc-

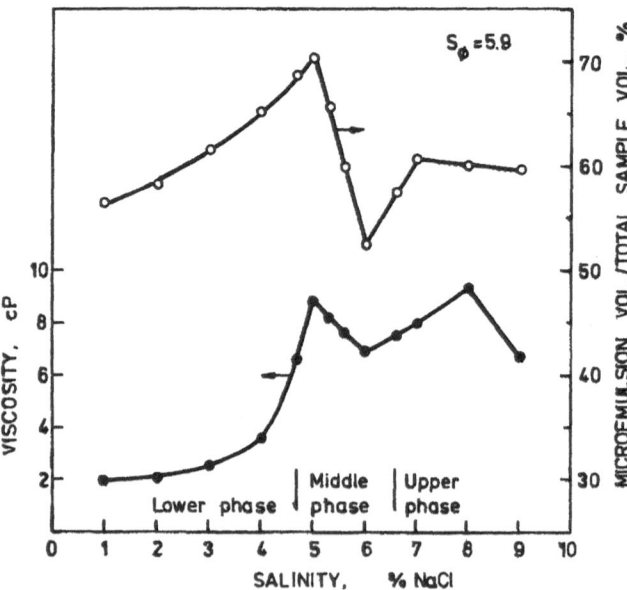

Fig. 10. Microemulsion viscosity and volume vs salinity

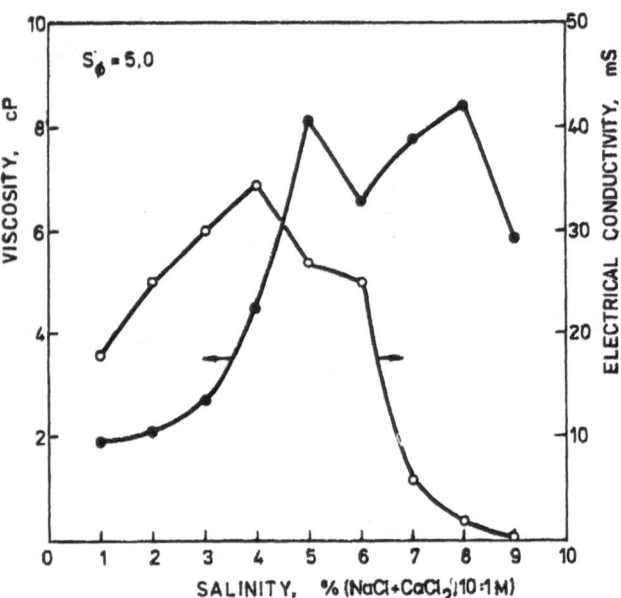

Fig. 11. Microemulsion viscosity and electrical conductivity vs salinity

tivity are related to optimal salinity ($S_\emptyset$ = 5.0% NaCl/CaCl$_2$). However, the electrical conductivity changes can afford speculations on the micellar structure of the microemulsion phases. In the lower-phase microemulsion the electrical conductivity is increasing with salinity. The micelles have

the spherical shape being dispersed in water. Since in the upper-phase microemulsion the conductivity decreases with increasing salinity, the micellar structure of the system is spherical oil-external. In terms of Winsor's intermicellar equilibrium concept [14], increasing salinity causes a gradual shift of intermicellar equilibrium from spherical water-external, S_1, to spherical oil-external, S_2. When microemulsion is lower-phase, the equilibrium favors S_1, and S_2 when it is upper-phase. In the middle-phase microemulsion both S_1 and S_2 are present and the micellar structure is represented by (S_1, S_2). However, this interpretation does not account for the abrupt changes in viscosity and phase volume recorded at boundary salinities for the middle-phase. These changes suggest that $l \rightarrow m$ and $m \rightarrow u$ are, in fact, sharp transitions. Unfortunately, it is very hard to distinguish between the micellar structure of the middle-phase microemulsion. Its properties may allow to conclude that the external phase is neither water nor oil, and if there are micelles they are not usually spherical, cylindrical or lamellar.

The recorded modifications of interfacial tension and phase behavior as a result of changing the parameters of the system are summarized in Table 3. Tabulated results show that whenever phase behavior changes in the direction $l \rightarrow m \rightarrow u$, γ_{mo} and V_w/V_s decrease, while γ_{mw} and V_0/V_s increase. When phase behavior changes in the opposite direction ($u \rightarrow m \rightarrow l$), opposite trends are observed.

The proposed concept relating changes in any variable of interest to directional changes in phase behavior ($l \rightarrow m \rightarrow u$ or $u \rightarrow m \rightarrow l$), together with the general conclusion that these phase behavior changes are associated with definite interfacial tension trends, suggests a rapid method for studying interfacial tension behavior.

The method consists of preparing a set of samples corresponding to a fixed overall composition of surfactant, cosurfactant, oil, and brine, and to change the variable of interest in a systematic, montonic manner. The samples are thoroughly mixed and allowed to equilibrate at constant temperature until the initial opaque emulsion completely disappears, and distinct translucent phases remain. If the direction of phase-behavior change is $l \rightarrow m \rightarrow u$, it follows that γ_{mo} decreases and γ_{mw} increases, and the reverse is true as well.

This approach finds application in determining, for a given surfactant, the effects of salinity, oil

Table 3. Influence of some variables on phase behavior, solubilization parameters, interfacial tensions, and optimal salinity

Increasing variable	Resulting trends					
	Phase behavior	V_0/V_s	V_w/V_s	γ_{mo}	γ_{mw}	$S_\varnothing$
Salinity	$l \rightarrow m \rightarrow u$	+	−	−	+	
Temperature	$u \rightarrow m \rightarrow l$	−	+	+	−	+
Electrolyte anion radius	$l \rightarrow m \rightarrow u$	+	−	−	+	+
Electrolyte cation radius	$l \rightarrow m \rightarrow u$	+	−	−	+	−
Surfactant concentration	$l \rightarrow m \rightarrow u$	+	−	−	+	−
HLB of nonionic surfactant	$l \rightarrow m \rightarrow u$	+	−	−	+	+
Molecular wt. of alcohol cosurfactant	$l \rightarrow m \rightarrow u$	+	−	−	+	−
Oil aromaticity	$l \rightarrow m \rightarrow u$	+	−	−	+	−
Carbon number of alkyl chain	$l \rightarrow m \rightarrow u$	+	−	−	+	+

l = lower phase 　　　(−) indicates a decrease
m = middle phase 　　　(+) indicates an increase
u = upper phase

composition or temperature on interfacial tension. Of special interest is the commonly occuring circumstances wherein salinity, oil composition, and temperature are fixed for a particular reservoir application and the method is used to screen surfactants for domains of requisite low interfacial tension.

Detailed study of multiphase aspects of microemulsion systems is useful for understanding microemulsions and surfactants in general and should prove of value in the design of effective flooding systems for tertirary oil recovery.

References

1. Healy RN, Redd RL (1977) Soc Pet Eng J 17:129
2. Foster WR (1973) J Pet Tech 25:205
3. Healy RN, Reed RL (1974) Soc Pet Eng J 14:491
4. Cayias JL, Schechter RS, Wade WJ (1975) "Adsorption at Interfaces", A.C.S. Symposium Series No 8, p 234
5. Winsor PA (1948) Trans Faraday Soc 44:376
6. Knickerbocker BM, Pesheck LE, Scriven LE, Davis HT (1979) J Phys Chem 83:1984
7. Kahlweit M, Strey R, Schomäcker R, Haase D (1989) Langmuir 5:305
8. Healy RN, Reed RL, Stenmark DG (1976) Soc Pet Eng J 16:147
9. Puerto MC, Gale WW (1977) Soc Pet Eng J 17:193
10. Dauben DL, Froning HR (1971) J Pet Tech 23:614
11. Minssieux L (1987) SPE Reserv Eng 2:605
12. Bansal VK, Shah DO (1978) Soc Pet Eng J 18:167
13. Bansal VK, Shah DO (1978) J Amer Oil Chem Soc 55:367
14. Winsor PA (1954) "Solvent Properties of Amphiphilic Compounds", Butterworths, London

Author's address:

D. F. Anghel
Institute of Chemical Physics
Department of Colloids
Spl. Independentei 202
79611 Bucharest, Romania

Progress in Colloid & Polymer Science   Progr Colloid Polym Sci 83:196—199 (1990)

Mesophases in petroleum

B. Paczyńska-Lahme

Consultingbüro, Osterode am Harz, FRG
Aus dem Sonderforschungsbereich 134 Erdöltechnik — Erdölchemie

Abstract: The formation of liquid crystalline mesophases in petroleum resins and interfacial films in petroleum water emulsions was observed. Structures are lamellar and hexagonal. The order of the structure grows with increasing temperature. We observed water in oil, oil in water, and multiple petroleum emulsions, as well as biliquid foams, as referred to by Sebba.

Key words: Petroleum emulsions; petroleum resins; mesophases; biliquid foams; petroleum treatment

Introduction

Petroleum is always recovered together with water; therefore, it generally forms water-in-oil and oil-in-water emulsions. The demulsification of those emulsions is an important problem for the petroleum industry.

Observations

During the demulsification of recovered petroleum emulsions, we have observed interlayers between the lower water phase and the upper oil phase, in which we are particularly interested. In such an interlayer, we found a polyhedral emulsion containing large amounts of water. This emulsion was optically anisotropic in polarized light. This is a surprising finding, because petroleum is a very complex mixture of many hydrocarbons and hetero-components [1].

In our further investigations, we observed the birefringence effect of polarized light on the volume phase of petroleum resins as being the interfaces of the polyhedral emulsion in the above-mentioned interlayers.

Petroleum resins [2] make up one of the two groups of petroleum colloids [3], the other being the asphaltenes. The petroleum resins and asphaltenes together are called the asphaltoids [4],

both are colloidally dispersed in all petroleum crude oils.

Petroleum emulsions are especially stable against coalescence [5], because of elastic interfacial films formed by petroleum resins and asphaltenes between the two phases, oil and water. They are not stable against flocculation.

Figure 1 shows a water-in-oil petroleum emulsion with spherical droplets of different sizes. Several droplets are in contact on the interface; they are flocculated. Nevertheless, there is seldom coalescence.

Investigations on petroleum resins

We isolated interfacial films of petroleum emulsions, and precipitated resins from petroleum. The following figures show microscopic pictures of those substances in polarized light. Figure 2 shows a water-in-petroleum polyhedral emulsion. In polarized light, we see the birefringence of interfacial films.

If we separate the petroleum resins out of a crude petroleum oil, we obtain a resin-volume phase with birefringence in several areas of mesophases [6, 7] in polarized light (Fig. 3). The order of structures in the mesophases grows with increasing temperature. Figure 3 shows double-refracting areas of mesophases at room temperature.

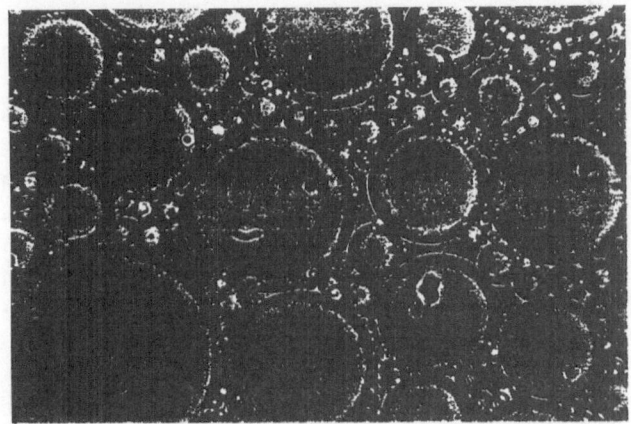

Fig. 1. Water-in-petroleum emulsion

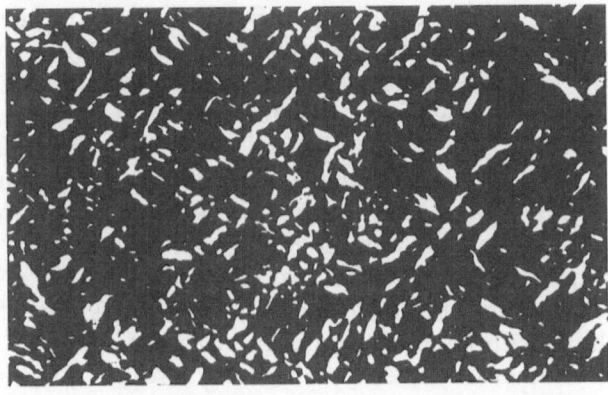

Fig. 4. Petroleum resins with a lamellar structure in polarized light at 50°C

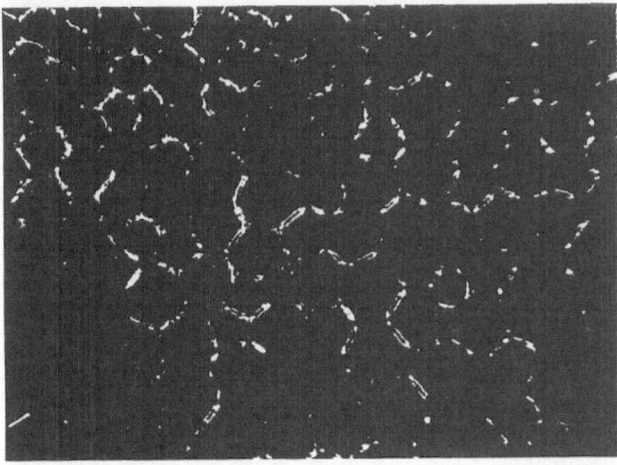

Fig. 2. Water-in-petroleum polyhedron emulsion in polarized light

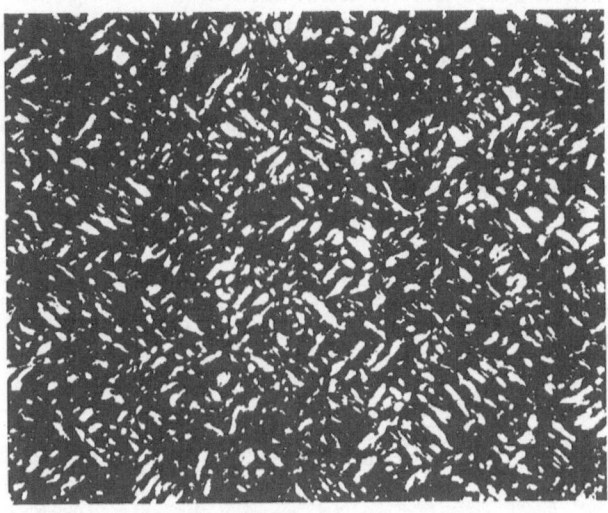

Fig. 5. Petroleum resins with a highly ordered lamellar structure in polarized light at 70°C

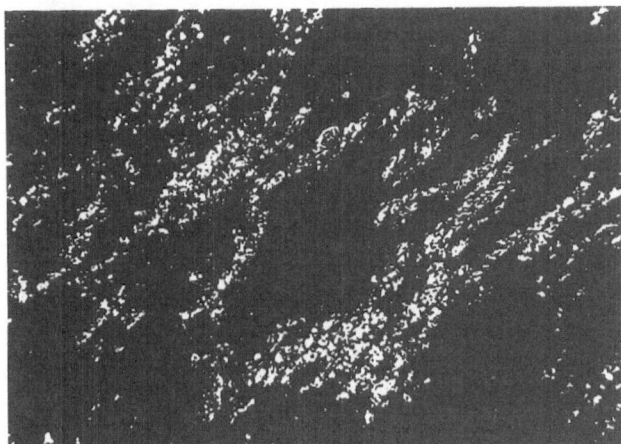

Fig. 3. Petroleum resins in polarized light at room temperature

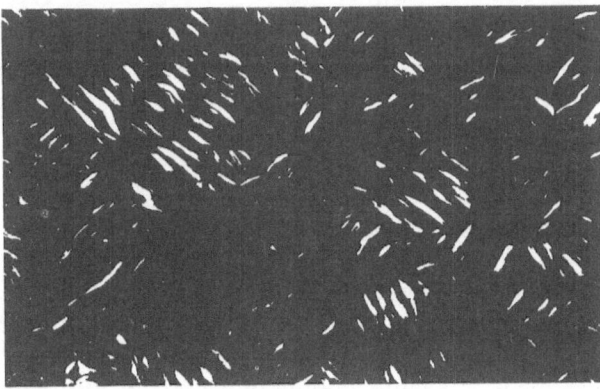

Fig. 6. Petroleum resins with a trace of water in polarized light at 70°C

After heating the system to 50 °C, we saw a higher range of order in the mesophase fields than in Fig. 4, where there is a lamellar structure.

After heating to a temperature of 70 °C, we obtained (Fig. 5), a marked lamellar structure with a higher order than that of the structure in Fig. 4. If we put a trace of water into the system shown in Fig. 5, the range of order grows again. The result is shown in Fig. 6, which has the highest range of order of the examples shown in Figs. 3—6, and of those with lamellar structures in Figs. 4—6.

Another example of a high range order is shown in Fig. 7, showing a hexagonal structure of a mesophase in a volume phase of another petroleum resin.

We can assume, according to temperature, water content, and ion concentration, that the range and the kind of order of the mesophases structure in petroleum resins vary. There are lamellar and hexagonal structures.

Up to now we have investigated the petroleum resins of 15 crude petroleums and have found mesophases in all of the examples.

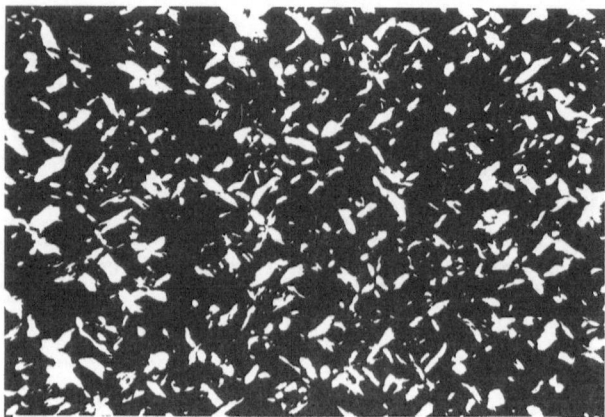

Fig. 7. Petroleum resins with a hexagonal structure in polarized light

Petroleum emulsions

Petroleum emulsions form spherical emulsions and polyhedral emulsions, especially the latter if one of either phase predominates in volume. In systems with a large amount of water, we observed, besides large drops, smaller droplets having a polyhedron shape.

In Fig. 8 we see the microscopic picture of a water-in-oil emuslion; we found very little coalescence in such a system.

In that system, we observed, beside the droplets of a water-in-oil emulsion, also droplets of a water-in-water emulsion with very thin oil films between the water droplets and the water-volume phase. That is a "biliquid foam", as referred to by Sebba [8]. Such an emulsion exists in an interlayer between the lower phase and the upper oil phase, for example during treatment of recovered petroleum emulsions.

Another example of a petroleum emulsion is the multiple emulsion (Fig. 9). In a continuous-water

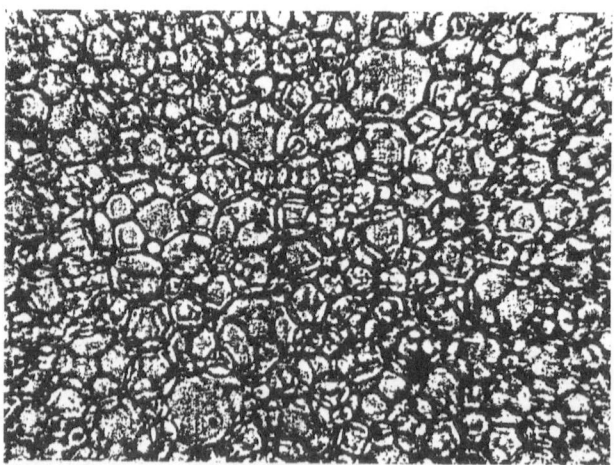

Fig. 8. Microscopic picture of a water-in-oil emulsion with a high ratio water: oil

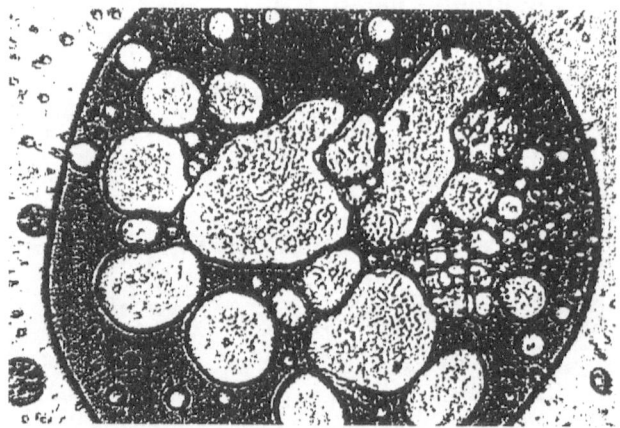

Fig. 9. Multiple petroleum water emulsion

phase there are oil droplets. In these oil droplets we see water droplets. Several of those droplets are spherical, but the smaller ones are irregularly shaped. In the water droplets there are again, oil droplets.

Conclusions

In petroleums we found different types of emulsions: water-in-oil, oil-in-water, multiple emulsions, and "biliquid foams" referred to by Sebba [8].

There are special, irregular, and polyhedral emulsions in petroleum water systems. These facts are important for demulsification.

The various petroleum emulsion types with different shapes of the dispersed droplets are functions of the type and the range of order of mesophases in the petroleum resins in the emulsion interfaces.

This is the hypothesis for our future investigations.

References

1. Neumann H-J, Rahimian I (1984) Petroleum refining. Ferdinand Enke, Stuttgart
2. Paczyńska-Lahme B (1979) Dissertation, University of Clausthal
3. Neumann H-J (1967) Habilitationsschrift Braunschweig
4. Speight HG (1980) The chemistry and technology of petroleum. Marcel Dekker, New York Basel
5. Neumann H-J, Paczyńska-Lahme B (1981) Chem-Ing Tech 53:911—916
6. Hoffmann H, Ulbricht W (1981) In: Stache H (ed) Tensidtaschenbuch. Hanser, München Wien
7. Friberg S (1971) J Colloid Interface Sci 40:291—295
8. Sebba F (1972) J Colloid Interface Sci 40:468—474

Author's address:

Dr.-Ing. Barbara Paczyńska-Lahme
Am Bergwäldchen 22
3360 Osterode am Harz, FRG

Progress in Colloid & Polymer Science   Progr Colloid Polym Sci 83:200—210 (1990)

Molecular mechanisms during the thermoreversible gelation of gelatin-water-systems *)

W. Borchard and B. Burg

Angewandte Physikalische Chemie der Universität — GH — Duisburg, FRG

Abstract: A dynamic viscometer is presented that enables the determination of the complex shear modulus and the optical rotation of systems during the process of thermoreversible gelation. It is demonstrated that the storage and loss modulus during an oscillating shear deformation can be obtained from the ratio of the amplitudes and the phase shift of the signals due to the driving external momentum and the response of the deformation. — Results of measurements of the optical and rheological properties of the system gelatin-water with a polymer content of 2.94% by wt. after a temperature step to the transformation range are reported. From the time dependence of the physical quantities it is deduced that the first process is followed by an increase of the viscosity and can be related to the formation of macromolecular clusters. Finally, the storage modulus develops as a function of time, indicating the formation of an infinite molecular network. Near the gel point the results agree well with the predictions of the percolation theory, whereas in the subsequent range of gelation a kinetic approach has been discussed to describe the results.

Key words: Thermoreversible gelation; gelatin/water; complex shear modulus; optical rotation; gelation kinetics

Introduction

Gelatin is a product of the native protein collagen which consists of long chains of linear poly-[α]-aminoacids. During the different processes, which are known to make the polymer soluble in water, the biopolymer, which is built up of three strands, is more or less partly decomposed [1—3]. Thus, the name gelatin is used for a large family of substances, some of which have polymodal molar mass distributions. If the origin is known, perhaps a more accurate information on the aminoacid composition is given. Only in a single case has the primary structure or the sequence of aminoacids been established of the two identical strands of the three-strand collagen molecule [4].

Gelatin has been long used in the photo and food industries, because it is soluble in water at high temperatures and easily forms thermoreversible gels on cooling. The gel state can be characterized by a fluid mixture having elastic properties. As the interlinking of the macromolecules is caused by physical bonding, the system can undergo the transition from the liquid to the gel state or vice versa by a change of the external variables like, for example, temperature.

Although a more detailed knowledge of the number and kind of the binding sites per single strand does not exist, arguments have been presented stating that during the gelation process three of the single strands are joined together (like in the biopolymer) in certain regions of temperature and concentration [5, 6]. This is only possible if there are forces keeping the coiled strands in the matching positions already before the process of helix formation starts or at the moment when the helices have been formed [7, 8]. It is known that the values of optical rotation are not completely recovered in gelatin solutions so that it is probable

---

*) Extended version of the oral presentation at the Kolloid-Tagung in Bochum 1989.

that only a part of the natural order of the biopolymer is restored [9]. In calorimetric studies it has been shown that the transition of gelatin-water systems is incomplete at polymer concentrations larger than 5% by weight [10], which is also derived from the mixed crystal formation of gelatin-water at the invariant eutectic conditions [11, 12].

The kinetics of the gelation process have been studied in several papers under various conditions of temperature, polymer concentration, and prehistory [5, 6, 13—18], and previously, by simultaneous measurements of the optical rotation and the complex shear modulus of gelatin-water systems [19, 20]. Up to now there is no general treatment for all properties of the gelled system up to the gel point and far beyond it at very long gelation times.

In this work our goal was to study the rheological behavior of the system close to the gel point and at times where nearly stationary values of the dynamic moduli are obtained.

The dynamic measurements

Rheological methods are preferred for the detection of the transition, because the gelation is combined with strong changes of the mechanical properties of the systems. Dynamic measurements are especially useful, because elastic and viscous parts of the modulus during the process can be separated. At times lower than the gelation time t_{gel} the viscosity of the polymer solution increases and goes to infinity at $t = t_{gel}$, which can be deduced from the frequency dependence of loss modulus [21, 19]. At the gel point the mechanical rigidity develops from zero values and finally reaches nearly stationary values after some days.

A new dynamic viscometer is presented that gives the opportunity to measure the simultaneous registration of the complex shear modulus and the optical rotation.

In Fig. 1 the measuring device for the quantities is shown. It consists of two concentric cylinders of which the inner (IZ) one is suspended between two torsional wires. The outer cylinder (GZ) is made of glass in the form of a double cylinder through which water of a constant temperature flows. The polymer solution or the gel is situated in the slit between IZ and GZ and in the optical channel (OK) which is connected to the rheometer chamber by bore-holes. In the slit the sample is forced to a sinusoidal shear oscillation. The ratio of the

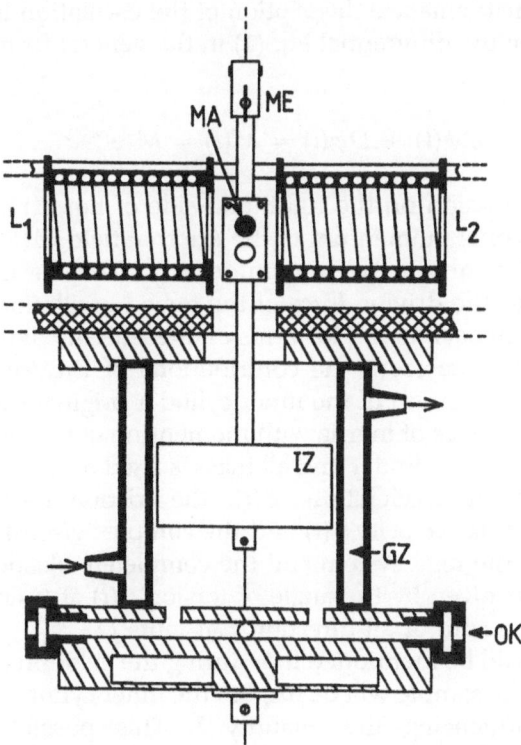

Fig. 1. Measuring device for the determination of the complex viscosity and of the optical rotation: ME = metal foil for the detection of the oscillation; L_1, L_2 = inducting coils; MA = magnetic bar; IZ = inner cylinder; GZ = double glass cylinder; OK = optical channel

amplitudes and the phase difference betwen the torsional momentum and the resulting distortion of the oscillating inner cylinder are characteristic of the sample in the slit. The torsion of the inner cylinder is achieved by an electro-dynamically induced magnetic field in the coils (L_1 and L_2) working vertically on a magnetic bar (MA) fixed to the inner cylinder (22).

In this apparatus frequencies lower than 1 Hz, below the resonance frequency, and torsional angles in the linear viscoelastic range of a maximum 2° are used.

The oscillating quantities like the external momentum of torsion and the resulting distortion of the inner cylinder are recorded by means of two proportional alternating voltages. By the Fourier analysis of the two harmonic vibrations the ratio of the amplitudes and the phase difference between the generating vibration and the response of the oscillating system is determined. The complex shear modulus G^* can be calculated from these two quantities.

The mathematical description of the oscillation is given by the differential Eq. (1) in the general form [23]:

$$I\ddot{\varphi}(t) + \eta^*\dot{\varphi}(t) + D\varphi(t) = M(t) = M_0 e^{i\omega t} . \quad (1)$$

The expression on the righthand side of Eq. (1) is the torsional momentum of the external field $M(t)$, with the maximum amplitude M_0. This term describes the driving force of the forced oscillation of the inner cylinder. The terms on the lefthand side of Eq. (1) represent the contributions of the torsional momentum of the inner cylinder origination from the forces of inertia with momentum of inertia I of the inner cylinder and all masses fixed to it and the angular acceleration, $\ddot{\varphi}(t)$, the friction term with angular velocity $\dot{\varphi}(t)$ and the complex viscosity η^* of the total system and the completely elastic response given by the angle of torsion $\varphi(t)$ at time t and the sum of all direction constants D.

It should be mentioned that during the gelation a part of the sample will be fixed to the inner cylinder thus influencing the quantity I. This possible change of I has been estimated to be of minor importance for frequencies ≤ 5 Hz and soft gels. The quantities η^* and D are sums of parts belonging to the apparatus and the sample characterized by the symbols ap and sa:

$$\eta^* = \eta'_{ap} + \eta'_{sa} - i\eta''_{sa} , \quad (2)$$

and

$$D = D_{ap} + D_{sa} . \quad (3)$$

The unusual contribution η'_{ap} of the apparatus in Eq. (2) stems from the not completely elastic support of the oscillating inner system by wires of stainless steel instead of corroding but Hookean steel. During the calibration measurements of the instrument it became evident that the viscous contribution of the apparatus that is represented by the real part η'_{ap} of η^* could not be neglected.

However, there is a repulsive torsional momentum of the sample after the gel point that is independent of the frequency of excitation and which is proportional to the torsional angle. This is considered by D_{sa} in Eq. (3).

The relation (4) leads to the solution of the differential Eq. (1):

$$\varphi(t) = \varphi_0 e^{i(\omega t - \delta)} , \quad (4)$$

where φ_0 is the maximum torsional amplitude and δ is the phase angle between the oscillation of the inner cylinder and the external torsional momentum. The first and second derivatives with respect time t lead to

$$\dot{\varphi}(t) = i\omega\varphi_0 e^{i(\omega t - \delta)} = i\omega\varphi(t) ,$$

and

$$\ddot{\varphi}(t) = -\omega^2\varphi_0 e^{i(\omega t - \delta)} = i\omega\dot{\varphi}(t) . \quad (6)$$

From Eqs. (1), (4), (5) and (6) it follows that

$$\frac{M(t)}{\dot{\varphi}(t)} = i\omega I + \eta^* + \frac{D}{i\omega} = \frac{M_0}{i\omega\varphi_0} e^{i\delta} . \quad (7)$$

Now, introducing the mechanical impedance Z for the lefthand side of Eq. (7), which, in principal, is a complex quantity given by the ratio of the external torsional momentum and the resulting angular velocity of the oscillating system, we get [23]

$$Z = \frac{M_0}{i\omega\varphi_0} e^{i\delta} = \frac{M_0}{i\omega\varphi_0} (\cos\delta + i\sin\delta) . \quad (8)$$

According to Eqs. (2), (3) and (7) the mechanical impedance of the empty apparatus (e) and that filled with a sample (f) is given by

$$Z_f = i\omega I + \eta'_{ap} - i\eta''_{sa} + \eta'_{sa} + \frac{1}{i\omega}(D_{ap} + D_{sa}), (9)$$

$$Z_e = i\omega I + \eta'_{ap} + \frac{1}{i\omega}D_{ap} . \quad (10)$$

The difference of Eqs. (9) and (10) leads to the mechanical impedance of the viscoelastic sample Z_{sa}

$$Z_{sa} = \eta'_{sa} - i\eta''_{sa} + \frac{1}{i\omega}D_{sa} = \eta'_{sa}$$

$$- i\left(\eta''_{sa} + \frac{D_{sa}}{\omega}\right) . \quad (11)$$

The quantity Z_{sa} can now be determined by the time analysis of the driving oscillation and the response of the oscillating system by means of Eq. (8), using the indices e and f for an empty and a filled device, respectively, as before. With the abbreviations:

$$K_e = \frac{M_0}{\varphi_{0,e}}, \quad K_f = \frac{M_0}{\varphi_{0,f}}, \tag{12}$$

and characterizing in the same way the phase angles by indices, it follows that

$$Z_{sa} = \frac{1}{i\omega}\left[(K_f\cos\delta_f - K_e\cos\delta_e)\right.$$

$$\left. + i(K_f\sin\delta_f - K_e\sin\delta_e)\right]. \tag{13}$$

The complex shear modulus G^* is calculated from the quantity Z by use of the relation $G^* = i\omega\frac{1}{b}Z$ and Eq. (11) where b is the form factor typical for the arrangement of the two coaxial cylinders with the radii R_i, R_0 for the inner and outer cylinder and the length L of the inner cylinder with $b = 4\pi L R_i^2 R_0^2/[R_0^2 - R_i^2]$ [23, 24]:

$$G^*_{sa} = G'_{sa} + iG''_{sa} = i\frac{\omega}{b}Z_{sa}$$

$$= [D_{sa} + \omega\eta''_{sa} + i\omega\eta'_{sa}]\frac{1}{b}, \tag{14}$$

where the parts of the complex shear modulus G^*_{sa} are given by the following equations.

Real part: $G'_{sa}(\omega) = [D_{sa} + \omega\eta''_{sa}]\dfrac{1}{b}; \tag{15a}$

Imaginary part: $G''_{sa}(\omega) = \dfrac{\omega}{b}\eta'_{sa}. \tag{15b}$

Both quantities are dependent on frequency, where $G'_{sa}(\omega)$ is proportional to the work stored by the sample during a single cycle of deformation and named storage modulus. The imaginary part $G''_{sa}(\omega)$ is proportional to the work dissipated in the sample during a single cycle of oscillation and named loss modulus. By means of Eq. (13) both quantities can be related to the phase angles δ_f, δ_e, and the quantities K'_f and K'_e given by $K'_f = K_f/b$ and $K'_e = K_e/b$:

$$G'_{sa}(\omega) = K'_f\cos\delta_f - K'_e\cos\delta_e \tag{16a}$$

$$G''_{sa}(\omega) = K'_f\sin\delta_f - K'_e\sin\delta_e. \tag{16b}$$

The ratios K'_f respectively K'_e are determined only indirectly by the ratio of two amplitudes of the voltage signals U_M and U_φ, where the indices refer to the measurables. Therefore, the proportionality constants k_M and k_φ are introduced:

$$K'_f = \frac{M_0}{b\varphi_{0,f}} = \frac{k_M U^0_M}{bk_\varphi U^0_{\varphi,f}} = E\frac{U^0_M}{U^0_{\varphi,f}} = EA_f, \tag{17a}$$

and correspondingly:

$$K'_e = \frac{M_0}{b\varphi_{0,f}} = \frac{k_M U^0_M}{bk_\varphi U^0_{\varphi,f}} = E\frac{U^0_M}{U^0_{\varphi,f}} = EA_e. \tag{17b}$$

The quantity E, the ratio of k_M and bk_φ is an electro-mechanical constant of the device that depends on the geometry of the measuring cell and the sensitivity of the electronic measuring devices for the momentum M and torsional angle φ. By the introduction of the ratio of the voltage amplitudes A_f, respectively A_e, finally the storage and loss modulus of the sample are calculated from Eqs. (16) and (17).

$$G'_{sa}(\omega) = E[A_f\cos\delta_f - A_e\cos\delta_e] \tag{18a}$$

and

$$G''_{sa}(\omega) = E[A_f\sin\delta_e - A_e\sin\delta_e]. \tag{18b}$$

Experimentally, the constant E of the measuring device can be determined by filling the system with an oil of known viscosity. At low frequencies the storage modulus of the liquid can be neglected, thus the loss modulus in Eq. (18b) is given by the Newtonian viscosity of the oil by $G''_{oil}(\omega) = \omega\eta_{oil}$.

During the experiments it became evident that an undamped oscillation of the empty device is very sensitive to external perturbations, leading to a low accuracy of the reference values. Therefore, water has been used as a reference fluid under the consideration of its viscosity. In this case, the resulting equations are

$$G'_{sa}(\omega) = E[A_f\cos\delta_f - A_{H_2O}\cos\delta_{H_2O}] \tag{19a}$$

$$G''_{sa}(\omega) = E[A_f\sin\delta_f - A_{H_2O}\sin\delta_{H_2O}]$$

$$+ \omega\eta_{H_2O}; \tag{19b}$$

η_{H_2O} is the Newtonian viscosity of the pure water.

The constant E has been determined by measuring the values of ratio of the amplitudes for an oil of known viscosity and water by use of Eq. (19b) with $G''_{oil}(\omega) = \omega \eta_{oil}$. The electro-mechanical constant has a typical value of $E = 3.31 \pm 0.04$ Pa at 293 K. Each time that the torsion wires or the amplification of the electrical system is changed, the calibration has to be repeated.

The system for determination of the optical rotation

The scheme of the optical system can be seen in Fig. 2. The laser beam with wave length of $\lambda = 632.8$ nm, the polarisation of which is stabilized, passes through the optical channel (OK) in the lower part of the cell. The windows are of thin glass plates. The beam passes the solution or gel and is split. At right angles the reference beam passes through an intensity attenuating wedge and is detected by a silicon-photodiode after having passed the analyzer of the polarizing filter. Additional to the polarization of the laser, a polarizing filter is installed at the beam exit of the laser. Both polarizer and analyzer are fixed in a 90°-position with respect to the planes of polarization. If an optically inactive substance is now introduced into the optical channel, the attenuation of the reference beam is chosen so that the ratio of both intensities is unity. In case of an optical rotation of the medium there is a rotation of the plane of polarization which gives an intensity increase on the detector for the transmitted beam. By use of beam-splitting all influences stemming from intensity variations of the laser or the increase of the turbidity of the samples can be eliminated, because the ratio of both signals is amplified which remains constant and depends only on the optical rotation of the medium.

The two polarizing filters, P_0 and P_1, which generate linearly polarized light are mounted on a common optical axis. The light (having passed the first filter P_0, called the polarizer) is characterized by the plane of polarization in which the electrical vector oscillates. If the direction of polarization of the second polarizing filter P_1 (called the analyzer), which can be rotated perpendicularly in the direction of the light propagation, makes an angle a with the plane of polarization of the beam, the amplitudes of the light waves just behind P_0 and P_1 (given by A_0 and A_1) are interrelated by

$$A_1 = A_0 \cos a . \tag{20}$$

For crossed polarizors ($a = 90°$) the light is completely absorbed by both filters, if the filters are ideal with respect to the generation of polarized light. Therefore, the intensity of the light waves, which is proportional to the square of the amplitude of the electrical field vector, is given by

$$\frac{I_1}{I_0} = \frac{(A_1)^2}{(A_0)^2} = \cos^2 a . \tag{21}$$

In this case, I_0 is characterized by the intensity that can be measured when both filters are in a parallel position, corresponding to $a = 0$. In order to have comparable intensities of light for both detectors in the position $a = 90°$ the reference beam has to be attenuated by a grey filter. Instead of I_0 the intensity I_{ref} was taken as reference, which is the intensity at the poisition of the analyzer P_1 at $a = 90°$. Introducing the attenuation factor F by

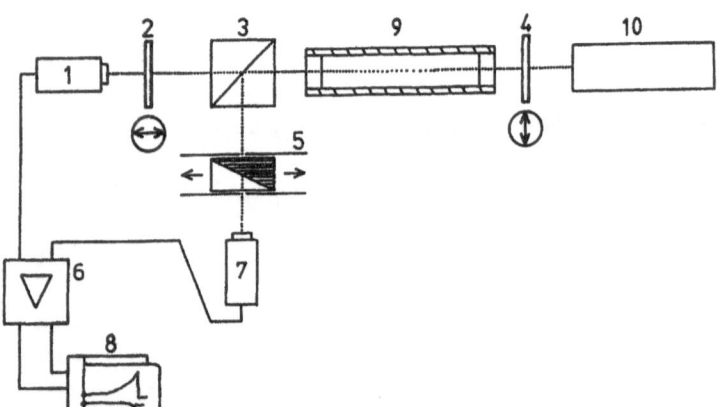

Fig. 2. Schematic representation of the continuous detection of the optical rotation.
1,7 = Si-photodiodes; 2,4 = polarization filters; 3 = beam splitter; 5 = attenuater; 6 = amplifier; 8 = recorder; 9 = sample; 10 = laser

$(I_{ref}/I_0) = F$, which is of the order 10^{-4}, we get from Eq. (21) by introduction of the complentary angle $\beta = 90 - a$:

$$F \frac{I_1}{I_{ref}} = \cos^2(90 - a) = \sin^2\beta . \tag{22}$$

The introduction of β has the advantage that β corresponds to the angle of rotation a_λ of an optically active sample placed between polarizer and analyzer.

The optical equipment can be calibrated by means of Eq. (22) using pure water as substance and changing the angle a from 75° to 105° and registrating the intensities. The result is presented in Fig. 3. The linear regression leads to

$$\sin^2\beta = (7.08 \cdot 10^{-4} \pm 2.210^{-6}) \frac{I_a}{I_{ref}}$$

$$- (8.3 \cdot 10^{-4} \pm 8.4 \cdot 10^{-5}) . \tag{23}$$

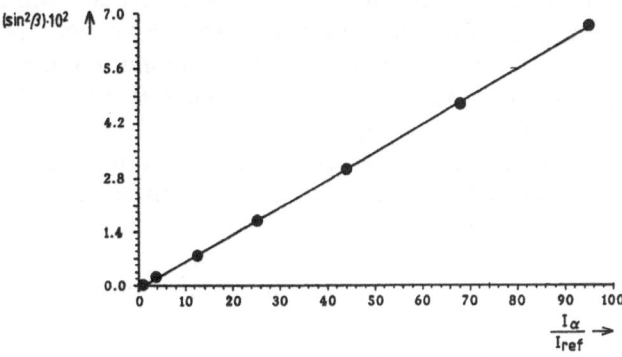

Fig. 3. Calibration curve for the determination of the angle of optical rotation (see text)

The small amount in the second bracket of Eq. (23) corresponds to the non vanishing intensity in the analyzer position for complete extinction. Therefore, the angle a_λ at wavelength λ is given by:

$$a_\lambda = \arcsin \left[(7.08 \cdot 10^{-4} \cdot \frac{I_a}{I_{ref}} - 8.3 \cdot 10^{-4})^{1/2} \right] . \tag{24}$$

The ratio of the signals is amplified logarithmically, where in the range from 10^{-8} to 10^{-5} W the ratio of the intensities per decade to the base 10 corresponds to a voltage of $U_{st} = -98.7$ mV. The ratio of the intensities is obtained from

$$\frac{I_a}{I_0} = 10^{\left(\frac{U_m - U_r}{U_{st}}\right)} , \tag{25}$$

where U_m is the voltage of the photodiode, which measures the transmitted light of the analyzer and U_r is the voltage of the reference beam. In this equipment it is not possible to detect the direction of the optical rotation, which is of no importance for our investigations as the direction is not changed during the thermoreversible gelation of the systems gelatin/water.

Materials

The solution have been prepared from a photo-gelatin containing 13% by wt. water. It is an acid-treated pigskin gelatin that has been freed from low molecular compounds by means of dialysis. Gelatin with a polymer content of 87% by wt. of the polymer and pure water have been weighed into a flask and allowed to swell overnight. For the conservation of the material against bacterial attack 0.15 ml of a 5% by wt. solution of 4-chloro-3-methylphenol in methanol per gram gelatin has been added to the aqeous gelatin solution of 250 ml. Solutions are obtained by heating the system to 50 °C under stirring for at least 1 h. These solutions are used only once for filling the viscometer.

Results and discussion

It has been shown that the thermoreversible gelation of aqueous gelation solutions starts with a fast change of the angle a_λ which leads to the specific rotation $[a]_\lambda$ by

$$[a]_\lambda = \frac{a_\lambda}{cd} \tag{26}$$

where c is the concentration of the solute in grams per 100 cm³ solution, and d is the thickness of the layer, which is the length of the optical channel in 0.1 m. The fast change of $[a]_\lambda$ is due to the coil-helix transformation of the polymer which occurs immediately within the change of the temperature,

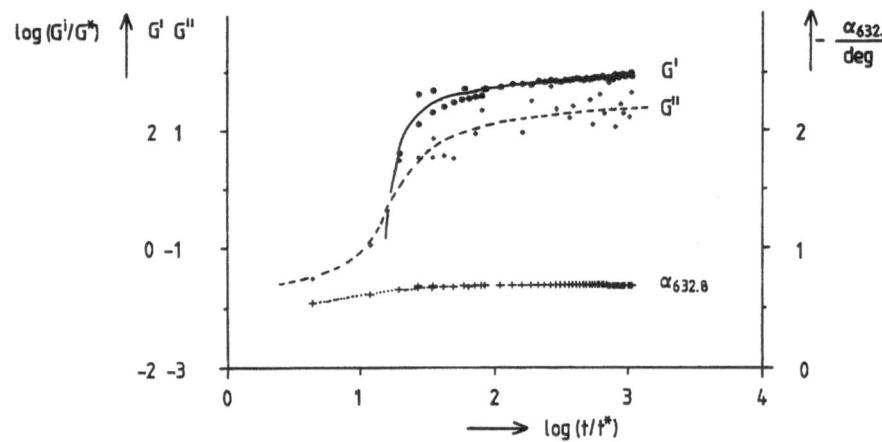

Fig. 4. Logarithm of the storage and loss modulus, G'/G^* and G''/G^* and $a_{632.8}$ vs logarithm of the time t/t^* of the system gelatin-water with a polymer content of 2.94% by wt, $a_{632.8}$ is the angle of the optical rotation in degrees at a wavelength of 632.8 nm, $G^* = 1$ Pa, $t^* = 1$ min

as soon as the range of transformation is reached during the cooling to isothermal conditions [19, 20]. The second step in the gelation process is the increase of the loss modulus G'' which is related to the change of the viscosity of the solution. At low polymer concentrations and high temperatures just below the end of the transformation range it is clearly resolved by experiments that the increase of $G''(t)$ starts much later than the specific optical rotation $[a]_\lambda(t)$. This demonstrates that, in addition to the stiffening of the chains, an aggregation of helices or partly helicated polymer molecules takes place, which is the reason why the viscosity tends to infinity at zero frequency [20, 21].

A typical time dependence of $a_{623.8}$ at a wavelength of $\lambda = 632.8$ nm and the quantities G'' and G' are represented in a double logarithmic plot in Fig. 4. The values have been measured after a temperature decrease from 313 to 293 K. The polymer concentration of the aqueous system is 2.94% by wt. It is clearly stated that the optical rotation has become stationary after a time of about 20 min, which is necessary sharply in the range from 10 to 60 min and then levels off after 300 min. At the time where the $G''(t)$-curve exhibits an inflexion point, a steep increase of the storage modulus is stated, indicating the existence of a physical network. From these and other results of this system it can be deduced that the aggregation of the polymer molecules finally leads to a gel, which is characterized by the appearance of a measurable storage modulus. After a total gelation time of 1000 min there is a change of G' over three decades. A further increase on the logarithmic time scale is rather low for this gel, but is clearly perceptible.

The mathematical description of the gelation curves

As the polymer chains aggregate under the condition of gelation it was tried to describe the time dependence of the storage modulus following a kinetic model in order to be able to predict the so-called equilibrium modulus G'_∞ obtained at infinite time of gelation. The problem is that the concentration of the junction zones of the polymer chains is unknown in reality. In case of gelation this is the number of chain sequences that is able to form a crosslink, requiring at least two of them per macromolecule with a functionality larger than two [25]. The optical rotation of the solution is not suited for this concentration, as it is a measure for all helices whether they are part of a junction zone or not.

A further attempt can be made to consider the number of the reacted chains being part of the aggregates and, later on, of the network. In this case the modulus of the network is a measure for the molecular processes taking place after the gel point. From the statistical theory or rubber elasticity it is known that the static equilibrium modulus of a network is proportional to the number of network chains for a given volume.

In earlier papers it was demonstrated that the modulus of gelatin gels is frequency independent for times larger than those necessary to reach the gel point t_{ind} [6, 19, 20, 22, 26]. This corresponds to the rubber plateau region in high molecular systems. Additionally, it is stated that G'' is very small in comparison to G' (1 to 5%). This means that most of the work done on a system corresponds to a change in the stored energy. From these considerations it follows that G' can be approximately used as a measure for the progress of the reaction. If we assume that n potential chain sequences of different chains K come together to form a network junction zone or a nodule N, we may formulate this as a "chemical reaction".

$$nK \rightarrow N . \tag{27}$$

From reaction kinetics we have for this reaction

$$-\frac{1}{n}\frac{d[K]}{dt} = k[K]^n , \tag{28}$$

where $d[K]/dt$ is the reaction rate of the different potential junction sections, k is the rate constant, and n is the stoichiometric coefficient of the crosslinking reaction. If all chain sequences that are able to crosslink have reacted, the concentration of the educt can be expressed by the concentration of the product so that

$$-\frac{1}{n}\frac{d[K]}{dt} = \frac{d[N]}{dt} = k\{[N_\infty] - [N(t)]\}^n , \tag{29}$$

where $[N_\infty]$ is the concentration of the chains finally being part of the network. It is possible that the crosslinking of the chains for a complete transfer of all chains to network chains is finished at very long times ($t \rightarrow \infty$), because the kinetics of gelation near the stationary values of G' are a rather slow process. $N(t)$ is the concentration of the chains built into the network at time t. As already mentioned, we may substitute $[N_\infty]$ and $[N(t)]$ by the values of the storage modulus G'_∞ and $G'(t)$, which leads to:

$$\frac{dG'(t)}{dt} = k'[G'_\infty - G'(t)]^n . \tag{30}$$

Integration of Eq. (30) for $n = 2$, which corresponds to a modified procedure first proposed by Guggenheim gives

$$G'(t) = G'_\infty \left[1 - \frac{1}{G'_\infty k' t - 1} \right] . \tag{31}$$

The modulus $G'(t)$ can only be evaluated if the induction time t_{ind} is considered, which is close to the time t_{gel} introduced in section 2. This means that we have to introduce in Eq. (31) $t - t_{ind}$ instead of t, which results in

$$G'(t) = G'_\infty \left[1 - \frac{1}{G'_\infty k'(t - t_{ind}) + 1} \right] . \tag{32}$$

With this equation the $G'(t)$-curves of the gelatin-water systems with a polymer content between 0.6 and 1.2% by wt are well described in the middle part of the gelation times, as can be seen in Fig. 5. The resulting quantities G'_∞, k', and t_{ind} are gathered in Table 1 for the different concentrations of the gelatin.

At concentrations higher than 1.2% by wt. the induction time is so low that it can be neglected. This means that the strong increase of the storage modulus takes place during the first minutes of the gelation process at which the temperature is not yet constant; this concerns the polymer concentrations of 2 and 4.25% by wt. It is stated that the dependence of $G'(t)$ on time cannot be described by Eq. (32), because the quantity G'_∞ calculated from experimental values at the beginning of the gelation is exceeded by the measured quantity $G'(t)$ at longer times. For a better mathematical description of the gelation curves a further term $k'' \cdot t$ has been added to the two terms in brackets in Eq. (32); it has no physical meaning. This concerns the two highest concentrations and the k''-values are represented in Table 1.

Considering the kinetic expression (28), it is assumed that the potential junction sections react according to their concentration without any hindrance of the reaction. But with proceeding time

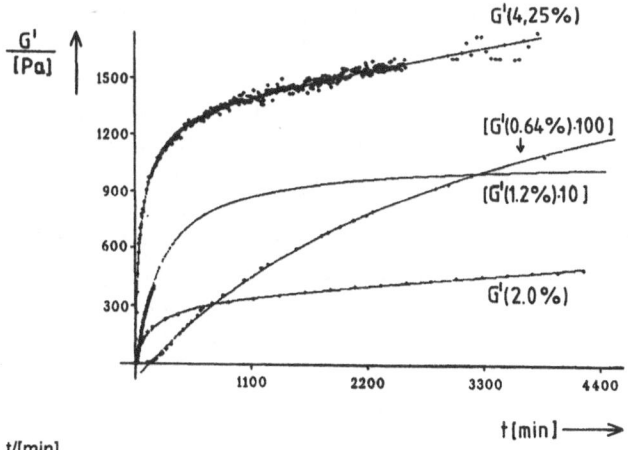

Fig. 5. Storage modulus G' in Pa vs gelation time in min of the system gelatin-water for different overall concentration of the polymers in % by wt. with different scaling

Table 1. Kinetic analysis of the G'-t-curve for different gelatin concentrations in the system gelatin-water at $T = 293$ K

w_2	$\dfrac{G'_\infty}{[\text{Pa}]}$		$\dfrac{k' \cdot 10^5}{[\text{Pa min}^{-1}]}$	$\dfrac{t_{\text{ind}}}{[\text{min}]}$	$\dfrac{k'' \cdot 10^4}{[\text{min}^{-1}]}$
0.004	3	± 0.1	11 ± 0.7	823.4 ± 11.7	0
0.0064	22	± 0.1	1.2 ± 0.02	138.9 ± 3.4	0
0.008	27.7	± 0.3	5.9 ± 0.14	56.5 ± 0.1	0
0.012	107	± 6.1	3.7 ± 0.5	18.6 ± 0.6	0
0.02	336	± 7.2	1.9 ± 0.1	0	1.18 ± 0.06
0.0425	1340	± 3.8	1.2 ± 0.02	0	0.78 ± 0.001

w_2 = weight fraction of the gelatin
G'_∞ = final or equilibrium modulus of the gels in Pa
k' = rate constant of Eq. (32); t_{ind} and k'', see text

after the gel point the network becomes much more dense so that the free chains have to move to the junction places inside an arrangement of fixed chains, which are obstacles. Thus, the condition for the reaction changes as soon as the network has been formed for the first time. Mainly, it is the diffusion of the free chain segments that will change continuously during the crosslinking. Therefore, it must be questioned if the values of G'_∞ listed in Table 1 are really the final achievable maximum values.

It is remarkable that the reaction constants k' have the same order of magnitude for all concentrations with slight differences out of the range of accuracy. The supposition that the order of the reaction changes during the reaction time can be tested. In this case the logarithm of the reaction rate is plotted vs the decreasing concentration of the educt [27]. If the order of the reaction changes, the dependence is non-linear. This test cannot be carried out directly as the concentration of the chains not yet built into the network is not known. This can only be obtained indirectly by assuming a reaction by integers like 2 or 3. For mixtures of low polymer content an exponent of about 2 has been found for the total measured gelation time [28]. This means that, on the average, two chains have to aggregate to form a junction point. From the remarks made up to now a good description is possible in the middle time range of gelation. If we look at gelation times just after the gel point it cannot be expected that the approach presented is suitable, because all experimentally determined $G'(t)$-curves start with a curvature corresponding to a convex shape of the curve with respect to the time axis. For large gelation times the shape is concave. Therefore, an inflexion point exists, which is not possible using the above description.

As with this apparatus, also the moduli at very low values could be detected, and it is possible to look for the network formation very close to the gel point. The theories treating the critical conditions for clusters in two, three and more dimensions have been developed by DeGennes, Stauffer and others [29—33]. If the fraction of the reacted bonds is proportional to the reaction time t and if the critical one for gel formation is proportional to the time t^*, the storage modulus $G'(t)$ for times $t > t^*$ is predicted to follow the relationship

$$G'(t) = K^*(t - t^*)^\varepsilon , \qquad (33)$$

where the exponent ε was derived from a random conductivity model in three dimensions to be $\varepsilon = 1.833$, where numerical estimates give $\varepsilon = 1.5$ [34].

The experimental results of the system gelatin-water with a polymer content of 0.8% by wt. are presented in Fig. 6 for a temperature of 25 °C. After an induction period or the time for gelation of $t^* = 445$ min the modulus G' increases. A non-linear regression analysis optimizing the curve fit of the first 59 non-vanishing $G'(t)$-values results in $K^* = 1.313 \cdot 10^{-5}$ Pa and an exponent $\varepsilon = 1.724$ shown by the full curve in Fig. 6.

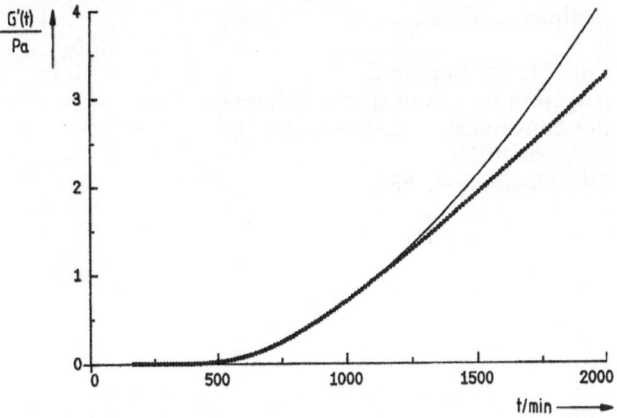

Fig. 6. Storage modulus $G'(t)$ of the system gelatin-water in Pa vs time t in min near the gel point at a temperature of T = 298 K. Dotted curve = experimental values, full curve = calculated (see text), w_2 = weight fraction of gelatin

All values near the inflexion point have been omitted. This kind of analysis close to the gel point has been made for different concentrations in the range $0.004 < w_2 < 0.04$ and various temperatures between 10° and 30°C. With exception of the lowest concentration and the highest temperature the same exponent $\varepsilon = 1.72 \pm 0.03$ has been found in all cases, which is clearly within the theoretically predicted range [36]. This is a remarkable result in favor of the percolation theory, which has been much debated in recent years [30, 31, 35]. But it should be stressed that the percolation theory is applicable only near the gel point, which is out of the technically important time range of gelation.

Acknowledgement

We thank the Deutsche Forschungsgemeinschaft for financial support. Dipl. Chem. A. Michalczyk has given help in computer calculations.

References

1. Veis A (1964) Macromolecular Chemistry of Gelatin, Academic Press, N.Y.
2. Ward AG, Courts A (1977) ed: The Science and Technology of Gelatin. Academic Press, N.Y.
3. Tomka I (1983) Chimia 37:33
4. Kühn KI (1974) Chemie in unserer Zeit 8:97
5. Peniche-Covacs C, Dev S, Gordon M, Judd M, Kajiware K (1987) In: Choempff A, Newman S (eds) Polymer Networks. Plenum Press, N.Y., 65
6. te Nijenhuis K (1981) Colloid & Polymer Sci 259:522
7. Bruckner P, Bächinger HP, Timpl R, Engel J (1978) Eur J Biochem 90:595
8. Bruckner P, Bächinger HP, Timpl R, Engel J (1978) Eur J Biochem 90:605
9. Engel J (1962) Leder 328:94
10. Meerson ST, Lipatov SM (1958) Colloid J USSR 20:336
11. Borchard W, Luft B, Reutner P (1986) The Journal of Photogr Sci 34:132
12. Borchard W, Luft B, Reutner P (1984) Ber Bunsenges 88:1010
13. Borchard W, Bergmann K, Rehage G (1976) In: Cox RJ (ed) Photogr Gelatin II. Academic Press, London, 60
14. Borchard W, Bergmann K, Emberger A, Rehage G (1976) Progr Colloid & Polymer Sci 60:120
15. Borchard W, Bremer W, Keese A (1980) Colloid & Polymer Sci 258:516
16. Godard P, Biebuyck, Daumerie M, Naveau H, Mercier JP (1978) J Polymer Sci Chem Ed 16:1817
17. Djabourov M, Leblond J, Papon P (1988) J Phys France 49:319
18. Djabourov M, Leblond J, Papon P (1988) J Phys France 49:333
19. Burg B, Borchard W (1989) In: Lemstra PJ, Kleintjens LA (eds) Integration of Fundamental Polymer Science and Technology III. Elsevier Applied Sci Publ, London, N.Y.
20. Borchard W, Burg B (1989) In: Baumgärtner A, Picot CE (eds) Molecular Basis of Polymer Networks. Springer Proceedings in Physics, Springer-Verlag, 42:162
21. Winter HH, Chambon F (1986) J of Rheology 30:367
22. Burg B, Dissertation, Duisburg 1988
23. Ferry JD (1980) Viscoelastic Properties of Polymers. J Wiley and Sons Inc, N.Y.
24. Pahl MH (1983) Prakt Rheology der Kunststoffschmelzen und Lösungen. VDI-Verlag GmbH, Düsseldorf
25. Flory PJ (1978) Principles of Polymer Chemistry. Cornell University Press, N.Y.
26. Ferry JD, Eldridge JE (1949) J Phys & Colloid Chem 53:184
27. Benson SW (1960) Foundation of Chemical Kinetics. Mac Graw-Hill, N.Y.
28. Croome RJ (1976) In: Cox RJ (ed) Photogr Gelatin II. Academic Press, London, 101
29. De Gennes PG (1976) J Phys Lett 37:L-1
30. De Gennes PG (1979) Scaling Concepts in Polymer Physics. Cornell University Press, Ithaca and London, 140
31. Stauffer D (1989) Physics Reports (Review Section of Physics Letters) 1,2:54
32. Stauffer D, Coniglio A, Adam M (1979) Adv Polym Sci 44:103

33. Ord G, Whittington SG (1982) J Phys A: Math Gen 15:L29
34. Kirkpatrick S (1973) Solid State Commun 12: 1279
35. Gordon M, Torkington A (1981) Pure Appl Chem 18:1461
36. Borchard W, Burg B in preparation

Authors' address:

Prof. Dr. W. Borchard
Angewandte Physikalische Chemie
der Universität — GH — Duisburg
Postfach 101503
4100 Duisburg 1, FRG

Progress in Colloid & Polymer Science

Progr Colloid Polym Sci 83:211—215 (1990)

NMR study of diffusion in protein hydration shells

K. Kotitschke, R. Kimmich, E. Rommel and F. Parak*)

Sektion Kernresonanzspektroskopie, Universität Ulm, Ulm, FRG
*) Institut für Molekulare Biophysik, Universität Mainz, Mainz, FRG

Abstract: Water diffusion coefficients have been measured in myoglobin single crystals and bovine serum albumin (BSA) solutions by the aid of the NMR field-gradient technique. The temperature and the concentration dependences have been determined. The diffusion coefficients in the myoglobin single crystals and in the hydration water of the BSA solutions indicate a strikingly high mobility of the water molecules. Percolation transitions have been observed with respect to the hydration shells and the free water phase as well.

Key words: Myoglobin single crystals; bovine serumalbumin; NMR field gradient technique; percolation transitions

Introduction

Water diffusion in aqueous protein systems is of interest because of several aspects. The interaction of water molecules with protein surfaces and the translational degrees of freedom in the hydration shells are important factors of nuclear magnetic relaxation (NMR) in such systems [1—3]. Recently, we proposed a corresponding model theory [3, 4], the crucial assumptions of which are first that water molecules are able to diffuse within the hydration shells along the rugged protein surface, and second that tumbling of the protein molecules requires a certain free-water volume, in analogy to the Cohen/Turnbull free-volume theory [5]. The diffusion experiments to be reported in the following are suitable to check predictions on this basis.

A quite different source of motivation for diffusion experiments with water/protein systems refers to modern theories of percolation transitions [6, 7]. We will show that such crossovers can be observed by the aid of the concentraion and temperature dependences of the water diffusion coefficient.

Experimental

The principle of the diffusion measurements is the NMR field-gradient method in the stimulated echo version [8]. Instead of rectangular gradient pulse shapes, we used a shifted cosine function as described in [9, 10]. The radio frequency pulses, as well as the echo signals, are positioned in "valleys" of the cosine function (Fig. 1). At these positions the oscillating gradient is negligible compared with a small stationary gradient which was applied additionally in order to stabilize the echo signal. The advantage of a continuous wave cosine function is that higher harmonics are filtered out by the gradient coil system itself and that operation in the steady state is possible including all phase shifts by eddy current fields from the pole caps. The stability of the system therefore is expected to be particularly high.

The total field gradient is

$$G_t(t) = G_0 + G_1[1 - \cos(\omega_g t)] , \qquad (1)$$

where G_0 is the statinoary gradient and ω_g is

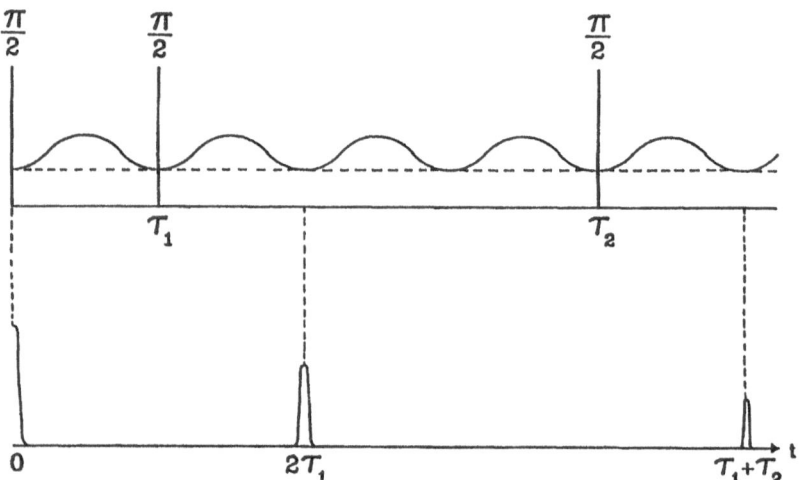

Fig. 1. Schematic representation of the pulse sequence for the measurement of diffusion coefficients. Three 90° RF pulses are irradiated at times 0, τ_1 and τ_2 in the "valleys" of the shifted cosine gradient function. As signals, either the Hahn echo at time $2\tau_1$ or the stimulated echo at time $\tau_1 + \tau_2$ can be evaluated as function of the amplitude of the field gradient. In this paper, only stimulated echoes have been recorded

the circular frequency of the oscillating gradient. With

$$\tau_2 = (m + 1)\tau_1 \quad (m = 1, 2, 3, \ldots) \tag{2}$$

and the expressions

$$F = \gamma^2 \tau_1^3 (k_1 G_1^2 + k_2 G_0^2 + k_3 G_0 G_1) \tag{3}$$

$$k_1 = \frac{2}{3} + \frac{5}{4\pi^2} + m \tag{4}$$

$$k_2 = \frac{2}{3} + m \tag{5}$$

$$k_3 = \frac{4}{3} + \frac{1}{\pi^2} + 2m \tag{6}$$

the echo decay function is given by

$$E = E_0 \exp(-DF) , \tag{7}$$

where D is the diffusion coefficient and γ the gyromagnetic ratio.

An early version of the home-built apparatus used has been described in [9]. Typical value ranges of the experimental parameters are

$$G_t \leqslant 930 \; G/cm, \; 33 \; Hz \leqslant \omega_g/2\pi \leqslant 625 \; Hz,$$
$$G_0 < 50 \; G/cm .$$

Single crystals of sperm whale myoglobin have been prepared according to the method of Kendrew

and Parrish [11]. Bovine serum albumin (BSA) was purchased from Behring-Werke, Marburg, FRG. Solutions of BSA with definite water contents were prepared by first lyophilizing the protein. The residual water content was about 1.6% [12]. Water contents up to 30% by weight were achieved by exposing the sample to a saturated vapor atmosphere for varying times. Contents above 30% were prepared by adding water with a pipette. Water contents are given as grams of water (including the residual water) per gram of solution.

Results

Figure 2 shows the temperature dependence of the water diffusion coefficient in myoglobin single crystals (water content about 40%). Data for bulk water have also been plotted for comparison. The fact that the myoglobin values are only a factor of 2 less then the bulk water data indicates that water in the crystals still has a very high mobility. The reduction can virtually be understood by the obstruction effect of the macromolecules alone [13, 14]. Note that water in the crystals does not freeze at zero degrees Celsius, so that the diffusion data are smoothly extended to minus temperatures. This finding is in complete agreement with the results obtained by dielectric relaxation [15]. The absence of freezing in the case of myoglobin may be due to the fact that the crystals have been grown in 3.75 m ammonium sulfate solution having, per se, a depression of freezing point by about 20°. Hydra-

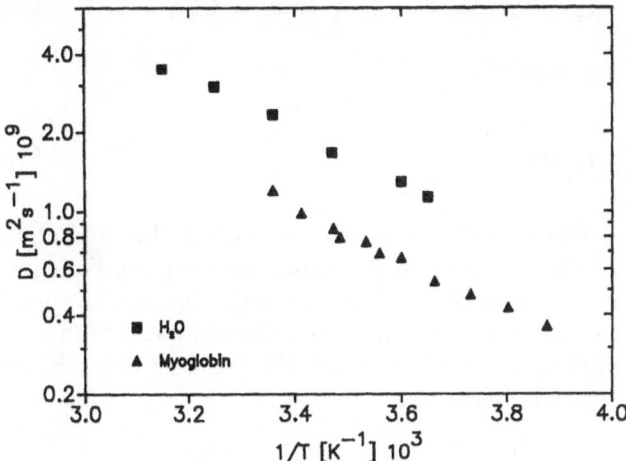

Fig. 2. Temperature dependence of the water diffusion coefficient in bulk water and in myoglobin single crystals

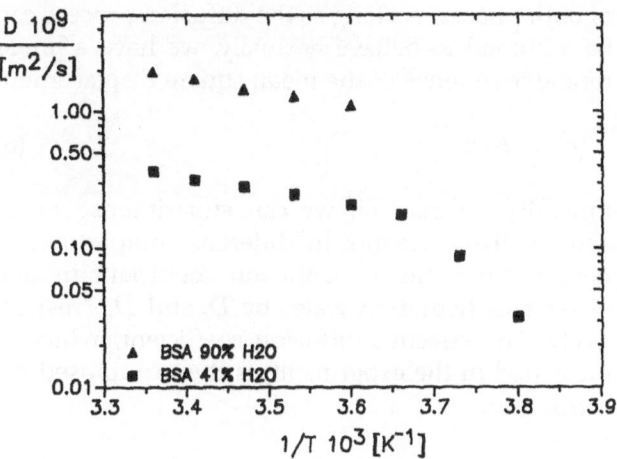

Fig. 3. Temperature dependence of the water diffusion coefficient in BSA solutions with water contents of 41% and 90%. Below the freezing temperature of free water, diffusion can only be measured if the hydration shells form "infinite" percolation clusters.

tion water of proteins, on the other hand, is known to remain generally liquid down to about −70 °C, as verified in the case of BSA solutions.

Figure 3 shows the temperature dependence of the water diffusion coefficient of two different BSA solutions. "Free water" freezes at the freezing temperature of ordinary water, while hydration water does not [4, 16]. In the 41% sample, diffusion coefficients could be measured even at −10 °C, and no abrupt change was visible at the freezing temperature, although the water content was above the saturation concentration of hydration water (about 30%).

Lastly, Fig. 4 shows the concentration dependences of the water diffusion coefficient in BSA solutions at different temperatures. Abrupt changes are visible at about 57% (b.w.) protein. Note that this concentration is less than the concentration corresponding to completely saturated hydration shells. This phenomenon must be considered to be indicative for a percolation transition; assuming that diffusion in the free water compartments is faster than in the hydration shells leads to the conclusion that diffusion becomes slower as soon as the percolation clusters of the free water become finite. One then has only "islands" of free water and the measurable displacements take place mainly in hydration water. At a protein content of about 70% the free-water islands finally disappear completely. In the following section, we present a model theory predicting this "free-water percolation" behavior of the data in Fig. 4 (fitted curves).

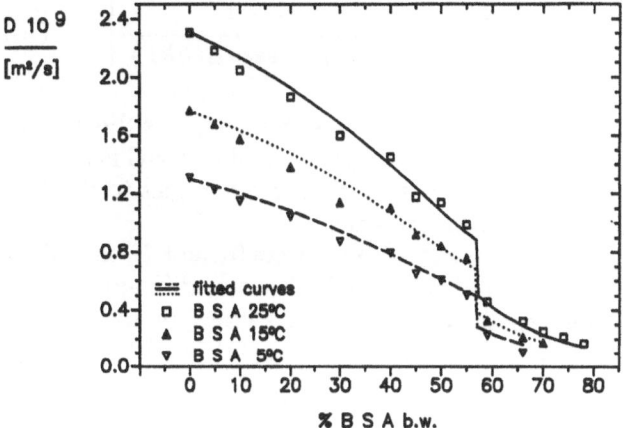

Fig. 4. Concentration dependence of the water diffusion coefficient in BSA solutions at various temperatures. "Free-water percolation transitions" are visible. The fitted curves have been calculated as described in the text. The parameters are $A_1 = 0.125$; $A_2 = 4.5$; $A_3 = 0.714$; $v_p = 0.75$; $a = 1.53$; $c_s = 0.27$; $c_c = 0.57$; $\Delta G/RT = 8$. The measured values of the diffusion coefficients in bulk water, D_0, are given by the first data points on the left

Free-water percolation transition

A water molecule in an unfrozen protein/water system is frequently exchanged between the free-water and the hydration water phases. Mean square displacements are thus composed of contributions of both phases. As long as the diffusion process can be assumed to behave normally, we have a linear time dependence of the mean square displacement

$$\langle r^2 \rangle = 6Dt . \tag{8}$$

Linearity means that we can superimpose mean square displacements in different compartments. Let us designate the diffusion coefficient in free water and hydration water by D_f and D_h, respectively. The effective diffusion coefficient, which is measured in the experiment, is then composed according to

$$D = p_f D_f + p_h D_h, \tag{9}$$

where p_f and $p_h = 1 - p_f$ are the time fractions a water molecules diffuses in the free and the hydration water, respectively. According to [4]

$$p_f = \frac{1}{1 - c_s} \left[\frac{c_w - c_s}{c_w} \right.$$

$$\left. + \frac{c_s}{c_w} (1 - c_w) \frac{1}{1 + \exp(\Delta G/RT)} \right], \tag{10}$$

where c_w is the water mass fraction; c_s is the saturation mass fraction of hydration shells; ΔG is the difference in the molar Gibbs free energies in the free and the hydration water.

The free-water diffusion coefficient D_f is reduced, compared with the bulk water diffusion coefficient D_0, by the obstruction effect of the macromolecules [13, 14]. Following [14], we have

$$D_f = D_0 \frac{1 - a\Phi}{1 - \Phi} , \tag{11}$$

with the volume fraction of the hydrated proteins being

$$\Phi = \frac{c_p v_p + c_h v_h}{c_p v_p + c_h v_h + c_f v_f} ; \tag{12}$$

c_p, c_h, c_f are the mass fractions, v_p, v_h, v_f are the specific volumes of proteins, hydration water and free water, respectively. a is a numerical parameter that takes into account the shape of the protein with respect to the obstruction effect [13, 14]. Above the saturation concentration of the hydration shells,

$$c_w = c_f + c_h = 1 - c_p \geqslant c_s \approx 0.3 , \tag{13}$$

we may use

$$c_h = c_p \frac{c_s}{1 - c_s} . \tag{14}$$

The diffusion coefficient within the hydration shells depends on the saturation degree. For $c_w < c_s$, water molecules increasingly become trapped on the surface and/or the diffusion paths along the surface become restricted [4]. We take this into account by assuming

$$D_h = D_0 \frac{A_1}{c_p} \exp[-A_2(c_p - c_c)] \quad (c_p \geqslant c_c) ; \tag{15}$$

A_1, A_2 are numerical constants, c_c is the threshold protein concentration of the free-water percolation process. Below the threshold protein concentration, the diffusion coefficient in the hydration shells is assumed to be constant

$$D_h = A_3 D_0 . \tag{16}$$

For fitting purposes, the denominator of Eq. (12) is approximated by 1 cm^3/g. We further assume that $v_f \approx v_h \approx 1$ cm^3/g. Equation (12) thus becomes, together with Eq. (14):

$$\Phi \approx c_p \left(\frac{v_p}{[1 \text{ cm}^3/\text{g}]} + \frac{c_s}{1 - c_s} \right) . \tag{17}$$

The fraction of water molecules in the free-water phase is given by Eq. (10). For c_p, this fraction cannot contribute to displacements large enough to be measurable, and we can set $p_f = 0$ in this case.

The fitted curves in Fig. 4 have been calculated by the aid of the above formalism. The good coincidence with the experimental data verifies the assumption of a free-water percolation mechanism.

Discussion

We have shown that hydration water in myoglobin single crystals, as well as in bovine serum albumin solutions, has translational degrees of freedom. The high mobility is maintained even below the freezing point of free water. These findings are crucial for the interpretation of nuclear magnetic relaxation in such systems.

It is known that water in such systems has orientational correlation times in the order of microseconds, what is longer than in bulk water by six orders of magnitude. A frequent conclusion, therefore, was that hydration water is irrotationally bound at the protein surface; this notion is ruled out by the present study. The high mobility of hydration water instead confirms a conclusion that we have drawn from deuterion field-cycling relaxation spectroscopy. Water molecules are locked within the hydration shells for longer periods, but are able to diffuse along the surface of the protein [3].

The percolating clusters either refer to free water or to hydration water compartments. Phenomena of that kind have a general aspect: it has been predicted that diffusion on percolating clusters behaves anomalously [7, 17, 18]. In principle, such behavior should be observable with the NMR method [19, 20]. In our own experiments, no deviation to normal diffusion was found within the experimental errors. This may be due to the fact that the expected effects in any case are small and are restricted to the critical density of the percolation network. Further investigations precisely at the percolation threshold are planned.

Problems of the sort we have discussed above also appear with systems other than proteins; examples are porous materials such as cement, zeolites, silica gels or condensed colloids. Analogous studies would be interesting and some unconventional findings have already been reported recently [21, 22].

*Acknowledgements*1

We thank F. Klammler, U. Zellhuber, J. Wiringer and W. Unrath for their cooperation in the course of this work. Financial support by the Deutsche Forschungsgemeinschaft and the Bundesministerium für Forschung und Technologie (grant no 01VF85203) is gratefully acknowledged.

References

1. Daszkiewicz OK, Hennel JW, Lubas B, Szczepkowski T (1963) Nature 200:1006—1007
2. Hallenga K, Koenig S (1976) Biochemistry 15:4255—4264
3. Schauer G, Kimmich R, Nusser W (1988) Biophys J 53:397—404
4. Kimmich R, Nusser W, Gneiting T (1990) Colloids and Surfaces 45:283—302
5. Cohen MH, Turnbull D (1959) J Chem Phys 31:1164—1169
6. de Gennes PG (1976) Recherche 7:919—927
7. Orbach R (1986) Science 231:814—819
8. Tanner JE (1970) J Chem Phys 52:2523—2526
9. Bachus R, Kimmich R (1983) Polymer 24:964—970
10. Kimmich R, Schnur G, Köpf M (1988) Progr NMR Spectr 20:385—421
11. Kendrew JC, Parrish RG (1956) Proc Roy Soc 238A:305—324
12. Andrew ER, Bryant DJ, Rizvi TZ (1983) Chem Phys Letters 95:463—466
13. Wang JH (1954) J Am Chem Soc 76:4755—4763
14. Clark ME, Burnell EE, Chapman NR, Hinke JAM (1982) Biophys J 39:289—299
15. Singh GP, Parak F, Hunklinger S, Dransfeld K (1981) Phys Rev Letters 47:685—688
16. Kuntz ID (1971) J Am Chem Soc 93:514—516
17. O'Shaughnessy B, Procaccia I (1985) Phys Rev Letters 54:455—458
18. Korb J-P, Gouyet J-F (1988) Phys Rev B 38:493—499
19. Kärger J, Vojta G (1987) Chem Phys Letters 141:411—413
20. Kärger J, Pfeifer H, Vojta G (1988) Phys Rev A 37:4514—4517
21. Blinc R, Lahajnar G, Zumer S, Pintar MM (1988) Phys Rev B 38:2873—2875
22. D'Orazio F, Bhattacharja S, Halperin WP (1989) Phys Rev Letters 63:43—46

Authors' address:

K. Kotitschke
Sektion Kernresonanzspektroskopie
Universität Ulm
7900 Ulm, FRG

Progress in Colloid & Polymer Science   Progr Colloid Polym Sci 83:216—221 (1990)

Thermal anomalies in the DSC-diagrams of a highly disperse silicic acid wetted by different aqueous electrolyte solutions

G. Peschel and U. Furchtbar

Institut für Physikalische und Theoretische Chemie, Universität Essen, Essen, FRG

Abstract: It is well established that many aqueous colloidal systems and even biological organisms exhibit more or less pronounced breaks in their physico-chemical properties at about the characteristic temperatures 15°, 30°, 45°, and 60°C. This phenomenon is ascribed to structural phase transitions of boundary water in the systems. In view of the gap in the understanding of the underlying mechanism, calorimetric tests employing differential scanning calorimetry (DSC-2, Perkin Elmer) were carried out with porous discs pressed from silica particles and wetted with various aqueous electrolyte solutions. The reference sample was a silica discs wetted with pure water. It could be demonstrated that pure water on a silica discs shows no thermal effect at the characteristic temperatures; but aqueous electrolyte solutions as the adsorbed phase display enthalpy changes at about these temperatures, according to water structure affecting ability of the ions.

Key words: Differential scanning calorimetry; structural transition of higher order; disperse silicic acid; alkali halides; thermal anomalies

Introduction

There is a large body of evidence that water near solid surfaces exhibits a modified structure and, hence, is different in many of its physical properties from those of the bulk phase [1—3]. But there is still much debate about the range of molecular ordering of water near interfaces. It seems, however, that the experimental findings up to now generally support the existence of structured water layers adjacent to interfaces with a thickness not exceeding the order of 10 nm.

Notwithstanding the question of what is the mechanism of the molecular ordering, additional difficulties in explaining these phenomena arise by the occurrence of thermal anomalies in the structure of aqueous boundary layers at about the characteristic temperatures 15°, 30°, 45°, and 60°C [4—6]. These anomalies, which have an extremely high importance in all vital processes, are commonly ascribed to a structural transition of higher order of small molecular entities of water [6]. Undoubtedly, interfaces are involved in this phenomenon,

since bulk water seems to not display any thermal anomalies [7—10]. According to our knowledge, no plausible explanations have been offered for these uncommon structural peculiarities, especially with regard to the influence of electrolytes. Of interest in this connection might be the detection of disjoining pressure maxima in thin water films between fused silica plates at just the characteristic temperatures [11]. Ling and Drost-Hansen [12] considered this problem in a DTA study of water in porous glass. Unfortunately, because of experimental difficulties they could not observe the boundary phase transitions with sufficient accuracy.

It is the purpose of this work to examine by aid of the DSC-method the thermal properties of aqueous electrolyte solutions in the pores of discs pressed from silica particles (Aerosil 200, Degussa, Hanau, FRG).

The hydration of silica and quartz surfaces

There is ample evidence in literature that silica hydrosol particles are subject to notable hydration effects which become, for example, evident by their contributing to hydrosol stability [13—18]. Peschel and Ludwig [19] succeeded in proving that hydration layers on silica hydrosol particles are apt to hamper the exchange of Li^+, Na^+ and K^+ in a concentration range below $3 \cdot 10^{-3}$ M.

In order to substantiate such findings on a more macroscopic scale experiments with samples or plates consisting of quartz or fused silica, respectively, were carried out by Peschel et al. [20, 21] by pressing two fused silica plates — one planar, the other spherically curved — against each other in aqueous solutions; they examined the hydration forces generated by the overlap of the hydration layers on both the surfaces.

Such tests are similar to those performed by Israelachvili and Adams [22] and Pashley [23], who used cylindrically bent mica platelets, and likewise, detected hydration forces. Infrared measurements carried out by Roberts and Zundel [24, 25] were indicative of the existence of a thick hydration sheath on silica surfaces. Similar conclusions were drawn by Belouschek and Suppa [26] when examining the thermal conductance of water in thin layers between macroscopically sized fused silica surfaces.

Experimental

The Method

The calorimetric investigations of the test samples were carried out by employing the method of differential scanning calorimetry (DSC) [27, 28], by which the heat flux through the test sample is measured as dependent on time or on temperature, since the temperature of the system is varied by a temporal program. The calorimeter (DSC 2, Perkin Elmer) is equipped with two analysis cells in twinned arrangement; one of the cells contains the test sample, the other contains the reference sample. Temperature differences that can occur by thermal transitions in the test sample are compensated by changing the heating power; this is indicated by a peak in the thermogram.

The change of the enthalpy of the sample per unit time is given by [29]:

$$\frac{dH}{dt} = -\frac{dq}{dt} + (C_S - C_R) \cdot \frac{dT}{dt} - R \cdot C_S \cdot \frac{d^2q}{dt^2} ;$$

(1)

dq/dt is the difference in heat production between the test and the reference sample. The second term on the right side of Eq. (1) describes the deviation of the basis line in the thermogram. This is dependent on the heat capacities

C_S of the test and C_R of the reference sample. From the peaks in the thermogram for exotherm or endotherm processes, respectively, the heats transferred can be derived by an integration procedure.

Measuring procedure

The adsorbent phase of the samples consisted of finely dispersed silica (Aerosil 200 from Degussa, Hanau) which was pressed into small discs under a pressure of $4 \cdot 10^8$ Pa. The surface area of the porous discs was determined to be about $1.90 \cdot 10^5$ $m^2 \cdot kg^{-1}$ (BET with nitrogen). Before each run, 20 mg of fragments of the discs were put into a small pan manufactured from stainless steel. A further 10 mg of the test solution was added to the silica sample, and then the pan was tightly sealed. The reference pan in most of the runs also contained 20 mg of the porous discs and, critically, triple-distilled water (10 mg), so that all thermal effects indicated by the peaks in the thermograms ought to be due to the influence of the electrolyte only. The heating rate was always 10 K/min, which implies an abscissa scale of 1 K/mm for the recorder connected to the calorimeter. The calibration of the transition enthalpies was based on the melting enthalpy of extremely pure indium (28.43 J/g).

Before starting a run the pans filled with the samples were heated up to the terminal temperature of 353 K, kept at this temperature for some minutes, and then cooled to the initial temperature (263 K) of each run.

Chemicals

The salts used (LiCl, NaCl, KCl, NaBr, NaI, NaF, $NaClO_4$, RbCl, CsCl, NaSCN, $MgCl_2 \cdot 6 H_2O$, $CaCl_2 \cdot 2 H_2O$, $BaCl_2 \cdot 2 H_2O$) were of p.a. grade and LiF and KF were of suprapure quality. Water as the solvent was triple distilled.

Results

Figure 1 reflects a thermogram in which the heat flow through the test sample (20 mg silica, 10 mg water) is compensated by that through the reference sample (20 mg silica, but without any liquid). There is every reason to state that the thermogram below 12 °C is governed by instrumental effects; above about 12 °C the thermogram represents, within the limits of error, a horizontal line displaying no notable anomalies at about the characteristic temperatures. Likewise, no effects concerning the heat capacity or the evaporation of water in the pan can be observed. In Figs. 2a—c the exothermic effects produced by LiCl, NaCl, and KCl for a concentration $c = 1$ M are shown. Obviously, we are faced

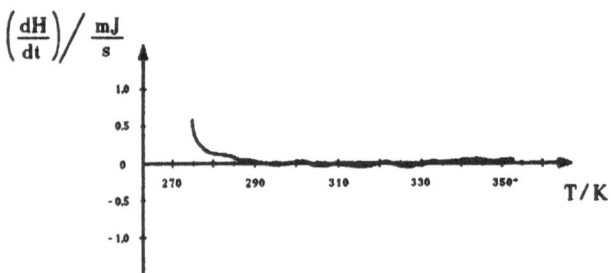

Fig. 1. Thermogram aerosil + H_2O vs aerosil

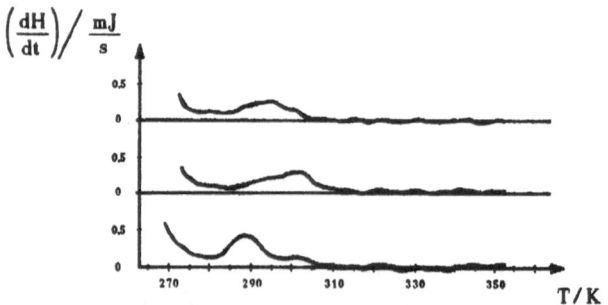

Fig. 2. Thermograms (from the top) for a) aerosil + LiCl-solution (1 M) vs aerosil + H_2O; b) aerosil + NaCl-solution (1 M) vs aerosil + H_2O; c) aerosil + KCl-solution (1 M) vs aerosil + H_2O

with surface hydration phenomena as dependent on temperature up to about 37°C. LiCl and NaCl display a broad peak and smooth peak in the range between about 12°C and 37°C, whereas KCl exhibits a significant peak at about 17°C. This value, however, must be considered with great precaution, because it lies rather close to instrumental effects.

In Table 1 the heats effectively transferred by the influence of the three salts are quoted. The thermograms for the remaining alkali chlorides (RbCl,

Table 1. Transition enthalpies (in $J \cdot mol^{-1}$ cations) for different aqueous alkali chloride solutions adsorbed on silica

Electrolyte	$\Delta H/J \cdot mol^{-1}$
LiCl	—4830
NaCl	—3505
KCl	—5678

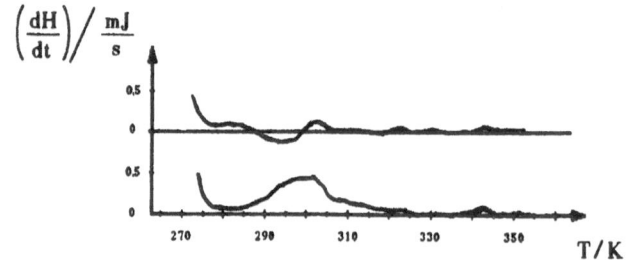

Fig. 3. Thermograms (from the top) for a) aerosil + RbCl-solution (1 M) vs aerosil + H_2O; b) aerosil + CsCl-solution (1 M) vs aerosil + H_2O

CsCl) are presented in Figs. 3a and b. RbCl shows a small endothermic peak at about 20°C and an additional exothermic one at about 30. CsCl brings about an extremely large exothermic effect which is also located at about 30°C.

A rather different picture emerges when testing LiF, NaF, and KF solutions (Figs. 4a—c). LiF (saturated solution) led to an endothermic peak at about 20°C and KF ($c = 1$ M) led to a similar but smaller one at about 25°C. NaF ($c = 1$ M), on the contrary, behaves like the alkali chlorides; it also shows an exothermic peak which is, however, located at about 15°C, that being one of the characteristic temperatures.

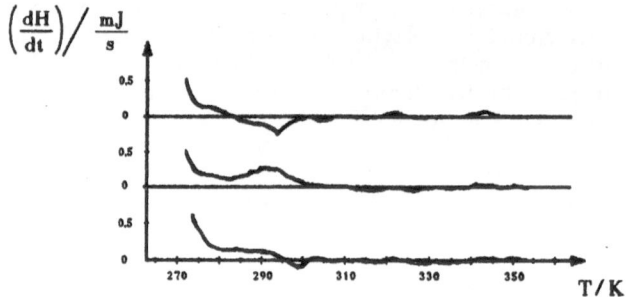

Fig. 4. Thermograms (from the top) for a) aerosil + LiF-solution (sat.) vs aerosil + H_2O; b) aerosil + NaF-solution (1 M) vs aerosil + H_2O; c) aerosil + KF-solution (1 M) vs aerosil + H_2O

Replacing, for example, the Cl^- ion by the water-structure-disrupting anions Br^- and I^-, respectively, yields results (Figs. 5a and b) dissimilar from those illustrated above in that a slight exothermic

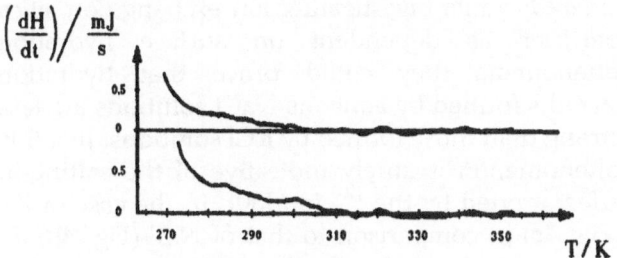

Fig. 5. Thermograms (from the top) for a) aerosil + NaBr-solution (1 M) vs aerosil + H$_2$O, b) aerosil + NaI-solution (1 M) vs aerosil + H$_2$O

effect can be observed up to about 40°C, but without exhibiting a typical peak.

A still more different picture appears if the earth alkali chlorides (MgCl$_2$, CaCl$_2$, SrCl$_2$, BaCl$_2$) are tested for the concentration $c = 0.5$ M (Figs. 6a—d). Here, attention must be drawn to the fact that all peaks indicating a structural transition are clearly endothermic and range according to the kind of the cation from about 20°C to about 30°C.

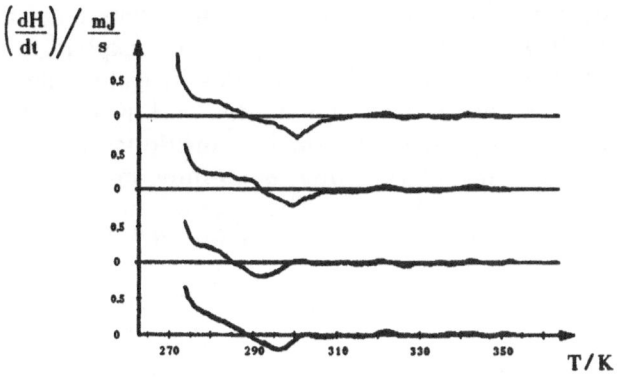

Fig. 6. Thermograms (from the top) for a) aerosil + MgCl$_2$-solution (0.5 M) vs aerosil + H$_2$O; b) aerosil + CaCl$_2$-solution (0.5 M) vs aerosil + H$_2$O; c) aerosil + SrCl$_2$-solution (0.5 M) vs aerosil + H$_2$O; d) aerosil + BaCl$_2$-solution (0.5 M) vs aerosil + H$_2$O

A further step in the calorimetric analysis concerns the concentration dependence of the heights of the exothermic peaks. This problem was treated in particular for differently concentrated KCl solutions (Figs. 7a—d in combination with Fig. 2c). It is clearly evident that the peak areas get larger with decreasing concentration.

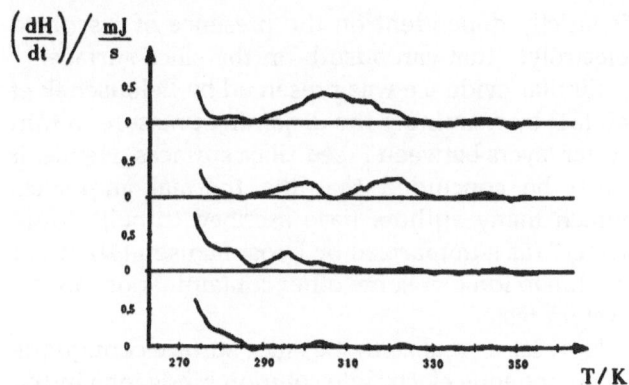

Fig. 7. Thermograms (from the top) for a) aerosil + KCl-solution (0.005 M) vs aerosil + H$_2$O; b) aerosil + KCl-solution (0.1 M) vs aerosil + H$_2$O; c) aerosil + KCl-solution (0.5 M) vs aerosil + H$_2$O; d) aerosil + KCl-solution (2 M) vs aerosil + H$_2$O

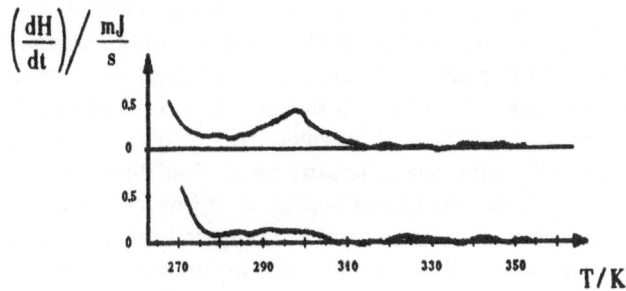

Fig. 8. Thermograms (from the top) for a) aerosil + LiCl-solution (0.1 M) vs aerosil + H$_2$O; b) aerosil + SrCl$_2$-solution (0.05 M) vs aerosil + H$_2$O

The concentration effect for LiCl (Fig. 8 in comparison with Fig. 2a) is insignificant; for the case of SrCl$_2$ (Fig. 8b in comparison with Fig. 6c) an inversion of the sign of the enthalpy effect even occurs for a concentration not larger than 0.05 M.

Tests with NaSCN and NaClO$_4$ (for both $c = 1$ M) yielded absolutely no heat effect; this is not surprising since SCN$^-$ and ClO$_4^-$ ions are well-known to strongly disrupt water structure.

Discussion

The findings in Fig. 1 for pure water, when compared with all the others presented in this paper demonstrate that the occurrence of notable thermic anomalies at about the characteristic temperatures is strictly dependent on the presence of dissolved electrolyte that can adsorb on the silica surface.

Similar evidence was presented by Belouschek et al. [31] by examining the disjoining pressure in thin water layers between fused silica surfaces. Hence, it must be concluded that the thermal anomalies which many authors have ascribed to bulk "pure water" (as summarized by Drost-Hansen [4]) might be due to ionic or some other contaminations in the system tested.

Now let us realize that a silica surface contiguous to an aqueous electrolyte solution binds ions in the adsorbed state. The surface hydration tendency exerted by these and the OH-groups on the silica surface [32] is counteracted by that of the nearby dissolved ions. Obviously, the latter mechanism is dominant for the alkali chlorides, since the peak area gets larger in the order from Li^+ to K^+ (Table 1), producing the most heat.

Just these experimental findings cannot be understood in the first instance since, with increasing temperature, H-bonded entities of water molecules should undergo an endothermic "melting" process. However, this process creates some disorder near the surface so that its ability to modify water structure nearby is improved. But the molecular orienting reaction is typically exothermic which means that, at a given temperature, exothermic and endothermic processes might be superimposed. What is effectively observed in experiment should then be the algebraic sum of both enthalpy changes. Therefore it is not surprising that the heat values in Table 1 are rather small. In the case of Figs. 2a—c the exothermic reaction obviously has priority. The situation is the same for Cs^+ (Fig. 3b) which, according to its water-structure-breaking ability [30], might have promoted the formation of high concentrated molecular entities of water and monomers near the surface.

LiF is well known as a potent water-structure-promoting electrolyte. Without question, the boundary region might be highly structured and separated from the bulk phase by a structural mismatch zone containing small molecular entities of water [6]. Because of the high structuredness of the solution the orienting influence of the silica surface appears to be quenched to such a degree that the endothermic process predominates (Fig. 4a). KF shows a similar but minor effect (Fig. 4c). But the result for NaF (Fig. 4b) does not fit into the picture of the lyotropic series. For interpreting this problem findings of Peschel and Ludwig [19] should be an-

sidered; while investigating ion exchange on silica particles as dependent on surface hydration phenomena they could prove that hydration sheaths formed by aqueous NaCl solutions are less strong than those formed by KCl solutions. Just this phenomenon is surely indicative of the salting-in effect exerted by the K^+ ion [30]. In the case of KF (Fig. 4c) in comparison to that of NaF (Fig. 4b) an analogous mechanism might work.

The salting-in effect produced by the big I^- ion surely exceeds that of the Br^- ion, so that the results in Figs. 5a and b (except the missing transition peaks) can readily be understood.

Earthalkali cations are subject to more favorable ion-water dipole interactions [30] than are alkali cations. One can well imagine that the thermograms shown in Figs. 6a—d are produced by a similar mechanism, as commented on in the case of LiF (Fig. 4a), where also the endothermic effect totally outweights the exothermic one.

When considering the concentration dependence of the calorimetric effect, we should bear in mind that, concerning the hydration process, the silica surface and the dissolved ions nearby compete for the water molecules available. As a consequence, the surface hydration is promoted with decreasing ion concentration, as is clearly demonstrated for KCl (Figs. 7a—d). This might generally explain the fact that traces of ionic impurities in water might generate notable surface hydration and, if the occasion arises, even thermal transitions at the characteristic temperatures as commented on by Drost-Hansen [4—6]. The same underlying mechanism is obviously at work for LiCl (Fig. 8a in comparison with Fig. 2a).

In the case of $SrCl_2$, hydration is likewise strengthened at a lower concentration (Fig. 8b in comparison to Fig. 6c). The hydration process, in our opinion, means evolution of heat which, in fact, counterbalances the endothermic effect necessary for distintegration of the molecular aggregates of water to a large extent (Fig. 8b).

In the present work it could be demonstrated that the thermal transitions found in many colloidal and biological systems are strongly dependent on a dissolved substance owing to its hydration tendency. It is, however, unclear why molecular entities of water near a solid surface exhibit a structural transition at just the characteristic temperatures.

It deserves attention that the measured heat effects are only large enough with respect to the limits of errors if porous discs pressed from silica particles are used; silica in its powdered form (likewise 10 mg) shows only minor effects that can hardly be analyzed.

Recent results that demonstrate that bulk aqueous electrolyte solutions can exhibit an exothermal structural transition of higher order in the range about 50 °C will be discussed in a future paper [33].

Appendix

Our recent work [34] in this particular field has revealed that the occurrence of thermal transitions is strictly associated with the presence of aqueous solutions in narrow pores in which, because of spatial restrictions, the formation of bulk water structure might be more or less prevented.

Acknowledgement

Thanks are due to the Deutsche Forschungsgemeinschaft for granting the DSC-equipment.

References

1. Derjaguin BV, Churaev NV (1981) Progr Surf Membr Sci 14:69
2. Churaev NV (1983) J Eng Phys 45:824
3. Derjaguin BV, Churaev NV (1986) In: Croxton CA (ed) Fluid Interfacial Phenomena, John Wiley & Sons, New York
4. Drost-Hansen W (1969) Ind Eng Chem 61:10
5. Drost-Hansen W (1969) Chem Phys Lett 2:647
6. Drost-Hansen W (1971) In: Brown HD (ed) Chemistry of the Cell Interface, Academic Press, New York, p 3
7. Senghaphan W, Zimmermann GO, Chase GE (1969) J Chem Phys 51:2543
8. Cini R, Lolglio G, Ficalbi A (1969) Nature (London) 233:1148
9. Rushe EW, Good WB (1966) J Chem Phys 45:4667
10. Korsen L, Drost-Hansen W, Millero FJ (1969) J Phys Chem 73:34
11. Peschel G (1976) In: Lange OL, Kappen L, Schulze E-D (eds) Ecological Studies, Analysis and Synthesis, Vol 19, Springer-Verlag, Berlin, Heidelberg, New York
12. Ling Chaur-Sun, Drost-Hansen W (1975) In: Mittal KL (ed) Adsorption at Interfaces, American Chemical Society Symposium Series 8, Washington
13. Iler RK, Dalton RL (1956) J Phys Chem 60:955
14. Fripiat JJ, Uytterhoeven J (1962) J Phys Chem 66:800
15. Fripiat JJ (1963) In: Clays and Clay Minerals Proc 12th Nat Conf Atlanta, Ga, p 327
16. Lange KR (1965) J Colloid Interf Sci 20:231
17. AllenLH, Matijevic (1969) J Colloid Interf Sci 31:287
18. Harding RD (1971) J Colloid Interf Sci 35:172
19. Peschel G, Ludwig P (1987) Ber Bunsenges Phys Chem 91:536
20. Peschel G, Belouschek P (1976) Prog Colloid Polym Sci 60:108
21. Peschel G, Belouschek P, Müller MM, Müller MR, König R (1982) J Colloid Polym Sci 260:444
22. Israelachvili JN, Adams GE (1978) J Chem Soc Faraday Trans I 74:975
23. Pashley RM (1981) J Colloid Interf Sci 80:153; 83:531
24. Roberts NK, Zundel G (1980) J Phys Chem 84:3655
25. Roberts NK (1981) J Phys Chem 85:2706
26. Belouschek P, Suppa M (1985) Z Physik Chem NF 146:77
27. Hemminger W, Höhne G (1979) Grundlagen der Kalorimetrie, Verlag Chemie, Weinheim, New York
28. Wendlandt WWM (1986) Thermal Analysis, John Wiley & Sons, New York, Chichester, Brisbane, Toronto, Singapore
29. Gray AP (1968) In: Analytical Calorimetry, Plenum, New York, p 209
30. Conway BE (1981) Ionic Hydration in Chemistry and Biophysics, Elsevier Scientific Publ Co., Amsterdam, Oxford, New York
31. Belouschek P, Suppa M, Maier S (1986) Colloid J USSR 981 (1986)
32. Davydov VY, Kiselev AV, Zhralev LT (1964) Trans Farad Soc 60:2254
33. Peschel G, Krämer R, Zmarsly being prepared
34. Peschel G, Krämer R (in preparation)

Authors' address:

G. Peschel
Institut für Physikalische und Theoretische Chemie
Universität Essen
4300 Essen, FRG

Author Index

Subject Index